Mechanical Science for Higher Technicians

D.H.Bacon, B.Sc., C.Eng., M.I.Mech. E.

and

R.C.Stephens, M.Sc(Eng.), C.Eng., M.I.Mech.E.

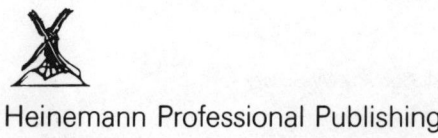

Heinemann Professional Publishing

Heinemann Professional Publishing Ltd
22 Bedford Square, London WC1B 3HH

LONDON   MELBOURNE   AUCKLAND

First published by Butterworth & Co. (Publishers) Ltd 1981
Reprinted 1983, 1985
First published by Heinemann Professional Publishing Ltd 1988

© Heinemann Professional Publishing Ltd 1988

British Library Cataloguing in Publication Data
Bacon, Dennis Harry,
   Mechanical science for higher technicians
   1. Mechanics
   I. Title   II. Stephens, Richmond Courtney
   531'.02'46   QC125.2   80-41531

ISBN 0 434 91871 7

Printed in Great Britain by
LR Printing Services Ltd, Manor Royal, Crawley, W Sussex

PREFACE

This book is primarily intended to cover the BTEC Level IV and Level V units in Mechanical science for the HNC and HND courses; the text is largely based on the author's earlier work, *Mechanical Technology*. The scope is greater than that required for the standard units to allow for variations in college-devised units and to provide background theory, where necessary. The book may also prove useful for degree and diploma students in other disciplines who require a fundamental outline of mechanical engineering principles.

The theory is presented concisely and is followed by worked examples which illustrate and amplify the theory. A large number of tutorial examples, with answers, are included with each chapter.

D.H. BACON
R.C. STEPHENS

CONTENTS

1 SIMPLE STRESS AND STRAIN   1

    Introduction. Direct stress and strain. Shear stress and strain. Elasticity, plasticity and Hooke's Law. Factor of safety. Stresses in compound bars.

2 SHEARING FORCE AND BENDING MOMENT   8

    Introduction. Sign convention. Relation between load intensity, shearing force and bending moment. Shearing force and bending moment diagrams. Graphical construction methods.

3 BENDING STRESSES   20

    Second moment of area. Stresses due to pure bending. Values of I and Z for common cases. Radius of curvature. Position of neutral axis. Combined bending and direct stresses.

4 TORSION   33

    Stress due to twisting. Angle of twist.

5 DEFLECTION OF BEAMS   37

    Introduction. Basic relations. Graphical method. Application of integration method to standard cases. Principle of superposition. Unsymmetrical loading—Macaulay's Method. Distributed loads. Built-in beams. Standard cases. Area-moment method. Application of area-moment method method to standard cases.

6 STRUTS   63

    Introduction. Euler's Theory. Validity of Euler's Theory. The Rankine-Gordon relation.

7 STRAIN ENERGY   71

    Introduction. Strain energy due to direct stress. Impact loading. Strain energy due to a uniform moment or torque. Strain energy due to a variable bending moment. Castigliano's Theorem. Application to deflection of beams and curved bars. Variable torque.

8 SHEAR STRESS IN BEAMS   83

    Shear stress distribution. Application to common cases.

9 COMPLEX STRESS AND STRAIN   90

    Stresses on an oblique section. Material subjected to two perpendicular direct stresses. Material subjected to shear stress. The general case of two-dimensional stress. Mohr's stress circle. Simple applications of principal stresses. Theories of elastic failure. Principal strains. Electric resistance strain gauges. Volumetric strain and bulk modulus. Rela- between E, G and ν.

CONTENTS

10 CYLINDERS — 112

Stresses in thin cylindrical shells. Lamé's Theory. Thick cylinder subjected to internal pressure only. Longitudinal and shear stress. The Lamé line. Compound cylinders.

11 DYNAMICS — 123

Introduction. Linear motion. Angular motion. Mass, force, weight, momentum. Newton's Laws of Motion. Impulse. Circular motion. Work, energy and power. Moment of inertia. Theorem of parallel axes. Values of I and k for common cases. Torque and angular acceleration. Angular momentum and angular impulse. Angular work, power and kinetic energy. Equivalent mass of a rotating body. Acceleration of geared systems. Maximum acceleration of vehicles.

12 VELOCITY AND ACCELERATION DIAGRAMS — 148

Introduction. Velocity of a rigid link. Velocity of a block sliding on a rotating link. Acceleration of a rigid link. Acceleration of a block sliding on a rotating link. Inertia force on a link.

13 RECIPROCATING MECHANISMS — 162

Introduction. Piston velocity. Piston acceleration. Analytical method for piston velocity and acceleration. Crankshaft torque due to piston force and mass. Crankshaft torque due to connecting rod inertia.

14 TURNING MOMENT DIAGRAMS — 172

Introduction. Crank effort diagrams

15 FRICTION CLUTCHES, BEARINGS AND BELT DRIVES — 178

Plate clutches. Cone clutches. Centrifugal clutches. Bearings. Journal bearings. Lubricated surfaces. Viscosity. Application to bearings. Belt drives. Centrifugal tension. Initial tension. V-belt drives.

16 GYROSCOPIC MOTION — 197

Introduction. Gyroscopic couple.

17 GEAR TRAINS — 204

Introduction. Gear teeth definitions. Simple gear trains. Compound gear trains. Epicyclic gear trains. Torques in gear trains.

18 FREE VIBRATIONS — 215

Introduction. Simple harmonic motion. Linear motion of an elastic system. Angular motion of an elastic system. Effect of mass of spring and inertia of shaft. Motion of a pendulum. Differential equation of motion.

19 TRANSVERSE VIBRATIONS AND WHIRLING SPEEDS — 228

Light beam with single load. Uniformly distributed load. Several loads. Whirling speed of shafts.

20 DAMPED AND FORCED VIBRATIONS — 237

Introduction. Damped vibrations. Forced vibrations. Forced damped vibrations. Periodic movement of support.

APPENDIX – DIFFERENTIAL EQUATIONS — 249

1 Simple stress and strain

1.1 Introduction When a material is loaded, the force may be resolved into components normal and parallel to any plane within the material. The normal component is termed a *tensile* or *compressive* force and the intensity of loading per unit area is called the *direct stress*, σ. The parallel component is termed a *shear* force and the intensity of loading per unit area is called the *shear stress*, τ.

The distortion of the material due to the direct and shear forces is measured by the *direct strain*, ϵ, and the *shear strain*, ϕ, respectively.

1.2 Direct stress and strain Fig. 1.1(a) shows a piece of material subjected to a tensile force P. If the cross-sectional area of the material is a, then the tensile stress on the cross-section

$$\sigma = \frac{P}{a} \tag{1.1}$$

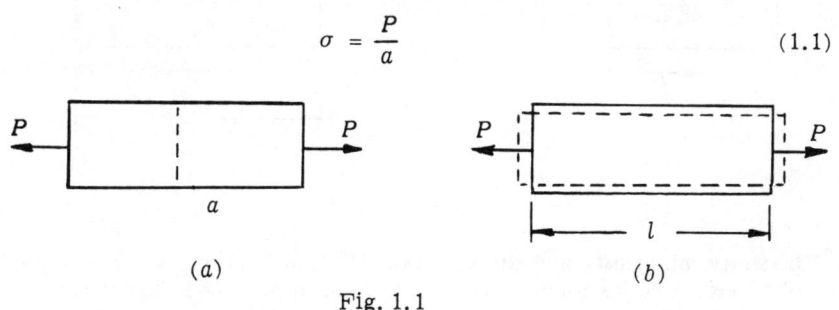

Fig. 1.1

If the original length of the bar in the direction of P is l, Fig. 1.1(b), and the extension due to the load is x, then the tensile strain

$$\epsilon = \frac{x}{l} \tag{1.2}$$

It will be seen in Fig. 1.1(b) that the extension in the direction of P is accompanied by a contraction in perpendicular directions and the ratio $\dfrac{\text{transverse strain}}{\text{longitudinal strain}}$ is called *Poisson's Ratio*, ν.

1.3 Shear stress and strain Fig. 1.2(a) shows a piece of material subjected to a shear force P. If the cross-sectional area of the material is a, then the shear stress

$$\tau = \frac{P}{a} \tag{1.3}$$

If the deformation in the direction of P is x and the distance between the opposite faces is l, then shear strain

$$\phi = \frac{x}{l} \tag{1.4}$$

ϕ is the angular distortion in radians since $\dfrac{x}{l}$ is very small.

Due to the shear forces, P, a clockwise couple $Pl = \tau a l$ is applied to the material. If the material is to remain in external equilibrium, an equal and opposite couple must be applied by shear stresses τ' induced on perpendicular faces, Fig. 1.2(b). Thus, for equilibrium,

$$\tau a l = \tau' a' l'$$

But
$$a l = a' l'$$

$$\therefore \tau = \tau'$$

This induced stress is called the *complementary shear stress*.

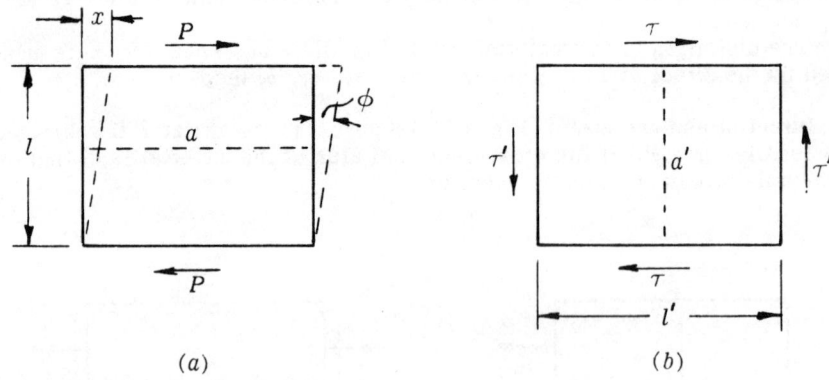

Fig. 1.2

1.4 Elasticity, plasticity and Hooke's law If a load is removed from a material and it returns to its former shape, it is said to be *elastic* but if it remains deformed, it is said to be *plastic*. Many engineering materials are elastic up to a certain stress, termed the *elastic limit*, after which they are partly elastic and partly plastic.

Elastic materials obey Hooke's Law, which states that the deformation of a material is directly proportional to the load, i.e., the ratio stress/strain is a constant.

For direct stress, the constant of proportionality is called the *Modulus of Elasticity* (or *Young's Modulus*), E.

Thus
$$E = \frac{\sigma}{\epsilon} = \frac{P/a}{x/l} = \frac{Pl}{ax} \tag{1.5}$$

For shear stress, the constant of proportionality is known as the *Modulus of Rigidity*, G.

Thus
$$G = \frac{\tau}{\phi} = \frac{P/a}{x/l} = \frac{Pl}{ax} \tag{1.6}$$

1.5 Factor of safety The stress at which a material will break is called the ultimate stress but the maximum design stress is considerably less than this to allow for overloading, non-uniformity of stress distribution, faulty workmanship and material, etc.

The ratio $\dfrac{\text{breaking stress}}{\text{design stress}}$ is called the *factor of safety* and may range from about 3 to 10, depending upon the material and nature of the loading.

SIMPLE STRESS AND STRAIN

1.6 Stresses in compound bars A compound bar is one which is made up of two different materials rigidly fixed together so that there is no relative movement between the ends.

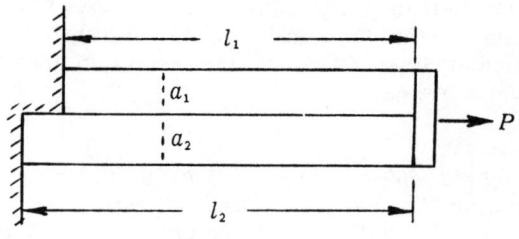

Fig. 1.3

If the bar shown in Fig. 1.3 is subjected to a force P, the extension of each part will be the same,

i.e. $\quad x_1 = x_2$

or $$\frac{\sigma_1 l_1}{E_1} = \frac{\sigma_2 l_2}{E_2} \qquad (1.7)$$

Also the load will be carried by the two parts such that

$$P = P_1 + P_2$$
$$= \sigma_1 a_1 + \sigma_2 a_2 \qquad (1.8)$$

σ_1 and σ_2 may be determined from equations (1.7) and (1.8).

If a compound bar undergoes a temperature change, differences in the coefficients of expansion, α, of the two materials coupled with the constraint of no relative movement between the ends will lead to stresses.

Let the bar shown in Fig. 1.4(a) be subjected to a temperature rise T. Then material (1) should extend an amount $l_1 a_1 T$ and material (2) by an amount $l_2 a_2 T$. Since they are constrained to extend the same amount, Fig. 1.4(b), material (1) is extended a distance x_1, inducing stress σ_1, and material (2) is compressed a distance x_2, inducing stress σ_2.

Hence $\quad l_1 a_1 T + x_1 = l_2 a_2 T - x_2$

or $$(l_2 a_2 - l_1 a_1)T = x_1 + x_2 = \frac{\sigma_1 l_1}{E_1} + \frac{\sigma_2 l_2}{E_2} \qquad (1.9)$$

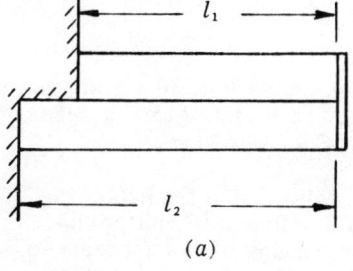
(a)

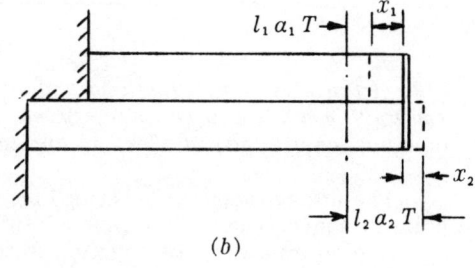
(b)

Fig. 1.4

If there is no external force on the bar, the internal forces due to the stresses must balance,

i.e.,
$$\sigma_1 a_1 = \sigma_2 a_2 \qquad (1.10)$$

σ_1 and σ_2 may be determined from equations (1.9) and (1.10).

If there is an external force and a change in temperature, the effects of each can be calculated separately and the results are then combined to obtain the resultant stresses.

1. *A mild steel rod, 600 mm long, is 25 mm diameter for 150 mm of its length and 50 mm for the rest of its length. It carries an axial tensile pull of 18 kN. With the axial pull applied, the ends of the rod are secured by rigid fixings. Find the temperature through which the rod must be raised to reduce the axial pull by two-thirds.* $a_{\text{steel}} = 11 \times 10^{-6}/\deg C$; $E_{\text{steel}} = 200$ GN/m^2.

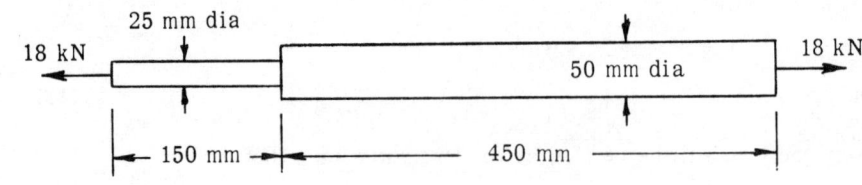

Fig. 1.5

$$x = \frac{Pl}{aE} \quad \ldots \ldots \ldots \text{from equation (1.5)}$$

$$\therefore \text{total extension} = \frac{18 \times 10^3 \times 0.15}{\frac{\pi}{4} \times 0.025^2 \times 200 \times 10^9} + \frac{18 \times 10^3 \times 0.45}{\frac{\pi}{4} \times 0.05^2 \times 200 \times 10^9}$$

$$= 48 \cdot 1 \times 10^{-6} \text{ m}$$

To reduce the axial pull by two-thirds, the natural extension under a temperature rise T must equal two-thirds of the extension due to the load,

i.e. $\qquad laT = \tfrac{2}{3} x$

i.e. $\qquad 0 \cdot 6 \times 11 \times 10^{-6} T = \tfrac{2}{3} \times 48 \cdot 1 \times 10^{-6}$

$$\therefore T = \underline{4 \cdot 86 \text{ deg C}}$$

2. *A flat steel bar, 10 m long and 10 mm thick tapers from 60 mm at one end to 20 mm at the other. Determine the change in length of the bar when an axial tensile load of 12 kN is applied to it.* $E = 200 \ GN/m^2$.

The arrangement is shown in Fig. 1.6 and, from similar triangles, the point of convergence of the tapered sides is at 5 m from the smaller end.

In problems where the section of the bar is varying, it is necessary to consider the extension of a small element of the bar and then obtain the overall extension by integration.

SIMPLE STRESS AND STRAIN 5

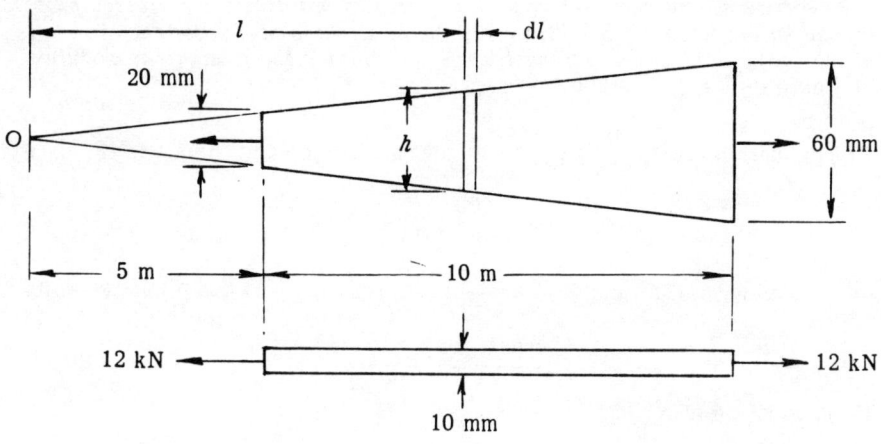

Fig. 1.6

From equation (1.5), $x = \dfrac{Pl}{aE}$.

Thus the extension of an element of the bar of length dl is given by

$$dx = \frac{P \, dl}{aE}$$

Taking the origin for l at the point O,

$$h = \frac{l}{15} \times 0.06 = 0.004l$$

$$\therefore dx = \frac{12 \times 10^3 \, dl}{0.01 \times 0.004l \times 200 \times 10^9}$$

$$= 0.0015 \frac{dl}{l}$$

$$\therefore x = 0.0015 \int_5^{15} \frac{dl}{l}$$

$$= 0.0015 \Big[\log_e l\Big]_5^{15}$$

$$= 0.0015 \log_e 3$$

$$= 0.001\,648 \text{ m} \quad \text{or} \quad \underline{1.648 \text{ mm}}$$

If, as an approximation, the bar is treated as a uniform bar of width 40 mm,

$$x = \frac{12 \times 10^3 \times 10}{0.01 \times 0.04 \times 200 \times 10^9}$$

$$= 0.0015 \text{ m} \quad \text{or} \quad \underline{1.5 \text{ mm}}$$

3. *A wire strand consists of a steel wire 2·7 mm diameter, covered by six bronze wires, each 2·5 mm diameter. If the working stress for the bronze is 60 MN/m², calculate the strength of the strand and the equivalent modulus of elasticity.* $E_{steel} = 200$ *GN/m²;* $E_{bronze} = 85$ *GN/m².*

From equation (1.8), $\dfrac{\sigma_s}{E_s} = \dfrac{\sigma_b}{E_b}$ since the length of each part is the same

$$\therefore \sigma_s = 60 \times \frac{200}{85} = 141 \cdot 2 \text{ MN/m}^2$$

$$\therefore \text{ strength of strand } = \frac{\pi}{4} \times \left(\frac{2 \cdot 7}{10^3}\right)^2 \times 141 \cdot 2 \times 10^6 + 6 \times \frac{\pi}{4} \times \left(\frac{2 \cdot 5}{10^3}\right)^2 \times 60 \times 10^6$$

$$= \underline{2\,575 \text{ N}}$$

From equation (1.5), $E = \dfrac{\sigma}{\epsilon} = \dfrac{P}{a\epsilon}$

For the equivalent modulus of the composite strand, the strain will be that of the steel alone or bronze alone, since both parts stretch by the same amount. Thus, using the given stress in the bronze,

$$E = \frac{2\,575}{\dfrac{\pi}{4} \times \left(\dfrac{2 \cdot 7}{10^3}\right)^2 + 6 \times \dfrac{\pi}{4} \times \left(\dfrac{2 \cdot 5}{10^3}\right)^2} \times \frac{85 \times 10^9}{60 \times 10^6}$$

$$= \underline{104 \times 10^9 \text{ N/m}^2}$$

4. *A copper tube of mean diameter 120 mm and 6·5 mm thick has its open ends sealed by two rigid plates connected by two steel bolts of 25 mm diameter, initially tensioned to 20 kN at a temperature of 30°C, thus forming a pressure vessel. Determine the stresses in the copper and steel at 0°C.*

$E_{steel} = 200$ *GN/m²;* $\alpha_{steel} = 11 \times 10^{-6}/\text{deg C}$

$E_{copper} = 100$ *GN/m²;* $\alpha_{copper} = 18 \times 10^{-6}/\text{deg C}$.

At 30°C, stress in steel, $\sigma_s = \dfrac{20 \times 10^3}{\dfrac{\pi}{4} \times \left(\dfrac{25}{10^3}\right)^2} = 40 \cdot 7 \times 10^6 \text{ N/m}^2$ (tensile)

and stress in copper, $\sigma_c = \dfrac{2 \times 20 \times 10^3}{\pi \times \dfrac{120}{10^3} \times \dfrac{6 \cdot 5}{10^3}} = 16 \cdot 32 \times 10^6 \text{ N/m}^2$ (comp)

When the temperature is reduced by 30°C,

$$\frac{\sigma_s}{E_s} + \frac{\sigma_c}{E_c} = (\alpha_c - \alpha_s)T \quad . \quad . \quad \text{from equation (1.9)}$$

i.e. $\dfrac{\sigma_s}{200 \times 10^9} + \dfrac{\sigma_c}{100 \times 10^9} = (18 - 11) \times 10^{-6} \times 30$

or $\sigma_s + 2\sigma_c = 4 \cdot 2 \times 10^6$ \hfill (1)

Also $\sigma_s a_s = \sigma_c a_c$ . . . . . from equation (1.10)

i.e. $\sigma_s \times 2 \times \dfrac{\pi}{4} \times \left(\dfrac{25}{10^3}\right)^2 = \sigma_c \times \pi \times \dfrac{120}{10^3} \times \dfrac{6 \cdot 5}{10^3}$

SIMPLE STRESS AND STRAIN

or $\qquad \sigma_s = 2\cdot 5 \sigma_c \qquad (2)$

Therefore, from equations (1) and (2),

$$\sigma_s = 23\cdot 31 \times 10^6 \text{ N/m}^2$$

and $\qquad \sigma_c = 9\cdot 33 \times 10^6 \text{ N/m}^2$

Since the coefficient of expansion of the copper is greater than that of the steel, a temperature rise will lead to a compressive stress in the copper and a tensile stress in the steel, and vice versa for a fall in temperature.

Thus resultant stress in steel $= 40\cdot 7 \times 10^6 - 23\cdot 31 \times 10^6$

$\qquad\qquad\qquad\qquad\qquad\quad = \underline{17\cdot 39 \times 10^6 \text{ N/m}^2}$ (tensile)

Resultant stress in copper $= 16\cdot 32 \times 10^6 - 9\cdot 33 \times 10^6$

$\qquad\qquad\qquad\qquad\qquad\quad = \underline{6\cdot 99 \times 10^6 \text{ N/m}^2}$ (comp)

5. A brass rod 6 mm diameter and 1 m long is joined at one end to a steel rod 6 mm diameter and 1·3 m long. The compound rod is placed in a vertical position with the steel rod at the top and connected at its ends to rigid fixings so that it carries a tensile load of 3·5 kN. A vertically downward load of 1·3 kN is then applied at the junction of the two metals. Calculate the stresses in the steel and brass.
$E_s = 200 \text{ GN/m}^2$; $E_b = 85 \text{ GN/m}^2$. $\qquad (Ans.: 153\cdot 5 \text{ MN/m}^2; 107\cdot 5 \text{ MN/m}^2)$

6. A steel bar 40 mm diameter and 4 m long is heated through 60 deg C, after which its ends are firmly fixed. After cooling to normal temperature again, the length of the bar is 1·2 mm less than when at its highest temperature. Determine the force and stress in the cold bar.
$E = 200 \text{ GN/m}^2$ and $\alpha = 11 \times 10^{-6}/\text{deg C}$. $\qquad (Ans.: 90\cdot 5 \text{ kN}; 72 \text{ MN/m}^2)$

7. A reinforced concrete column is 3 m high and of uniform cross-section 380 mm square. It is reinforced by four 25 mm diameter steel rods symmetrically placed. If the column carries an axial load of 600 kN, determine the stresses in the steel and concrete, the shortening of the column and the energy stored in the column.
$E_s = 200 \text{ GN/m}^2$; $E_c = 15 \text{ GN/m}^2$.
$\qquad\qquad\qquad (Ans.: 47\cdot 5 \text{ MN/m}^2; 3\cdot 56 \text{ MN/m}^2; 0\cdot 713 \text{ mm}; 214 \text{ J})$

8. A steel tube, 24 mm external diameter and 18 mm internal diameter, encloses a copper rod 16 mm diameter to which it is rigidly joined at each end. If there is no stress in the compound bar at $15°\text{C}$, determine the stresses in the rod and tube when the temperature is raised to $200°\text{C}$.
$E_s = 200 \text{ GN/m}^2$; $E_c = 90 \text{ GN/m}^2$; $\alpha_s = 11 \times 10^{-6}/\text{deg C}$; $\alpha_c = 18 \times 10^{-6}/\text{deg C}$
$\qquad\qquad\qquad\qquad\qquad\qquad (Ans.: 81\cdot 3 \text{ MN/m}^2; 80\cdot 0 \text{ MN/m}^2)$

9. A steel bar, 28 mm diameter and 400 mm long, is placed concentrically within a brass tube 40 mm outside diameter and 30 mm inside diameter. The length of the tube exceeds that of the bar by 0·12 mm. Rigid plates are placed on the ends of the tube through which an axial compressive force is applied to the compound bar. Determine the compressive stresses in the bar and tube due to a force of 60 kN.
$E_s = 200 \text{ GN/m}^2$; $E_b = 100 \text{ GN/m}^2$. $\qquad (Ans.: 48\cdot 2 \text{ MN/m}^2; 54\cdot 1 \text{ MN/m}^2)$

10. A steel rod of 320 mm² cross-sectional area and a coaxial copper tube of 800 mm² cross-sectional area are rigidly fixed together at their ends. An axial compressive load of 40 kN is applied to the composite bar and the temperature is then raised by 100 deg C. Determine the stresses in the steel and copper.
$E_s = 200 \text{ GN/m}^2$; $E_c = 100 \text{ GN/m}^2$; $\alpha_s = 12 \times 10^{-6}/\text{deg C}$; $\alpha_c = 16 \times 10^{-6}/\text{deg C}$
$\qquad\qquad\qquad\qquad\qquad\qquad (Ans.: 11\cdot 11 \text{ MN/m}^2; 45\cdot 55 \text{ MN/m}^2)$

2 Shearing force and bending moment

2.1 Introduction A structural member subject to transverse external forces is called a *beam* and these external forces cause shearing forces and bending moments at each cross-section of the beam. The external forces considered may be due to concentrated or distributed loads, beam mass and supporting reactions.

The *shearing force* at any cross-section of a beam is the algebraic sum of all the external forces to one side of the section. Referring to Fig. 2.1(a), the shearing force to the left of point C is $W - R_1$ downwards and to the right of point C, it is R_2 upwards. These are of the same magnitude, since $W = R_1 + R_2$.

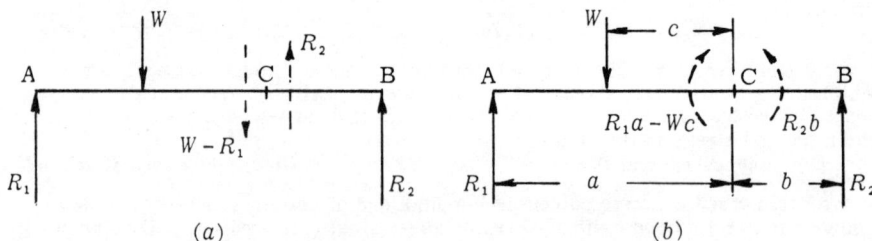

Fig. 2.1

The *bending moment* at any cross-section of a beam is the algebraic sum of the moments of all the external forces to one side of the section. Referring to Fig. 2.1(b), the bending moment to the left of point C is $R_1 a - Wc$ clockwise and to the right of point C, it is $R_2 b$ anticlockwise. These are the same magnitude since, taking moments about A,

$$R_2(a + b) = W(a - c)$$

or $$R_2 b = Wa - R_2 a - Wc = R_1 a - Wc$$

2.2 Sign convention If the cartesian coordinate system is chosen for beam deflection, Chapter 5, i.e., deflection positive upwards, then the sign convention for shearing force and bending moment is as follows:

(a) If the shearing force to the left of a section is upwards (or to the right of a section is downwards), it is regarded as positive.

(b) If the bending moment to the left of a section is clockwise (or to the right of a section is anticlockwise), it is regarded as positive.

This convention is illustrated in Fig. 2.2.

SHEARING FORCE AND BENDING MOMENT

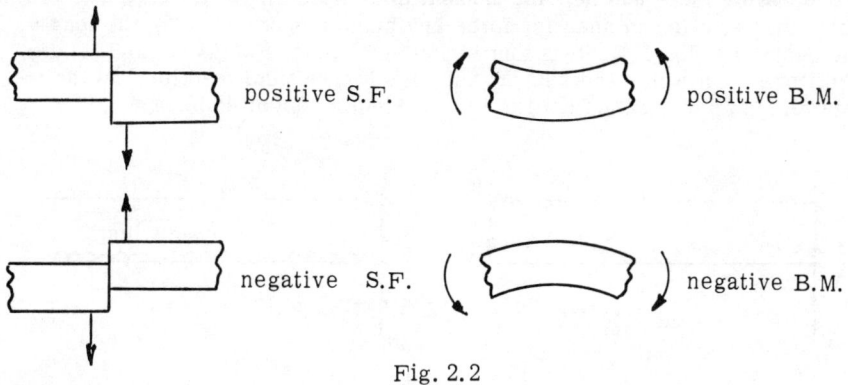

Fig. 2.2

2.3 Relation between load intensity, shearing force and bending moment

Consider a small length dx of beam, carrying a uniformly distributed load w per unit length, Fig. 2.3. Let the shearing force and bending moment at distance x from the origin be F and M and at distance $x + dx$, $F + dF$ and $M + dM$ respectively.

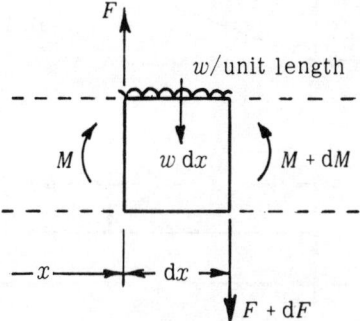

Fig. 2.3

For equilibrium of forces,

$$F = F + dF + w\,dx$$

$$\therefore w = -\frac{dF}{dx} \qquad (2.1)$$

For equilibrium of moments,

$$M + F\,dx + w\,dx \cdot \frac{dx}{2} = M + dM$$

$$\therefore F = \frac{dM}{dx}, \text{ neglecting second order terms} \qquad (2.2)$$

Equation (2.2) shows that a local maximum or minimum bending moment occurs when the shearing force is zero.

Combining equations (2.1) and (2.2) gives

$$w = -\frac{dF}{dx} = -\frac{d^2M}{dx^2} \qquad (2.3)$$

Bending moment may therefore be determined by integrating the shearing force and shearing force by integrating the loading. This may be performed mathematically where a suitable loading relation exists but otherwise it may be done graphically.

2.4 Shearing force and bending moment diagrams These are diagrams which show the variation in shearing force and bending moment along the length of the beam and Fig. 2.4 shows four simple examples. Cases (a) and (b) are cantilevers, in which the support supplies the vertical reaction and the restraining moment; cases (c) and (d) are simply supported beams.

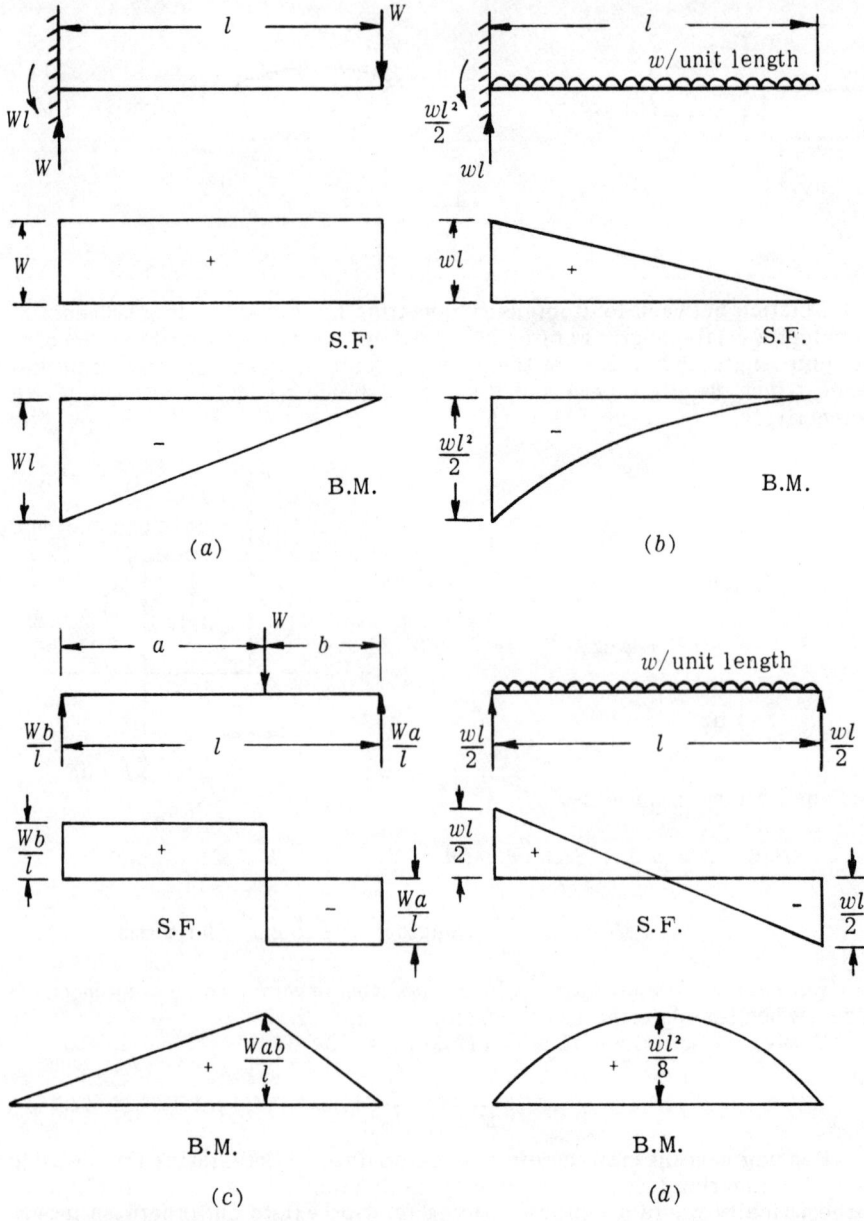

Fig. 2.4

SHEARING FORCE AND BENDING MOMENT

Concentrated loads are assumed to act at a point but in practice, they must be distributed over a short length of the beam.

In cases where the loading is arbitrary, the loading diagram may be divided into convenient strips (not necessarily of uniform width) and the load represented by each strip is replaced by a concentrated load of the same magnitude placed at the centroid of the strip. Approximate shearing force and bending moment diagrams are then drawn for the concentrated loads, Fig. 2.5, and finally, smooth curves are drawn through the points of intersection of the strip boundaries with these approximate diagrams.

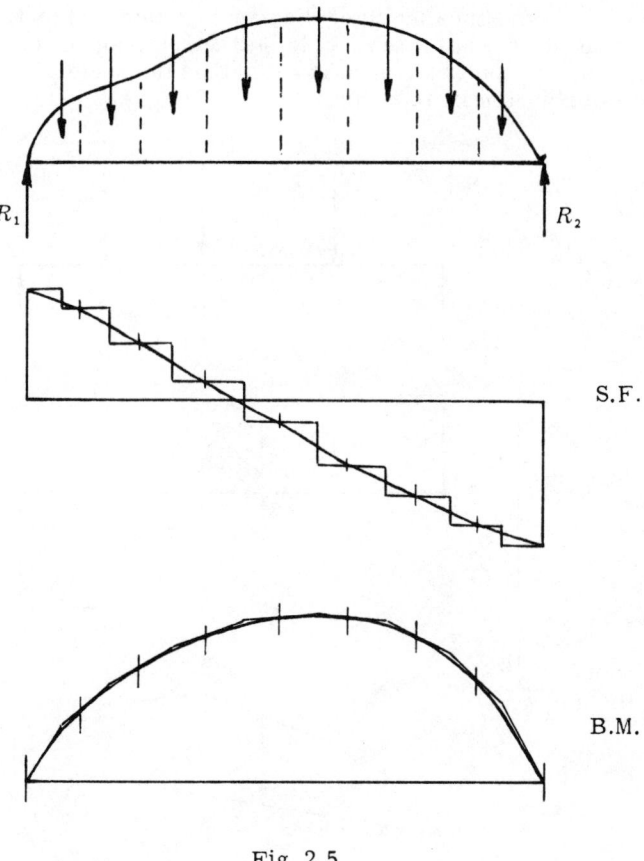

Fig. 2.5

2.5 Graphical construction methods Graphical integration may be performed by area measurement (counting squares) and plotting; to obtain the bending moment diagram from the loading diagram will require two stages and this may become tedious, so the funicular polygon method, Fig. 2.6, is often used.

SHEARING FORCE AND BENDING MOMENT

Distributed loads are first replaced by equivalent concentrated loads and the spaces on the load diagram are lettered, using Bow's notation. The load line *abcd* is drawn such that $ab = W_1$, $bc = W_2$, $cd = W_3$ and points *a*, *b*, *c* and *d* are projected across horizontally to form part of the shearing force diagram.

A pole O is chosen and is joined to *a*, *b*, *c* and *d*; lines *pq*, *qr*, *rs* and *st* are drawn parallel with O*a*, O*b*, O*c* and O*d* in spaces A, B, C and D respectively and the funicular polygon is closed with line *tp*. This then represents the bending moment diagram and the *vertical* ordinate shows the bending moment at any point.

The line O*e* is drawn on the polar diagram parallel with *pt* and the point *e* is then projected across horizontally to complete the shearing force diagram, whence *ea* and *ed* represent R_1 and R_2 respectively.

The scale of the shearing force diagram is determined by the choice of scale of the load line, say 1 mm = k_1 N. If the scale for the beam length is 1 mm = k_2 m and the polar distance is *l* mm, then the scale of the bending moment diagram is 1 mm = lk_1k_2 Nm.

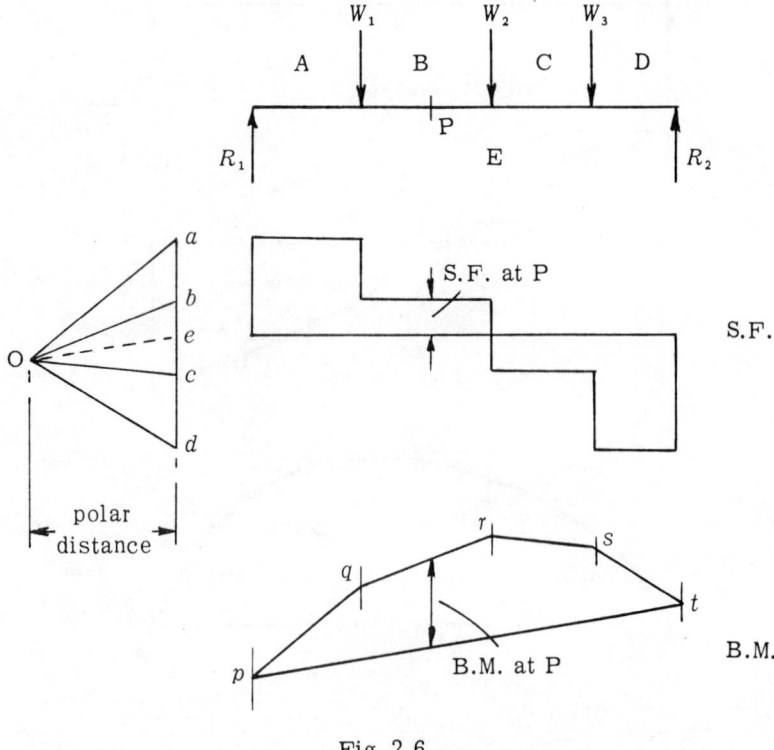

Fig. 2.6

1. *A beam 8 m long is freely supported over a span of 6 m and overhangs the right-hand support 2 m. It carries a uniformly distributed load of 20 kN/m together with loads of 160 kN and 130 kN concentrated at 3 m and 8 m from the left-hand support.*

Sketch the shearing force and bending moment diagrams, inserting principal numerical values.

Taking moments about B, Fig. 2.7(a):—

$$R_a \times 6 = 160 \times 3 - 130 \times 2 + 20 \times 8 \times 2$$
$$= 540$$
$$\therefore R_a = 90 \text{ kN}$$
$$R_b = 160 + 130 + 20 \times 8 - 90$$
$$= 360 \text{ kN}$$

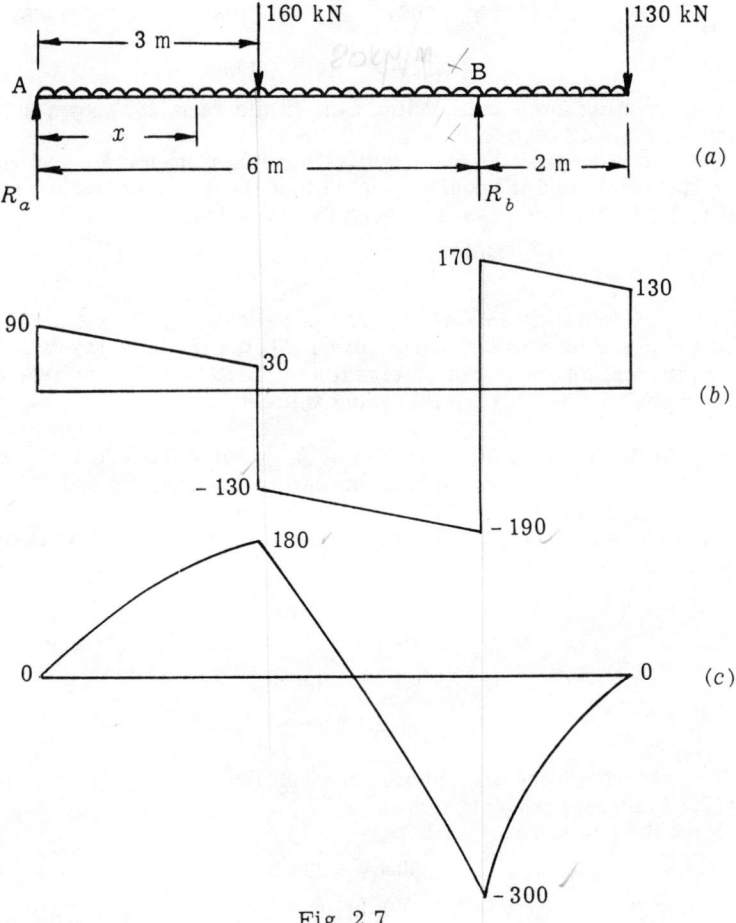

Fig. 2.7

Considering a section at a distance x from A, the shearing forces and bending moments at various points along the beam are as follows:

$\underline{0 < x < 3}$

$$F = 90 - 20x$$
$$M = 90x - 20x \cdot \frac{x}{2}$$
$$= 90x - 10x^2$$

$3 < x < 6$

$$F = 90 - 20x - 160$$
$$= -70 - 20x$$
$$M = 90x - 20x \cdot \frac{x}{2} - 160(x - 3)$$
$$= -70x - 10x^2 + 480$$

$6 < x < 8$

$$F = 90 - 20x - 160 + 360$$
$$= 290 - 20x$$
$$M = 90x - 20x \cdot \frac{x}{2} - 160(x - 3) + 360(x - 6)$$
$$= 290x - 10x^2 - 1680$$

The shearing force and bending moment diagrams are shown in Figs. 2.7 (b) and (c) respectively.

The maximum positive and negative bending moments are 180 and 300 kNm respectively and it should be noted that these values occur where the shearing force diagram passes through the zero line.

2. *Fig. 2.8 shows a beam ABC which is built in at A, has a hinge at B and is supported on a roller bearing at C. Sketch the shearing force and bending moment diagrams and determine the position and magnitude of the maximum positive and negative bending moments.*

Due to the hinge at B, there can be no bending moment at that point and the beam may therefore be split up into two beams, AB and BC, as shown in Fig. 2.9(a).

From the equilibrium of BC, the reaction R_b is obtained by taking moments about C,

i.e.
$$R_b \times 4 = 60 \times 2 \times 1$$
$$\therefore R_b = 30 \text{ kN}$$

and
$$R_c = 60 \times 2 - 30$$
$$= 90 \text{ kN}$$

The shearing force and bending moment diagrams are then as shown in Figs. 2.8(b) and (c) respectively.

From the equilibrium of AB,
$$\text{B.M. at A} = 30 \times 4 + 20 \times 2$$
$$= 160 \text{ kNm}$$

and this is the maximum negative bending moment on the beam.

On span BC, the shearing force will be zero at a point 1·5 m from C. Therefore, maximum bending moment between B and C

$$= 90 \times 1·5 - 60 \times 1·5 \times \frac{1·5}{2}$$
$$= 67·5 \text{ kNm}$$

and this is the maximum positive bending moment on the beam.

SHEARING FORCE AND BENDING MOMENT

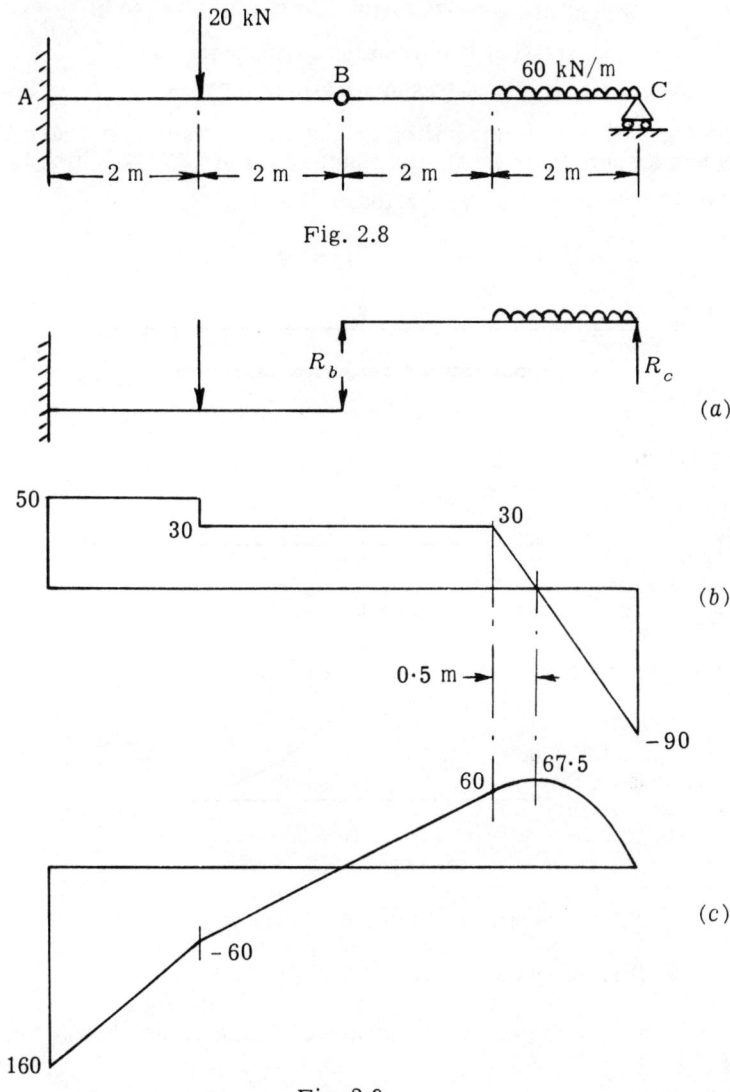

Fig. 2.8

Fig. 2.9

3. *A log of wood, 5 m long and of square cross-section 0·4 m by 0·4 m, floats in a horizontal position in fresh water. It is then loaded at the centre with a weight just sufficient to immerse it completely. Draw the shearing force and bending moment diagrams, stating maximum values. Take the specific gravity of wood as 0·8.*

Volume of log = $5 \times 0·4 \times 0·4$ = $0·8$ m³

When completely submerged,

upthrust of water = $0·8 \times 10^3 \times 9·81$ = 7850 N

since 1 m³ of water has a mass of 1 tonne or 10^3 kg.

Weight of log = $0.8 \times 10^3 \times 9.81 \times 0.8$ = 6 280 N

Therefore concentrated load required to submerge log

= 7 850 - 6 280 = 1 570 N

The log is therefore in equilibrium under a central concentrated load of 1 570 N and an upward uniformly distributed load of 1 570 N, which is equivalent to $\frac{1570}{5}$ = 314 N/m, Fig. 2.10(a).

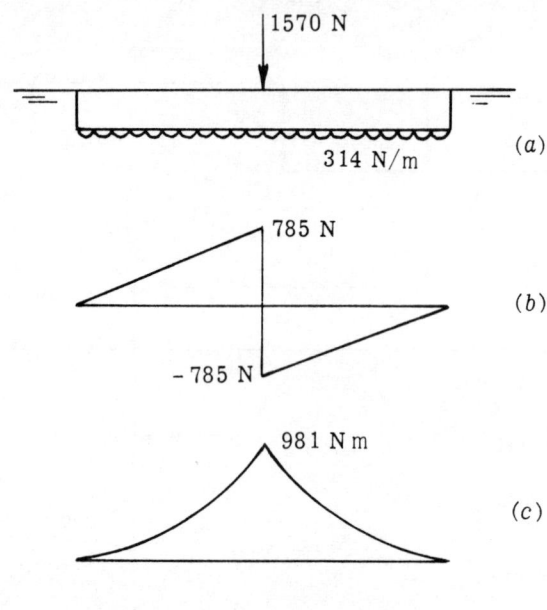

Fig. 2.10

S.F. at centre = 314×2.5 = 785 N

B.M. at centre = $314 \times 2.5 \times \frac{2.5}{2}$ = 981 Nm.

The shearing force and bending moment diagrams are shown in Figs. 2.10 (b) and (c) respectively.

4. *A beam 8 m long is supported at the ends and carries a distributed load which varies uniformly in intensity from zero at one end to 30 kN/m at a section 2 m from the other end and over the remaining length is constant at 30 kN/m.*

Derive equations for shearing force and bending moment at any section of the beam and sketch the shearing force and bending moment diagrams.

Method 1 For the purposes of calculating the reactions, the distributed load may be replaced by two concentrated loads acting at the centres of gravity of the respective parts. Thus the triangular load is equivalent to a load of ½ × 30 × 6 = 90 kN acting at a point 4 m from B and the rectangular load is equivalent to a load of 30 × 2 = 60 kN acting at a point 1 m from B, Fig. 2.11(a).

SHEARING FORCE AND BENDING MOMENT

Taking moments about B:—

$$R_a \times 8 = 90 \times 4 + 60 \times 1 = 420$$
$$\therefore R_a = 52\cdot 5 \text{ kN}$$
$$R_b = 90 + 60 - 52\cdot 5 = 97\cdot 5 \text{ kN}$$

At a section distant x from A, intensity of loading $= \frac{x}{6} \times 30 = 5x$ kN/m. The load to the left of this section is therefore $\frac{1}{2} \times 5x \times x = \frac{5x^2}{2}$, which may be considered to act at $\frac{x}{3}$ from the section. Thus, for $0 < x < 6$,

$$F = 52\cdot 5 - \frac{5x^2}{2}$$

and
$$M = 52\cdot 5 x - \frac{5x^2}{2} \cdot \frac{x}{3} = 52\cdot 5 x - \frac{5x^3}{6}$$

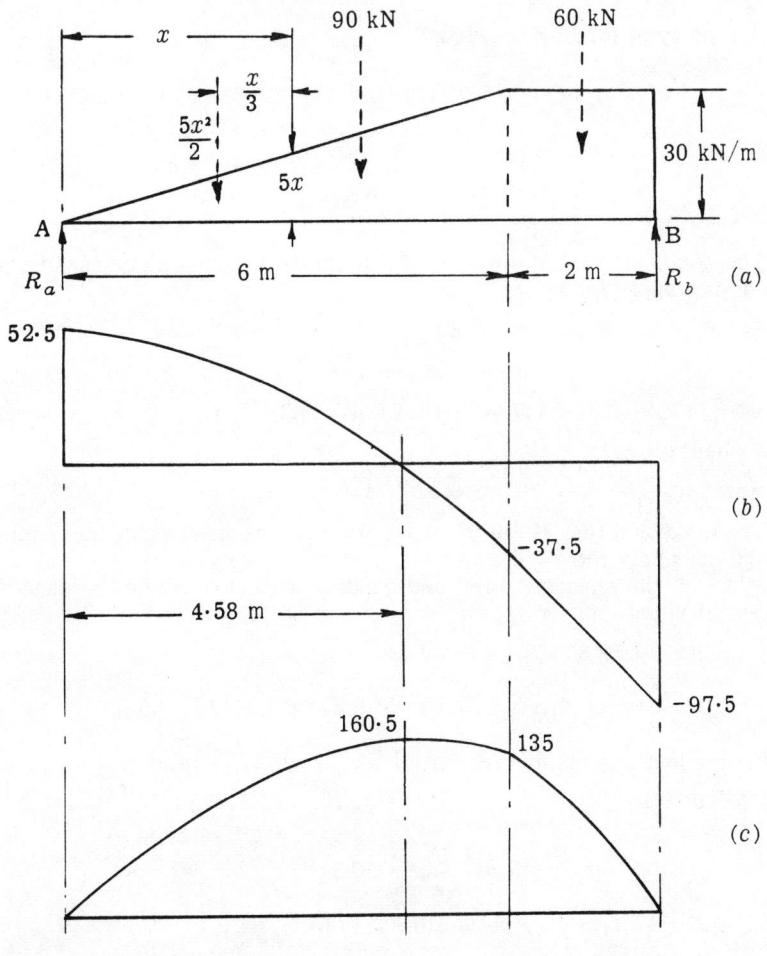

Fig. 2.11

For $6 < x < 8$, the triangular load is equivalent to 90 kN at 4 m from A and the rectangular load is equivalent to a concentrated load of $30(x-6)$ kN acting at a point $\frac{x-6}{2}$ m from the section.

Thus $\quad F = 52 \cdot 5 - 90 - 30(x-6) = 142 \cdot 5 - 30x$

and $\quad M = 52 \cdot 5x - 90(x-4) - 30(x-6) \cdot \frac{x-6}{2}$

$\quad\quad\quad = -180 + 142 \cdot 5x - 15x^2$

The shearing force and bending moment diagrams are shown in Figs. 2.11(b) and (c).

Method 2 For $0 < x < 6$,

intensity of loading $w = 5x$

$$\therefore F = -\int w\,dx = -\frac{5}{2}x^2 + A_1$$

and $\quad\quad M = \int F\,dx = -\frac{5}{6}x^3 + A_1 x + B_1$

$M = 0$ when $x = 0$, $\quad \therefore B_1 = 0$

The condition that $M = 0$ when $x = 8$ cannot be used since these equations do not apply for $x > 6$.

For $6 < x < 8$, $\quad w = 30$

$\quad\quad\quad \therefore F = -30x + A_2$

and $\quad\quad M = -15x^2 + A_2 x + B_2$

$M = 0$ when $x = 8$,

from which $\quad\quad B_2 = 8(120 - A_2)$

The condition that $M = 0$ when $x = 0$ cannot be used since these equations do not apply for $x < 6$.

At $x = 6$, the shearing force and bending moment must be the same from each set of equations.

Thus $\quad\quad -\frac{5}{2} \times 6^2 + A_1 = -30 \times 6 + A_2$ $\quad\quad\quad\quad$ (1)

and $\quad -\frac{5}{6} \times 6^3 + A_1 \times 6 = -15 \times 6^2 + A_2 \times 6 + 8(120 - A_2)$ $\quad\quad$ (2)

From equations (1) and (2), $A_1 = 52 \cdot 5$ and $A_2 = 142 \cdot 5$

Then, for $0 < x < 6$, $\quad F = -\frac{5}{2}x^2 + 52 \cdot 5$

and $\quad\quad\quad M = -\frac{5}{6}x^3 + 52 \cdot 5x$

and for $6 < x < 8$, $\quad F = -30x + 142 \cdot 5$

and $\quad\quad\quad M = -15x^2 + 142 \cdot 5x - 180 \quad$, as before.

SHEARING FORCE AND BENDING MOMENT

5. A beam 6 m long is freely supported at each end and carries a point load of 70 kN at a distance of 1·5 m from the left-hand end, together with a uniformly distributed load of 15 kN/m from the centre of the span to the right-hand end. Determine the position and magnitude of the maximum B.M.

(*Ans*.: 95·6 kNm at 1·5 m from L.H. end)

6. A beam 6 m long overhangs its two supports symmetrically and is loaded with three equal loads, one at each end and one at mid-span. Determine the distance between the supports in order that the B.M. at mid-span shall be equal to that at each support.

Draw the S.F. and B.M. diagrams and find the points of contraflexure.

(*Ans*.: 4·8 m; 1·8 m from each end)

7. A beam 9 m long is freely supported over a span of 7 m and overhangs the right-hand support 2 m. It carries a uniformly distributed load of 25 kN/m together with loads of 150 kN, 70 kN and 120 kN concentrated at 3 m, 5 m and 8 m respectively from the left-hand support.

Draw the S.F. and B.M. diagrams and calculate the value of the maximum bending moment. (*Ans*.: 395·5 kNm under 150 kN load)

8. A beam ABCD is simply supported at B and C, 6 m apart, and the overhanging parts AB and CD are 2 m and 4 m long respectively. The beam carries a uniformly distributed load of 60 kN/m between A and C and there is a concentrated load of 40 kN at D.

Draw the S.F. and B.M. diagrams. Calculate the position and magnitude of the maximum bending moment between B and C. (*Ans*.: 130 kNm at 2·89 m from B)

9. A horizontal beam AD, 10 m long, carries a uniformly distributed load of 200 N/m, together with a concentrated load of 500 N at the left-hand end A. The beam is supported at a point B, 1 m from A, and at C which is in the right-hand half of the beam and x m from the end D.

Determine the value of x if the mid-point of the beam is a point of contraflexure and for this arrangement, draw the B.M. diagram. Locate any other points of contraflexure. (*Ans*.: 3 m; B.M. at supports, 600 and 900 Nm; B.M. at 3·75 m from A, 156·25 Nm; 2·5 m from A)

10. Two slings are used to raise a concrete pile of length l and uniform cross-section, the pile remaining horizontal during the lift. Determine the most suitable positions for the slings if the B.M. due to its own weight is to be kept as small as possible. For this arrangement, draw the S.F. and B.M. diagrams.

(*Ans*.: 0·207 l from ends)

11. A beam 7 m long rests on two supports A and B, 4 m apart, with a 1 m length of beam overhanging the right-hand support B. The beam carries a uniformly distributed load w from A to the right-hand end and a uniformly distributed load nw on that part of the beam to the left of A. If there is a point of contraflexure 0·5 m to the left of B, find the value of n and the other point of contraflexure. Find also the maximum B.M. between the points of contraflexure. Sketch the S.F. and B.M. diagrams. (*Ans*.: 1·75; 2 m from A; $9w/32$ at 2·75 m from A)

12. Determine the position and magnitude of the maximum B.M. for a simply supported beam of length l carrying a distributed load which increases uniformly from zero at one end to w N/m at the other.

(*Ans*.: $l/\sqrt{3}$ m from end of zero load; $wl^2/9\sqrt{3}$ Nm)

13. A horizontal beam 4 m long is freely supported at its ends and carries a distributed load which increases uniformly from zero at each end to a maximum of 60 kN/r at the centre. Draw the S.F. and B.M. diagrams.

(*Ans*.: Max S.F., 60 kN at each end; max B.M., 80 kNm at centre)

14. A simply supported beam with a span of 4 m carries a distributed load which varies uniformly from 30 kN/m at one end to 90 kN/m at the other. Determine the position and magnitude of the maximum bending moment. Sketch the S.F. and B.M. diagrams. (*Ans*.: 121 kNm at 2·16 m from lightly loaded end)

3 Bending stresses

3.1 Second moment of area The second moment of an element of an area about an axis in the same plane is the product of the area and the square of its distance from that axis. Thus, in Fig. 3.1, the second moment of area of the element da from the axis OO is $da.l^2$ and the total second moment of the area about OO is therefore $\int da.l^2$. This quantity is denoted by I and whenever a beam is bent, the stresses and curvature are inversely proportional to I.

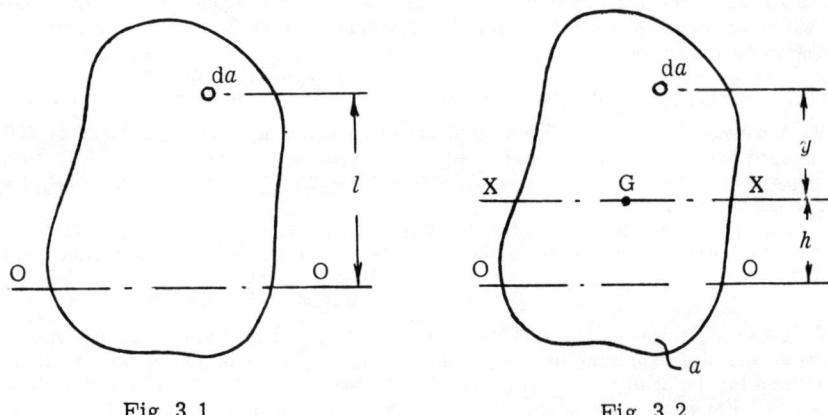

Fig. 3.1   Fig. 3.2

If XX is an axis parallel to OO and passing through the centroid G, Fig. 3.2, then

$$I_{OO} = \int da\,(y+h)^2 = \int da\,(y^2 + 2yh + h^2)$$
$$= \int da.y^2 + 2h\int da.y + h^2\int da$$

But $\int da.y^2 = I_{XX}$, $\int da.y$ is the first moment of the area about XX, which is zero since this axis passes through the centroid G, and $\int da$ is the total area a.

Thus $\qquad I_{OO} = I_{XX} + ah^2 \qquad\qquad (3.1)$

This is known as the *theorem of parallel axes*.

For triangular figures, it can be shown that for the determination of first and second moments of area, the area a of the triangle is equivalent to three areas, each $a/3$, considered concentrated at the mid-points of the three sides. Since these areas are imagined concentrated at points, they will have no second moment about their own axes.

BENDING STRESSES

Thus, for the triangle of area a shown in Fig. 3.3,

first moment of area about OO

$$= \frac{a}{3}.\alpha + \frac{a}{3}.\beta + \frac{a}{3}.\gamma$$

and
second moment of area about OO

$$= \frac{a}{3}\alpha^2 + \frac{a}{3}\beta^2 + \frac{a}{3}\gamma^2$$

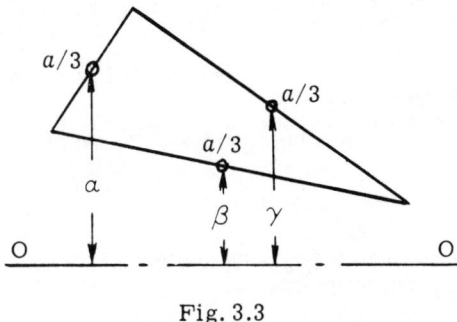

Fig. 3.3

This equivalent system, known as an *equimomental system*, may be extended to any area which can be divided into triangles (see Example 4).

3.2 Stresses due to pure bending If a straight, elastic, homogeneous beam is subjected to a pure bending moment (i.e., one which is not accompanied by shear or direct forces), the beam will bend, causing some layers of the beam to be extended and others to be compressed. The stresses resulting from these strains will produce an internal moment of resistance, equal and opposite to the applied bending moment.

If the radius of curvature of the beam is large in comparison with the depth of the beam, plane transverse sections will remain plane after bending and, assuming that the cross-section is symmetrical about the plane of bending, the strain will vary linearly from a maximum at the top face to another maximum at the bottom face, changing from tension to compression as it does so, Fig. 3.4.

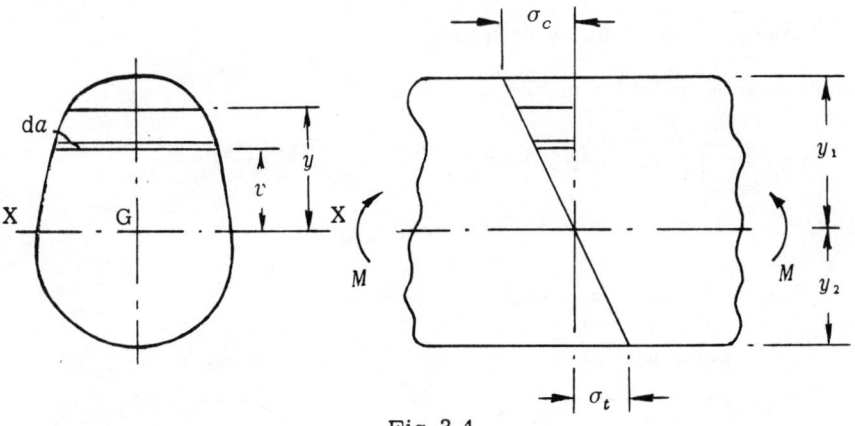

Fig. 3.4

Between the areas of tension and compression, there must be a layer, normal to the plane of bending which is unstrained; this layer is called the *neutral plane* and its intersection with the cross-section is termed the *neutral axis*; this is shown as **XX** in Fig. 3.4. The strain at any point, and by Hooke's Law the stress, is directly proportional to the distance from the neutral plane, so that, if the stress at distance y from **XX** is σ, the stress on an elementary strip of area da, distance v from **XX**, is $\frac{v}{y}\sigma$.

$$\therefore \text{ force on area } da = \frac{v}{y}\sigma . da$$

$$\therefore \text{ moment of force about XX} = \frac{v}{y}\sigma da . v$$

For the whole section, the moment of resistance,

$$M = \int \frac{\sigma}{y} v^2 \, da = \frac{\sigma}{y} . I$$

where I is the second moment of area of the cross-section of the beam about the neutral plane.

Thus
$$\frac{M}{I} = \frac{\sigma}{y} \qquad (3.2)$$

The maximum stresses occur at the top and bottom faces of the beam, so that

$$\text{maximum compressive stress, } \sigma_c = \frac{M}{I} . y_1$$

and

$$\text{maximum tensile stress, } \sigma_t = \frac{M}{I} . y_2$$

In general,
$$\sigma_{max} = \frac{M}{I} . y_{max}$$

$$= \frac{M}{I/y_{max}} = \frac{M}{Z} \qquad (3.3)$$

where Z is the *modulus of section* of the beam.

3.3 Values of I and Z for common cases

(a) Rectangle, breadth b and depth d, Fig. 3.5(a)

$$I_{XX} = \frac{bd^3}{12} \quad ; \quad I_{OO} = \frac{bd^3}{3} \quad ; \quad Z_{XX} = \frac{bd^2}{6}$$

(b) Circle, diameter d, Fig. 3.5(b)

$$I_{XX} = I_{YY} = \frac{\pi d^4}{64} \quad ; \quad Z_{XX} = Z_{YY} = \frac{\pi d^3}{32}$$

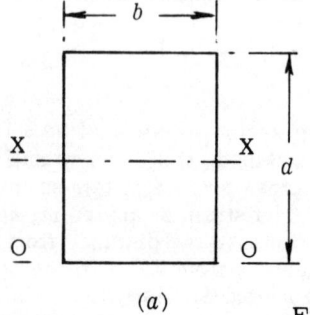
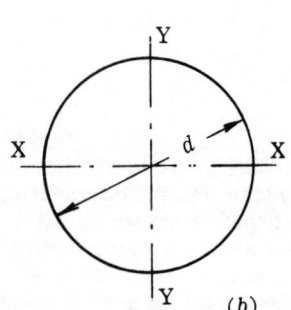

Fig. 3.5

BENDING STRESSES

3.4 Radius of curvature If the radius of curvature of the neutral axis is R, Fig. 3.6, the length of the neutral plane is $R\theta$. The extension of a layer distant y from the neutral plane is $(R+y)\theta - R\theta$ and the strain at this plane,

$$\epsilon = \frac{(R+y)\theta - R\theta}{R\theta} = \frac{y}{R}$$

Thus the stress

$$\sigma = E\epsilon = \frac{Ey}{R}$$

so that

$$\frac{\sigma}{y} = \frac{E}{R} \quad (3.4)$$

From equations (3.2) and (3.3)

$$\frac{M}{I} = \frac{\sigma}{y} = \frac{E}{R} \quad (3.5)$$

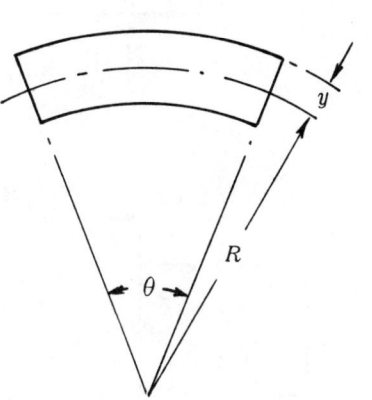

Fig. 3.6

3.5 Position of neutral axis Since there are no external longitudinal forces acting on the beam, the sum of the internal longitudinal forces must be zero. On the elementary area da, Fig. 3.4, the longitudinal force is $(\sigma/y)v\,da$ and the total longitudinal force on the section is $\int \frac{\sigma}{y} v\,da$. Thus $\int \frac{\sigma}{y} v\,da = 0$ or $\int v\,da = 0$ since $\frac{\sigma}{y}$ is a constant, but $\int v\,da$ is the first moment of area of the section about the neutral axis and this can only be zero if the neutral axis passes through the centroid of the section.

3.6 Combined bending and direct stresses Materials are frequently subject to both bending and direct stresses and the resultant stresses are obtained by the algebraic addition of the stresses due to the two causes considered separately.

If the material shown in Fig. 3.7(a) is subjected to a bending moment M and an axial load P, the stress along the edge AB,

$$\sigma_1 = \frac{P}{a} + \frac{My_1}{I}$$

and the stress along the edge CD,

$$\sigma_2 = \frac{P}{a} - \frac{My_2}{I}$$

Fig. 3.7(b) shows the stress diagrams due to the direct load and the bending moment and Fig. 3.7(c) shows the resultant stress diagrams; the stress σ_1 is tensile but the stress σ_2 may be tensile, zero or compressive, depending on whether the bending stress is less than, equal to or greater than the direct stress.

BENDING STRESSES

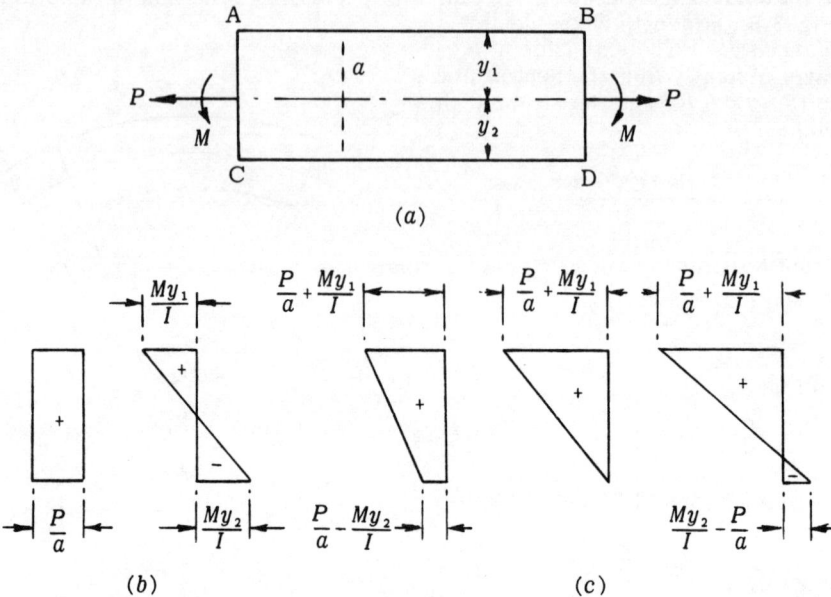

Fig. 3.7

A common example of combined bending and direct stresses occurs when a short member is subjected to an eccentric load. If the load is P and its eccentricity to the longitudinal axis is e, the bar is effectively subjected to an axial load P and a bending moment Pe, as shown in Fig. 3.8. The maximum stresses are then given by

$$\sigma = \frac{P}{a} \pm \frac{Pe}{Z}$$

If there is to be no change in the nature of the stress in the section

$$\frac{P}{a} \geqslant \frac{Pe}{Z}$$

or

$$e \leqslant \frac{Z}{a}$$

Fig. 3.8

BENDING STRESSES

For a rectangular section, $b \times d$, $Z = \dfrac{bd^2}{6}$, so that $e \leqslant \dfrac{d}{6}$, i.e., the load must be within the middle third of the bar. For a circular section, diameter d, $Z = \dfrac{\pi}{32} d^3$, so that $e \leqslant \dfrac{d}{8}$.

Examples arise in the case of structures built of brittle materials such as concrete, masonry, etc., where the compressive load at any section must not be sufficiently eccentric to the axis of that section to give rise to tensile stress.

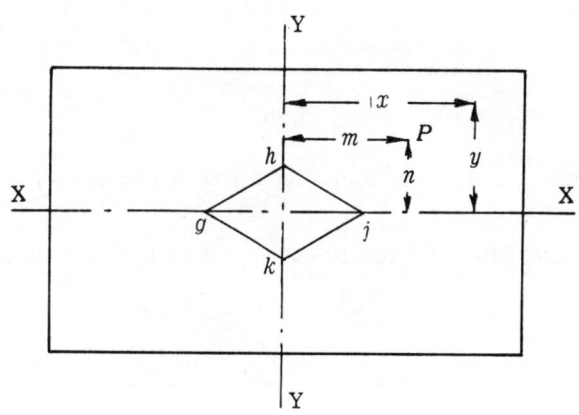

Fig. 3.9

If the load is eccentric to both axes of symmetry, it may be moved to the axis in two stages, each involving the addition of a moment equal to the force multiplied by the distance moved. Thus the load shown in Fig. 3.9 is equivalent to a load on the axis, together with moments Pm in plane XX and Pn in plane YY. The stress at any point, coordinates x and y is then given by

$$\sigma = \frac{P}{a} + \frac{Pm}{I_{YY}} \cdot x + \frac{Pn}{I_{XX}} \cdot y$$

If there is to be no change in the nature of the stress in the section, the load must be within the rhombus $ghjk$ formed by joining the middle third points, gj being one-third of the horizontal axis and hk one-third of the vertical axis.

1. *Fig. 3.10 shows a T-section beam which is used as a cantilever, 2 m long, which supports a concentrated load of 100 N at the free end.*

Determine the maximum stress in the beam and calculate also the width of flange required if the maximum stresses in compression and tension are to be in the ratio of 2 to 1.

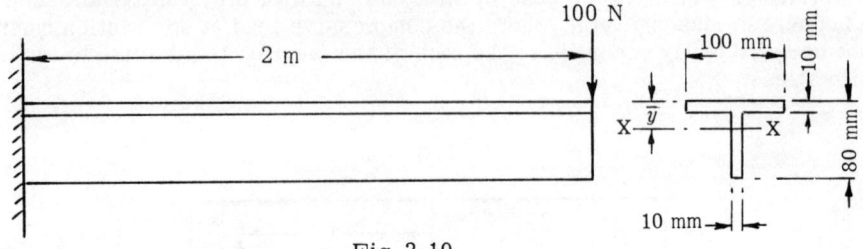

Fig. 3.10

Taking moments of area about the top of the flange to find the position of the neutral axis:

$$100 \times 10 \times 5 + 70 \times 10 \times 45 = (100 \times 10 + 70 \times 10)\bar{y}$$

$$\therefore \bar{y} = 21 \cdot 5 \text{ mm}$$

$$I_{XX} = \frac{100 \times 10^3}{12} + 100 \times 10 \times 16 \cdot 5^2 + \frac{10 \times 70^3}{12} + 10 \times 70 \times 23 \cdot 5^2$$

$$= 953\,600 \text{ mm}^4 = 0 \cdot 953\,6 \times 10^{-6} \text{ m}^4$$

The maximum B.M. occurs at the wall and the maximum stress will be at the bottom of the web, this being the farthest point from the neutral axis.

$$\frac{M}{I} = \frac{\sigma}{y} \quad \text{from equation (3.2)}$$

$$\therefore \sigma = \frac{100 \times 2}{0 \cdot 953\,6 \times 10^{-6}} \times 58 \cdot 5 \times 10^{-3}$$

$$= \underline{12 \cdot 26 \times 10^6 \text{ N/m}^2}$$

For the maximum compressive stress to be twice the maximum tensile stress, the neutral axis must be one-third of 80 mm from the upper surface, as shown in Fig. 3.11.

Taking moments about the top of the flange:

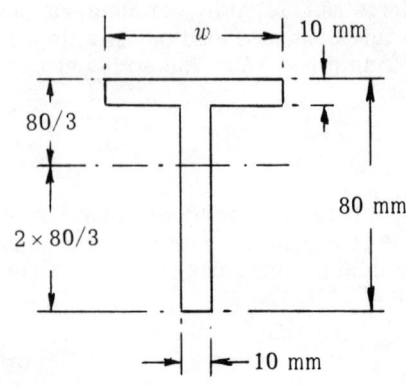

Fig. 3.11

$$w \times 10 \times 5 + 70 \times 10 \times 45 = (w \times 10 + 70 \times 10) \times \frac{80}{3}$$

BENDING STRESSES

2. *Fig. 3.12 shows the section of a beam. Find the maximum stress due to bending if a moment of 8 kN m is applied (a) in the plane YY and (b) in the plane XX.*

(a) The second moment of area about XX is that of a rectangle, 70 mm wide by 100 mm deep, less that of a circle of diameter 60 mm,

i.e. $I_{XX} = = \dfrac{70 \times 100^3}{12} - \dfrac{\pi}{64} \times 60^4$

$= 5 \cdot 197 \times 10^6 \text{ mm}^4$

$= 5 \cdot 197 \times 10^{-6} \text{ m}^4$

$\dfrac{M}{I} = \dfrac{\sigma}{y}$

$\therefore \sigma = \dfrac{8 \times 10^3}{5 \cdot 197 \times 10^{-6}} \times 50 \times 10^{-3}$

$= \underline{77 \times 10^6 \text{ N/m}^2}$

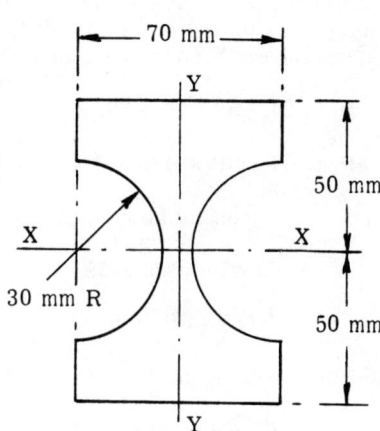

Fig. 3.12

(b) The second moment of area about YY is that of a rectangle 100 mm wide by 70 mm deep, less that of the two semicircles. The second moment of area of a semi-circle about YY is that about the diameter (half the value for a full circle), *less* area × (distance from diameter to centroid)², *plus* area × (distance from centroid to YY)².

Distance of centroid of semicircle from diameter $= \dfrac{4r}{3\pi} = \dfrac{4 \times 30}{3\pi}$

$= 12 \cdot 74$ mm

$\therefore I_{YY} = \dfrac{100 \times 70^3}{12} - 2\left\{\dfrac{\pi}{128} \times 60^4 - \dfrac{\pi}{8} \times 60^2 \times 12 \cdot 74^2 + \dfrac{\pi}{8} \times 60^2 \times [35 - 12 \cdot 74]^2\right\}$

$= 1 \cdot 28 \times 10^6 \text{ mm}^4$

$= 1 \cdot 28 \times 10^{-6} \text{ m}^4$

$\dfrac{M}{I} = \dfrac{\sigma}{y}$

$\therefore \sigma = \dfrac{8 \times 10^3}{1 \cdot 28 \times 10^{-6}} \times 35 \times 10^{-3}$

$= \underline{218 \cdot 5 \times 10^6 \text{ N/m}^2}$

3. *A brick chimney, 30 m high is 2·3 m outside diameter at the top and tapers uniformly to 3·3 m outside diameter at the bottom. The chimney has a mass of 224 tonnes and is 0·675 m thick at the base. If a horizontal wind pressure of 1 kN/m² acts on the projected area of chimney, determine the maximum and minimum stresses on the base.*

The projected area may be divided into a rectangle and a triangle, Fig. 3.13, the forces on the two parts, P_1 and P_2, acting at the respective centroids.

$P_1 = 1 \times 10^3 \times 2\cdot3 \times 30 = 69 \times 10^3$ N

$P_2 = 1 \times 10^3 \times \tfrac{1}{2} \times 30 = 15 \times 10^3$ N

Therefore moment about base,

$M = 69 \times 10^3 \times 15 + 15 \times 10^3 \times 10$

$ = 1\,185 \times 10^3$ N m

Second moment of area of base section,

$I = \dfrac{\pi}{64}(3\cdot3^4 - 1\cdot95^4) = 5\cdot1$ m⁴

Therefore bending stress at base,

$\sigma_b = \dfrac{M}{I} \times y = \dfrac{1\,185 \times 10^3}{5\cdot1} \times 1\cdot65$

$ = 383\cdot5 \times 10^3$ N/m²

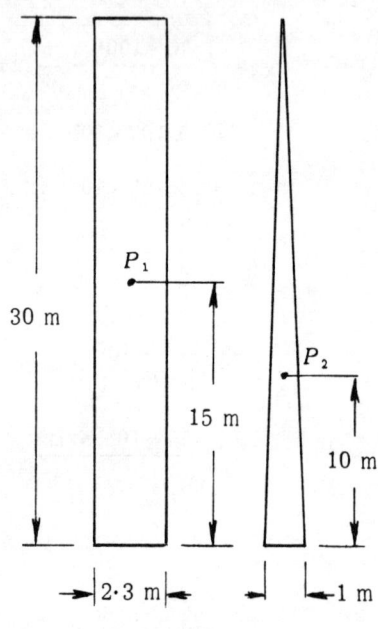

Fig. 3.13

Direct stress, $\sigma_d = \dfrac{W}{a} = \dfrac{224 \times 10^3 \times 9\cdot81}{\dfrac{\pi}{4}(3\cdot3^2 - 1\cdot95^2)} = 395 \times 10^3$ N/m²

∴ maximum compressive stress $= (395 + 383\cdot5) \times 10^3 = \underline{778\cdot5 \times 10^3 \text{ N/m}^2}$

and minimum compressive stress $= (395 - 383\cdot5) \times 10^3 = \underline{11\cdot5 \times 10^3 \text{ N/m}^2}$

4. *A vertical flagstaff 10 m high is of square section throughout, tapering uniformly from 150 mm × 150 mm at the base to 75 mm × 75 mm at the top. A horizontal pull of 300 N is applied at the top, the direction being along a diagonal of the section. Calculate the maximum stress due to bending.*

For a square of side s, the value of I about a diagonal may be obtained most simply by use of the equimomental system, Art. 3.1. If the square is divided into two triangles and one third of the area of each triangle is imagined concentrated at the mid-points of the three sides, the resulting system is as shown in Fig. 3.14(b). The mid-points of the sides are each at $s/2\sqrt{2}$ from the diagonal so that

$$I = 4 \times \dfrac{s^2}{6} \times \left(\dfrac{s}{2\sqrt{2}}\right)^2 + \dfrac{s^2}{3} \times 0 = \dfrac{s^4}{12}$$

BENDING STRESSES 29

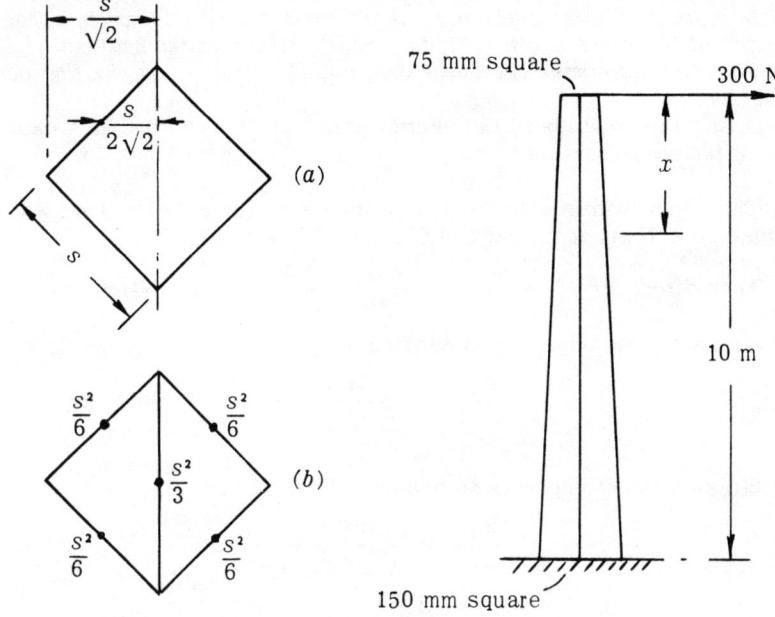

Fig. 3.14   Fig. 3.15

At a section distance x metres below the top, Fig. 3.15,

$$\text{side of square, } s = 75 + \frac{x}{10} \times 75 = 7 \cdot 5(10+x) \text{ mm}$$

and B.M. at that section $= 300x$ N m

$$\sigma = \frac{M}{I} \times y = \frac{300x}{\frac{s^4}{12}} \times \frac{s}{\sqrt{2}}$$

$$= \frac{3\,600x}{\sqrt{2}\,s^3}$$

$$= \frac{3\,600x}{\sqrt{2}\,[7 \cdot 5(10+x) \times 10^{-3}]^3}$$

$$= \frac{6 \cdot 03 \times 10^9\,x}{(10+x)^3}$$

For maximum stress $\quad \dfrac{d\sigma}{dx} = 0$

from which $\quad x = 5$ m

$$\therefore \sigma_{\max} = \frac{6 \cdot 03 \times 10^9 \times 5}{15^3} = \underline{8 \cdot 93 \times 10^6 \text{ N/m}^2}$$

BENDING STRESSES

5. *A rectangular bar, 80 mm × 40 mm cross-section, is subjected to a pull of W N applied parallel to the axis of the bar but offset 10 mm from each of the axes of the cross-section. If the tensile stress in the material is limited to 150 MN/m², determine the value of W and the stresses at the four corners of the section.*

Obtain the equation of the neutral axis and show its position on a diagram of the cross-section.

The given load is equivalent to a load W on the axis together with moments $10W$ N mm about each of the axes XX and YY, Fig. 3.16.

Then direct stress $= \dfrac{W}{a} = \dfrac{W}{80 \times 40 \times 10^{-6}} = 312 \cdot 5 W$ N/m²

Stress at long edge due to bending about XX

$$= \dfrac{M_{XX}}{Z_{XX}} = \dfrac{10W \times 10^{-3}}{\dfrac{80 \times 40^2}{6} \times 10^{-9}} = 469 W \text{ N/m}^2$$

Stress at short edge due to bending about YY

$$= \dfrac{M_{YY}}{Z_{YY}} = \dfrac{10W \times 10^{-3}}{\dfrac{40 \times 80^2}{6} \times 10^{-9}} = 234 \cdot 5 W \text{ N/m}^2$$

The maximum tensile stress occurs at corner A,

$$\therefore \; 312 \cdot 5 W + 469 W + 234 \cdot 5 W = 150 \times 10^6$$

$$\therefore \; W = \underline{147 \cdot 5 \times 10^3 \text{ N}}$$

Stress at corner B $= 312 \cdot 5 W - 469 W + 234 \cdot 5 W = 78 W = 11 \cdot 5$ MN/m²
Stress at corner C $= 312 \cdot 5 W - 469 W - 234 \cdot 5 W = -391 W = -57 \cdot 5$ MN/m²
Stress at corner D $= 312 \cdot 5 W + 469 W - 234 \cdot 5 W = 547 W = 80 \cdot 7$ MN/m²

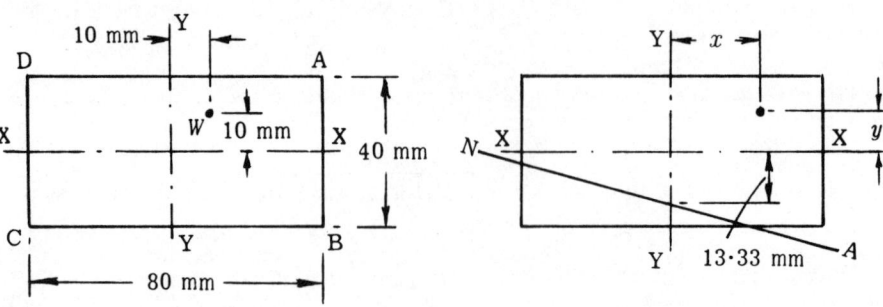

Fig. 3.16   Fig. 3.17

At any point whose coordinates are x and y relative to the axes of the section, Fig. 3.17,

$$\sigma = \dfrac{W}{a} + \dfrac{M_{XX}}{I_{XX}} \cdot y + \dfrac{M_{YY}}{I_{YY}} \cdot x$$

BENDING STRESSES

$$= \frac{W}{80 \times 40 \times 10^{-6}} + \frac{10W \times 10^{-3}}{\frac{80 \times 40^3}{12} \times 10^{-12}} \cdot y + \frac{10W \times 10^{-3}}{\frac{40 \times 80^3}{12} \times 10^{-12}} \cdot x$$

$$= W\{312 \cdot 5 + 23\,450\,y + 5\,860\,x\}$$

Along the neutral axis, $\sigma = 0$

i.e. $\qquad\qquad 312 \cdot 5 + 23\,450\,y + 5\,860\,x = 0$

from which $\qquad\qquad\qquad\qquad y = -0 \cdot 25\,x - 0 \cdot 013\,33$

This is the equation of the neutral axis, which is shown in Fig. 3.17.

6. A cast iron pipe 8 m long, 300 mm outside diameter and 250 mm inside diameter, is simply supported at its ends and is full of water at atmospheric pressure. If the density of water is 1 Mg/m^3 and that of cast iron is 7·8 Mg/m^3, what is the maximum tensile stress in the pipe? $\qquad$ (Ans.: 12·4 MN/m^2)

7. A cantilever 4 m long is of T-section with the web vertical. The section is 150 mm wide by 150 mm deep and the web and flange are 12 mm thick. The cantilever carries a concentrated load of 1·25 kN at a distance of 3 m from the fixed end. Determine the maximum stress due to bending, neglecting the effect of the weight of the cantilever. $\qquad$ (Ans.: 54 MN/m^2)

8. A horizontal girder of I-section 350 mm by 150 mm is simply supported at the ends of a span of 6 m. The flange thickness is 18 mm whilst that of the web is 10 mm. Calculate the uniformly distributed load which the beam can support including its own weight, if the maximum allowable tensile stress is 120 MN/m^2. $\qquad$ (Ans.: 26·6 kN/m)

9. The section of a steel beam is an inverted channel, outside dimensions 220 mm wide by 80 mm deep, thickness of web 10 mm, thickness of vertical flanges 12 mm. The beam is simply supported over a span of 3 m and carries two equal concentrated loads at points distant 0·5 m from each support. Find the value of these loads if the maximum tensile stress is not to exceed 100 MN/m^2. $\qquad$ (Ans.: 7·725 kN)

10. A wooden beam is 80 mm wide and 120 mm deep with a semi-circular groove of 25 mm radius planed out in the centre of each side. Calculate the maximum stress in the section when simply supported on a span of 2 m and loaded with a concentrated load of 400 N at a distance of 0·7 m from one end and a uniformly distributed load of 750 N/m over the whole span. $\qquad$ (Ans.: 2·825 MN/m^2)

11. A 250 mm by 150 mm I-section column, thickness of web and flanges 12 mm, is subjected to a thrust of 400 kN at a point on its longer axis 85 mm from its centroid. Determine the maximum tensile and compressive stresses in the column.
$\qquad$ (Ans.: 4·4 and 131·3 MN/m^2)

12. A tie-bar 75 mm wide and 25 mm thick sustains an axial load of 100 kN. What depth of metal may safely be removed from one of the narrow sides if the maximum stress over the reduced width may not exceed 100 MN/m^2? $\qquad$ (Ans.: 12·05 mm)

13. A chimney, outside diameter 3 m, inside diameter 2·5 m, is to be subjected to a horizontal wind force equal to a uniform load of 2 kN/m of its height. If the density of the masonry is 2·2 Mg/m^3, determine the maximum permissible height of the chimney if there is to be no tensile stress at the base. $\qquad$ (Ans.: 29·6 m)

14. A steel chimney is 30 m high, 1 m external diameter and 10 mm thick. It is acted upon by a horizontal wind pressure which is of a uniform intensity of 1 kN/m² of projected area for the lower 15 m and then varies uniformly from 1 kN/m² to 2 kN/m² over the upper 15 m. Calculate the maximum stress in the plates at the base. Steel has a density of 7·8 Mg/m³. *(Ans.:* 85·9 MN/m²)

15. A horizontal cantilever 3 m long is of rectangular cross-section 60 mm wide throughout, the depth varying uniformly from 60 mm at the free end to 180 mm at the fixed end. A load of 4 kN acts at the free end. Find the position of the most highly stressed section and the magnitude of the stress at that section.
(Ans.: Centre section ; 41·6 MN/m²)

16. A short cast iron column is of hollow section of uniform thickness, 200 mm outside diameter and 125 mm inside diameter. A vertical compressive load acts at an eccentricity of 50 mm from the axis of the column. If the maximum permitted stresses are 80 MN/m² in compression and 16 MN/m² in tension, calculate the greatest allowable load. *(Ans.:* 627·5 kN)

17. A short hollow pier, 1·2 m square outside and 0·75 m square inside, supports a vertical point load of 120 kN on a diagonal, 0·69 m from the vertical axis of the pier. Calculate the stresses at the four outside corners of a horizontal section of the pier.
(Ans.: 616·8 kN/m², comp ; 136·8 kN/m², comp ; 343·2 kN/m², tensile)

18. A tie-bar of rectangular section 75 mm × 25 mm carries an axial tensile load of 100 kN. Metal of thickness 5 mm is now removed from two adjacent faces while the load remains acting along the axis of the original section. Determine the value to which the load must be reduced if the maximum stress in the section is not to exceed the original stress. *(Ans.:* 38 kN)

19. Fig. 3.18 shows the cross-section of a short column which carries a load, the line of action of which is parallel with the axis of the column but eccentric to both XX and YY. The stresses at A, B and C in MN/m², are respectively 46 compressive, 75 compressive and 12 tensile.

Find the magnitude of the load, the position of its line of action with respect to XX and YY and the stress at point D.

$I_{XX} = 0\cdot2 \times 10^{-6}$ m⁴; $I_{YY} = 5 \times 10^{-6}$ m⁴; $A = 1300 \times 10^{-6}$ m².
(Ans.: 40·95 kN ; 1·7 mm from XX ; 40·7 mm from YY ; 17 MN/m² comp)

20. A short column has a solid cross-section in the form of an equilateral triangle of side 50 mm. Determine the shape and dimensions of the area within which a compressive load must be applied if there is to be no tensile stress in the cross-section.

Calculate the stress at each of the three corners when a compressive load of 80 kN acts at a point 2·5 mm eccentric to each of the axes of the section passing through the centroid, as shown in Fig. 3.19.
(Ans.: Equilateral triangle of side 12·5 mm ; 140·5, 51·9 and 23·6 MN/m²)

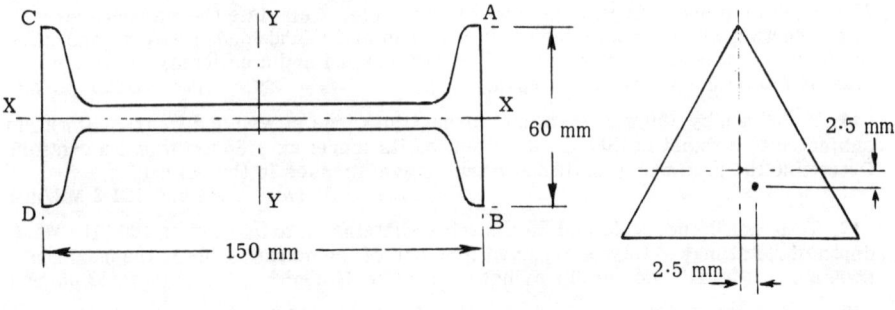

Fig. 3.18   Fig. 3.19

4 Torsion

4.1 Stress due to twisting When a shaft is subjected to a torque, every section is in a state of shear. The shaft will twist and the shear stress resulting from this strain will produce a moment of resistance, equal and opposite to the applied torque.

In the simple theory applicable to straight, circular, homogeneous, elastic shafts, it is assumed that radial lines remain radial after twisting. The shear strain is therefore proportional to the radius and it therefore follows from Hooke's Law that the stress is also proportional to the radius.

Thus, if the stress at radius r is τ, Fig. 4.1,

the stress on an element da at radius $v = \dfrac{v}{r}\tau$

$\therefore$ shear force on element $= \dfrac{v}{r}\tau\, da$

$\therefore$ moment of force about O $= \dfrac{v}{r}\tau\, da \cdot v$

$\therefore$ total moment of resistance $= \displaystyle\int \dfrac{v^2}{r}\tau\, da$

$= \dfrac{\tau}{r}\displaystyle\int v^2\, da$

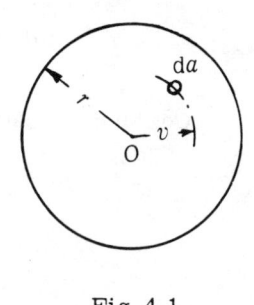

Fig. 4.1

$\int v^2\, da$ is the polar second moment of area about the axis of the shaft and is denoted by J.

Thus $\qquad T = \dfrac{\tau}{r}J$

or $\qquad \dfrac{T}{J} = \dfrac{\tau}{r}$ \hfill (4.1)

Equation (4.1) gives the stress at the surface of the shaft, where it is a maximum, but the stress at any other radius is proportional to that radius.

For a solid shaft of diameter d, $J = \dfrac{\pi}{32}d^4$ and for a hollow shaft of external diameter D and internal diameter d,

$$J = \dfrac{\pi}{32}(D^4 - d^4).$$

Equation (4.1) can be written as $\tau = \dfrac{T}{J/r}$; the quantity J/r is termed the *modulus of section* and is denoted by Z.

Thus $\qquad \tau = \dfrac{T}{Z}$ \hfill (4.2)

For a solid shaft, $\qquad Z = \dfrac{\pi}{32}d^4 / \dfrac{d}{2} = \dfrac{\pi}{16}d^3$

and for a hollow shaft, $\qquad Z = \dfrac{\pi}{32}(D^4 - d^4)/\dfrac{D}{2} = \dfrac{\pi}{16}\dfrac{D^4 - d^4}{D}$

It should be noted that values of the section modulus for twisting are not the same as those for bending.

4.2 Angle of twist
The shear strain in the shaft is the angle ϕ, Fig. 4.2.

$$\phi = \frac{AB}{l}$$

$$= \frac{r\theta}{l}$$

Also, from equation (1.6),

$$\phi = \frac{\tau}{G}$$

$$\therefore \frac{\tau}{G} = \frac{r\theta}{l}$$

or $\quad \dfrac{\tau}{r} = \dfrac{G\theta}{l} \quad$ (4.3)

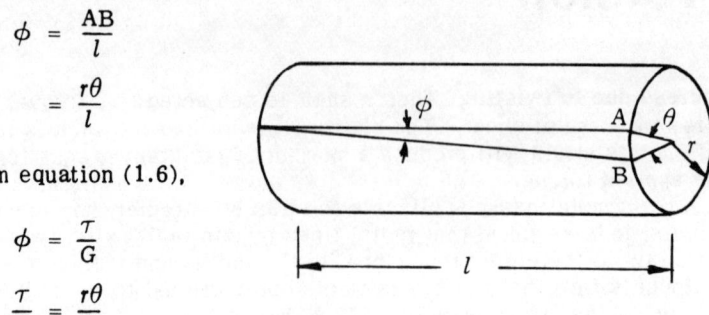

Fig. 4.2

Combining equations (4.1) and (4.3)

$$\frac{T}{J} = \frac{\tau}{r} = \frac{G\theta}{l} \tag{4.4}$$

1. *A hollow shaft of diameter ratio 3 : 5 is required to transmit 600 kW at 110 rev/min, the maximum torque being 12 per cent greater than the mean. The shearing stress is not to exceed 60 MN/m² and the twist in a length of 3 m is not to exceed 1°. Calculate the minimum external diameter to satisfy these conditions. G = 80 GN/m².*

Power = $T\omega$

i.e. $\quad 600 \times 10^3 = T \times \dfrac{2\pi}{60} \times 110$

$\therefore T = 52 \cdot 1 \times 10^3$ Nm

$\therefore$ maximum torque = $52 \cdot 1 \times 10^3 \times 1 \cdot 12 = 58 \cdot 3 \times 10^3$ N m

For the stress condition, $T = \tau Z$ . . . . . from equation (4.2)

i.e. $\quad 58 \cdot 3 \times 10^3 = 60 \times 10^6 \times \dfrac{\pi}{16} \dfrac{D^4 - (\tfrac{3}{5}D)^4}{D}$

from which $\quad D = 0 \cdot 178\,3$ m

For the twist condition, $\dfrac{T}{J} = \dfrac{G\theta}{l}$ . . . . . from equation (4.4)

i.e. $\quad \dfrac{58 \cdot 3 \times 10^3}{\dfrac{\pi}{32}\{D^4 - (\tfrac{3}{5}D)^4\}} = \dfrac{80 \times 10^9 \times 1 \times \dfrac{\pi}{180}}{3}$

from which $\quad D = 0 \cdot 195\,5$ m

The minimum external diameter to satisfy both conditions is therefore $0 \cdot 195\,5$ m and the internal diameter is $\underline{0 \cdot 117\,4 \text{ m}}$.

TORSION

2. *A shaft runs at 300 rev/min and transmits power from a pulley A at one end to two pulleys B and C which each drive a machine in a workshop. The distance between pulleys A and B is 3 m and that between B and C is 2·4 m. The shaft between A and B is 50 mm diameter and between B and C it is 40 mm diameter.*

If the maximum permissible shear stress in the shaft is 80 MN/m², calculate the maximum power which may be supplied from each of the pulleys B and C, assuming that both machines are in operation at the same time. Also calculate the total angle of twist of one end of the shaft relative to the other when running on full load. G = 80 GN/m².

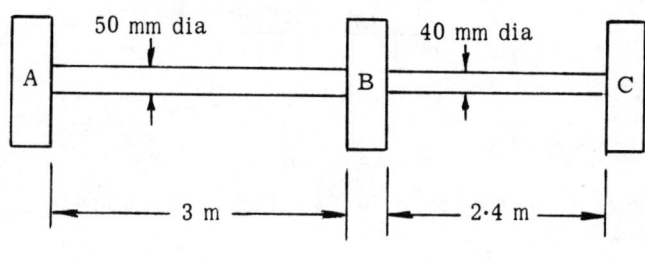

Fig. 4.3

The arrangement is shown in Fig. 4.3.

$$T = \tau Z \quad \ldots \quad \text{from equation (4.2)}$$

Therefore maximum torque transmissible by AB $= 80 \times 10^6 \times \dfrac{\pi}{16} \times \dfrac{50^3}{10^9}$

$$= 1964 \text{ Nm}$$

and maximum torque transmissible by BC $= 80 \times 10^6 \times \dfrac{\pi}{16} \times \dfrac{40^3}{10^9}$

$$= 1006 \text{ Nm}$$

Therefore maximum power transmissible by AB $= 1964 \times \dfrac{2\pi}{60} \times 300$

$$= 61\,700 \text{ W}$$

and maximum power transmissible by BC $= 1006 \times \dfrac{2\pi}{60} \times 300$

$$= 31\,600 \text{ W}$$

Therefore maximum power which can be taken from pulley C = __31·6 kW__

and maximum power which can be taken from pulley B = 61·7 − 31·6

= __30·7 kW__

$$\theta = \Sigma \dfrac{Tl}{Gr} \quad \ldots \quad \text{from equation (4.3)}$$

$$= \dfrac{80 \times 10^6}{80 \times 10^9}\left(\dfrac{3}{25 \times 10^{-3}} + \dfrac{2\cdot 4}{20 \times 10^{-3}}\right)$$

= 0·24 rad   or   __13·75°__

3. *Part of a steel tube 24 mm external diameter and 6 mm thick is enlarged to an external diameter of 36 mm. Find the diameter of the bore of the enlarged section so that when the tube is twisted, the maximum shear stresses in both sections are equal.*

If the total length of tube is 1 m, find the length of each part when the total angle of twist is 4° and the maximum shear stress is 75 MN/m².
G = 80 GN/m².

For equal stresses under the same torque, the modulus of section of each part must be the same,

i.e. $$\frac{24^4 - 12^4}{24} = \frac{36^4 - d^4}{36}$$

from which $\quad d = \underline{33 \cdot 2 \text{ mm}}$

From equation (4.3) $\quad \theta = \dfrac{\tau l}{Gr}$

Therefore, if the lengths of the 24 mm and 36 mm diameter parts are l_1 and l_2 respectively,

$$\frac{75 \times 10^6}{80 \times 10^9}\left(\frac{l_1}{0 \cdot 012} + \frac{l_2}{0 \cdot 018}\right) = 4 \times \frac{\pi}{180}$$

from which $\quad 1 \cdot 5\, l_1 + l_2 = 1 \cdot 34$

Also $\quad l_1 + l_2 = 1$

$\therefore\ l_1 = \underline{0 \cdot 68 \text{ m}} \text{ and } l_2 = \underline{0 \cdot 32 \text{ m}}$

4. A torque of 50 kN m is to be transmitted by a hollow shaft of internal diameter half the external diameter. If the maximum shear stress is not to exceed 80 MN/m², calculate the outside diameter of the shaft. What would be the angle of twist over a length of 3 m of the shaft under the above torque? $G = 80 \text{ GN/m}^2$.
(*Ans.*: 150 mm; 2°20′)

5. One quarter of the mass of a solid round shaft, 25 mm diameter, is removed by axial boring. Determine (*a*) by what percentage the strength of the shaft in torsion is reduced by this boring, (*b*) the maximum power that could be transmitted by the shaft before and after boring if the maximum allowable shear stress is 80 MN/m² and the speed is 150 rev/min. (*Ans.*: 6·25%; 3·86 kW; 3·62 kW)

6. A hollow shaft having external and internal diameter of 88 and 50 mm respectively transmits 1·5 MW at 2 000 rev/min. The shaft is connected to another shaft by a flanged coupling having eight bolts on a pitch circle of 180 mm diameter. Determine the minimum permissible diameter of the bolts if the shear stress is not to exceed 50 MN/m².

Calculate also the maximum shear stress in the shaft.
(*Ans.*: 15·9 mm; 59·8 MN/m²)

7. A hollow shaft is 50 mm outside diameter and 30 mm inside diameter. An applied torque of 1·6 kN m is found to produce an angular twist of 0·4°, measured on a length of 0·2 m of the shaft. Calculate the value of the modulus of rigidity. Calculate also the maximum power which can be transmitted by the shaft at 2 000 rev/min if the maximum allowable shearing stress is 65 MN/m². (*Ans.*: 86 GN/m²; 292 kW)

8. A shaft 50 mm in diameter and 0·75 m long has a concentric hole drilled for a portion of its length. Find the maximum length and diameter of the hole so that when the shaft is subjected to a torque of 1·67 kN m, the maximum shearing stress will not exceed 75 MN/m² and the total angle of twist will not exceed 1½°.
$G = 80 \text{ GN/m}^2$. (*Ans.*: 27·7 mm; 0·19 m)

5 Deflection of beams

5.1 Introduction The deflection of a beam depends upon the loading, the dimensions and material of the beam and the method of support. The principal types of beams to be considered are cantilevers, simply supported beams, in which rotation of the ends is unrestrained, and built-in (or *encastré*) beams, where the ends are rigidly built in to walls which restrain the ends to be horizontal but free from longitudinal tension.

5.2 Basic relations From equation (3.5)

$$\frac{M}{I} = \frac{E}{R}$$ which may be re-arranged to give

$$\frac{1}{R} = \frac{M}{EI} \tag{5.1}$$

The radius of curvature of any function $y = f(x)$ is expressed by

$$\frac{1}{R} = \frac{\frac{d^2y}{dx^2}}{\left[1 + \left(\frac{dy}{dx}\right)^2\right]^{3/2}}$$

When applied to a deflected beam, $\frac{dy}{dx}$ is very small, so that this equation reduces to

$$\frac{1}{R} = \frac{d^2y}{dx^2} \tag{5.2}$$

Combining equations (5.1) and (5.2) yields

$$\frac{d^2y}{dx^2} = \frac{M}{EI} \tag{5.3}$$

If the beam is of uniform cross-section and M may be expressed mathematically as a function of x, then the slope of the beam

$$\frac{dy}{dx} = \frac{1}{EI} \int M \, dx + A \tag{5.4}$$

and the deflection,

$$y = \frac{1}{EI} \int \int M \, dx \, dx + Ax + B \tag{5.5}$$

Further, from equations (2.1) and (2.2),

$$F = \frac{dM}{dx} \quad \text{and} \quad w = -\frac{dF}{dx}$$

so that
$$\frac{d^3y}{dx^3} = \frac{F}{EI} \qquad (5.6)$$

and
$$\frac{d^4y}{dx^4} = -\frac{w}{EI} \qquad (5.7)$$

Equation (5.7) is useful when the load is a function of x, as it may then be difficult to write down an expression for M in terms of x. In such cases, the deflection is obtained by integrating equation (5.7) four times.

Whichever equation is integrated, the resulting deflection will be in accordance with the Cartesian convention used, i.e., y positive upwards. This means that gravitational loading on a horizontal beam will usually produce negative deflections.

Slopes and deflections are inversely proportional to EI, which is termed the *flexural rigidity* of the beam.

5.3 Graphical method In practical cases, it may be impossible to express w or M in terms of x. In such cases, the bending moment diagram may be obtained from the loading diagram by the procedure described in Art. 2.5 and the deflection diagram may then be obtained from the bending moment diagram in a similar manner.

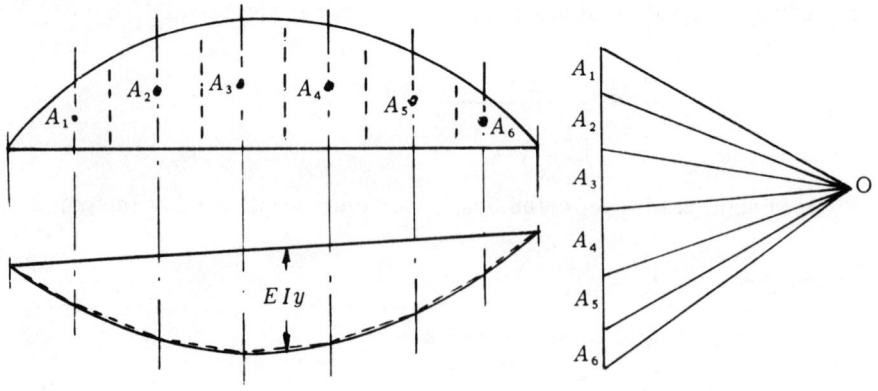

Fig. 5.1

Fig. 5.1 shows the bending moment diagram for a simply supported beam. The area is divided into convenient strips and the area of each strip is imagined concentrated at the centroid of that strip. The funicular polygon is then constructed, as in Art. 2.5, from which the deflection diagram (multiplied by EI) is obtained.

If the cross-section of the beam changes, then integration of the M/I diagram is made, leading to a deflection diagram with ordinates multiplied by E.

DEFLECTION OF BEAMS

5.4 Application of integration method to standard cases In the following examples, the sign of the bending moment is determined by the convention of Art. 2.2.

(a) Cantilever with concentrated end load, Fig. 5.2

Taking the origin at the fixed end,

B.M. at P $= -W(l-x)$

$$\therefore \frac{d^2y}{dx^2} = -\frac{W}{EI}(l-x)$$

$$\therefore \frac{dy}{dx} = -\frac{W}{EI}\left(lx - \frac{x^2}{2}\right) + A$$

When $x = 0$, $\frac{dy}{dx} = 0$, so that $A = 0$

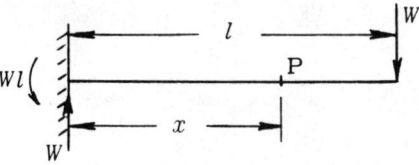

Fig. 5.2

$$\therefore y = -\frac{W}{EI}\left(\frac{lx^2}{2} - \frac{x^3}{6}\right) + B$$

When $x = 0$, $y = 0$, so that $B = 0$

The maximum slope and deflection occur at the free end, where $x = l$,

i.e.
$$\frac{dy}{dx} = -\frac{Wl^2}{2EI} \qquad (5.8)$$

and
$$y = -\frac{Wl^3}{3EI} \qquad (5.9)$$

(b) Cantilever with uniformly distributed load, Fig. 5.3

Taking the origin at the fixed end,

B.M. at P $= -w(l-x)\frac{(l-x)}{2}$

$$\therefore \frac{d^2y}{dx^2} = -\frac{w}{2EI}(l^2 - 2lx + x^2)$$

$$\therefore \frac{dy}{dx} = -\frac{w}{2EI}\left(l^2x - lx^2 + \frac{x^3}{3}\right) + A$$

When $x = 0$, $\frac{dy}{dx} = 0$, so that $A = 0$

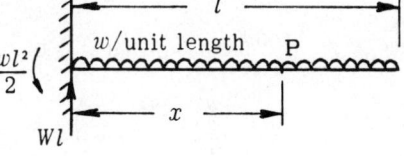

Fig. 5.3

$$\therefore y = -\frac{w}{2EI}\left(\frac{l^2x^2}{2} - \frac{lx^3}{3} + \frac{x^4}{12}\right) + B$$

When $x = 0$, $y = 0$, so that $B = 0$.

The maximum slope and deflection occur at the free end, where $x = l$,

i.e.
$$\frac{dy}{dx} = -\frac{wl^3}{6EI} \qquad (5.10)$$

and
$$y = -\frac{wl^4}{8EI} \qquad (5.11)$$

40　　　　　　　　　　　　　　　　　　　　　DEFLECTION OF BEAMS

(c) *Cantilever with end couple, Fig. 5.4*

Taking the origin at the fixed end,

B.M. at P $= -M$

$$\therefore \frac{d^2y}{dx^2} = -\frac{M}{EI}$$

$$\therefore \frac{dy}{dx} = -\frac{Mx}{EI} + A$$

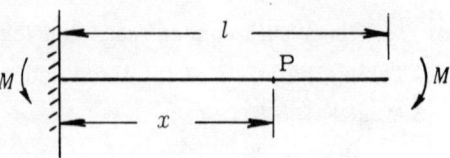

Fig. 5.4

When $x = 0$, $\frac{dy}{dx} = 0$, so that $A = 0$

$$\therefore y = -\frac{Mx^2}{2EI} + B$$

When $x = 0$, $y = 0$, so that $B = 0$.

The maximum slope and deflection occur at the free end, where $x = l$,

i.e. $$\therefore \frac{dy}{dx} = -\frac{Ml}{EI} \qquad (5.12)$$

and $$y = -\frac{Ml^2}{2EI} \qquad (5.13)$$

(d) *Simply supported beam with central concentrated load, Fig. 5.5*

Taking the origin at the centre,

B.M. at P $= \frac{W}{2}\left(\frac{l}{2} - x\right)$

$$\therefore \frac{d^2y}{dx^2} = \frac{W}{2EI}\left(\frac{l}{2} - x\right)$$

$$\therefore \frac{dy}{dx} = \frac{W}{2EI}\left(\frac{lx}{2} - \frac{x^2}{2}\right) + A$$

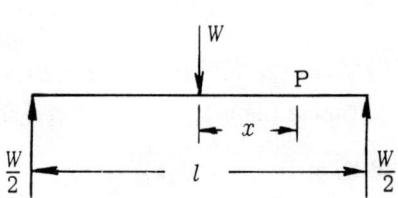

When $x = 0$, $\frac{dy}{dx} = 0$, so that $A = 0$

Fig. 5.5

$$\therefore y = \frac{W}{2EI}\left(\frac{lx^2}{4} - \frac{x^3}{6}\right) + B$$

When $x = \frac{l}{2}$, $y = 0$, so that $B = -\frac{W}{2EI} \cdot \frac{l^3}{24}$

$$\therefore y = \frac{W}{2EI}\left(\frac{lx^2}{4} - \frac{x^3}{6} - \frac{l^3}{24}\right)$$

The maximum slope occurs at the ends, where $x = \frac{l}{2}$,

i.e. $$\frac{dy}{dx} = \frac{Wl^2}{16EI} \qquad (5.14)$$

The maximum deflection occurs at the centre, where $x = 0$

i.e. $$y = -\frac{Wl^3}{48EI} \qquad (5.15)$$

DEFLECTION OF BEAMS

(e) Simply supported beam with uniformly distributed load, Fig. 5.6

Taking the origin at the centre,

$$\text{B.M. at P} = \frac{wl}{2}\left(\frac{l}{2} - x\right) - w\left(\frac{l}{2} - x\right)\frac{\left(\frac{l}{2} - x\right)}{2}$$

$$\therefore \frac{d^2y}{dx^2} = \frac{w}{2EI}\left(\frac{l^2}{4} - x^2\right)$$

$$\therefore \frac{dy}{dx} = \frac{w}{2EI}\left(\frac{l^2 x}{4} - \frac{x^3}{3}\right) + A$$

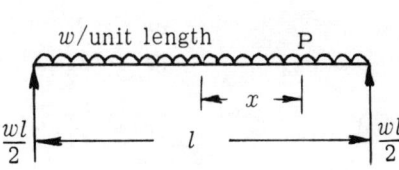

Fig. 5.6

When $x = 0$, $\frac{dy}{dx} = 0$, so that $A = 0$

$$\therefore y = \frac{w}{2EI}\left(\frac{l^2 x^2}{8} - \frac{x^4}{12}\right) + B$$

When $x = \frac{l}{2}$, $y = 0$, so that $B = -\frac{w}{2EI} \cdot \frac{5l^4}{192}$

$$\therefore y = \frac{w}{2EI}\left(\frac{l^2 x^2}{8} - \frac{x^4}{12} - \frac{5l^4}{192}\right)$$

The maximum slope occurs at the ends, where $x = \frac{l}{2}$,

i.e. $\quad \dfrac{dy}{dx} = \dfrac{wl^3}{24EI}$ \hfill (5.16)

The maximum deflection occurs at the centre, where $x = 0$,

i.e. $\quad y = -\dfrac{5wl^4}{384EI}$ \hfill (5.17)

5.5 Principle of superposition The principle of superposition states that if the bending moment, slope, deflection, etc., is directly proportional to the loading, the effect of a combination of loading may be obtained by the algebraic addition of the effects of each load considered separately. This may lead to the rapid solution of problems using the standard cases derived in Art. 5.4.

Thus for the loading shown in Fig. 5.7, the total deflection at the centre is the sum of the deflections due to the concentrated and distributed loads considered separately,

i.e. $\quad y = -\left[\dfrac{Wl^3}{48EI} + \dfrac{5}{384}\dfrac{wl^4}{EI}\right]$

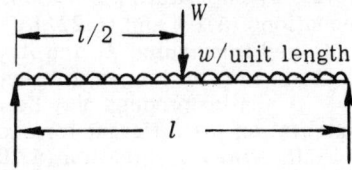

Fig. 5.7

5.6 Unsymmetrical loading – Macaulay's Method

For the loaded beam shown in Fig. 5.8, the bending moment at a point between A and C, distance x from A, is $R_1 x$ so that

$$\frac{d^2y}{dx^2} = \frac{R_1}{EI} \cdot x$$

$$\therefore \frac{dy}{dx} = \frac{R_1}{EI} \cdot \frac{x^2}{2} + A_1 \quad (5.18)$$

and $$y = \frac{R_1}{EI} \cdot \frac{x^3}{6} + A_1 x + B_1 \quad (5.19)$$

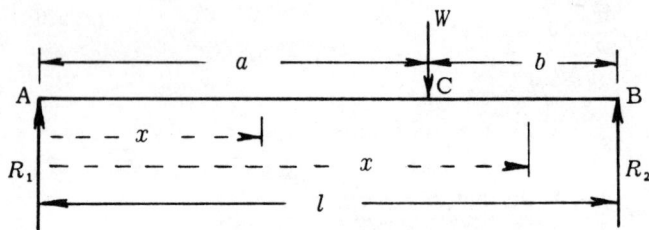

Fig. 5.8

When $x = 0$, $y = 0$, so that $B_1 = 0$. No other conditions are available within the range AC so that A_1 must remain unknown at present.

For a point between C and B, keeping the origin at A,

bending moment $= R_1 x - W[x - a]$

$$\therefore \frac{d^2y}{dx^2} = \frac{R_1}{EI} \cdot x - \frac{W}{EI}[x - a] \quad (5.20)$$

$$\therefore \frac{dy}{dx} = \frac{R_1}{EI} \cdot \frac{x^2}{2} - \frac{W}{EI}\left[\frac{x^2}{2} - ax\right] + A_2 \quad (5.21)$$

and $$y = \frac{R_1}{EI} \cdot \frac{x^3}{6} - \frac{W}{EI}\left[\frac{x^3}{6} - a\frac{x^2}{2}\right] + A_2 x + B_2 \quad (5.22)$$

When $x = l$, $y = 0$, so that B_2 can be determined. No other conditions are available within the range CB so that A_2 must also remain unknown at this stage.

However, the slope at C must be the same from equations (5.18) and (5.21) by substituting $x = a$ and the deflection at C must be the same from equations (5.19) and (5.22) by substituting $x = a$. Hence two equations are obtained from which A_1 and A_2 may be calculated. The slopes and deflections in the two ranges are then obtained by using the appropriate equations.

A similar process may be used for several loads but the constants of integration are different for each range and much laborious work is entailed.

If, however, equation (5.20) is integrated to give

$$\frac{dy}{dx} = \frac{R_1}{EI} \cdot \frac{x^2}{2} - \frac{W}{EI}\frac{[x - a]^2}{2} + A_2 \quad (5.23)$$

DEFLECTION OF BEAMS

and
$$y = \frac{R_1}{EI} \cdot \frac{x^3}{6} - \frac{W}{EI} \frac{[x-a]^3}{6} + A_2 x + B_2 \qquad (5.24)$$

then equating (5.18) and (5.23) at $x = a$ gives $A_1 = A_2$, and equating (5.19) and (5.24) at $x = a$ gives $B_1 = B_2$.

Thus, by this method of integration, the constants are the same for each range of the beam. Further, equations (5.23) and (5.24) may be regarded as applying to the whole beam if the terms involving $[x - a]$ are ignored when $[x - a]$ is negative, i.e., when $x < a$, since these then give the corresponding equations for the range AC. This method, known as *Macaulay's Method*, may be applied to any number of loads and it is conventional to use square brackets for terms such as $[x - a]$ which have to be treated in this special manner. The bending moment equation must always be written down at a point within the range remote from the origin so as to embrace all the loads.

For the simple case considered above, $R_1 = \dfrac{Wb}{l}$, so that the general deflection equation for the beam becomes

$$y = \frac{1}{EI} \left\{ \frac{Wb}{l} \cdot \frac{x^3}{6} - \frac{W}{6} [x - a]^3 + Ax + B \right\}$$

the constants A and B applying to the whole beam and the term $\dfrac{W}{6}[x-a]^3$ being ignored for values of x which make $[x - a]$ negative.

When $x = 0$, $y = 0$, so that $B = 0$, $[x - a]$ being negative for this value of x.

When $x = l$, $y = 0$, so that $A = -\dfrac{Wab}{6l}(l + b)$.

The deflection under the load is obtained by putting $x = a$, when

$$y = -\frac{Wa^2b^2}{3EIl} \qquad (5.25)$$

At the point of maximum deflection, $\dfrac{dy}{dx} = 0$ but it is first necessary to estimate which range of the beam will contain this point so that terms involving square brackets may be included or ignored as appropriate (see Ex. 6).

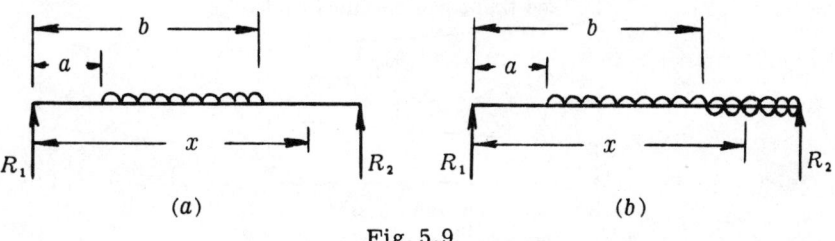

Fig. 5.9

5.7 Distributed loads A distributed load covering the entire span can be included in the bending moment equation for concentrated loads and is valid for all values of x. If, however, the load is discontinuous, it must extend to the end of the beam remote from the origin, adding, if necessary, a negative load to compensate for an additional load on top of the beam. Thus the load system shown in Fig. 5.9(a) must be converted into that in Fig. 5.9(b) in order that a continuous equation can be written down for the bending moment at any point, in accordance with the requirements of Macaulay's Method.

Considering a section within the range of the beam remote from the origin and taking moments of forces to the left of that section,

$$M = R_1 x - \frac{w}{2}[x - a]^2 + \frac{w}{2}[x - b]^2$$

from which $y = \dfrac{1}{EI}\left\{R_1 \dfrac{x^3}{6} - \dfrac{w}{24}[x - a]^4 + \dfrac{w}{24}[x - b]^4 + Ax + B\right\}$

The terms in square brackets are treated in exactly the same way as for concentrated loads; they must be integrated with respect to the quantities within the brackets and must be ignored when the quantities within the brackets become negative

5.8 Built-in beams If the ends of a beam are rigidly built-in to walls or similar fixings at each end, these fixings exert moments on the beam to maintain the ends horizontal. The bending moment diagram for the beam then consists of the 'free' bending moment diagram (i.e. the bending moment diagram due to the transverse loads only, as if on a simply supported beam) and the bending moment diagram due to the end fixing moments. These are of opposite sign and the resultant bending moment diagram is then given by the difference of the two diagrams, as shown in Fig. 5.10.

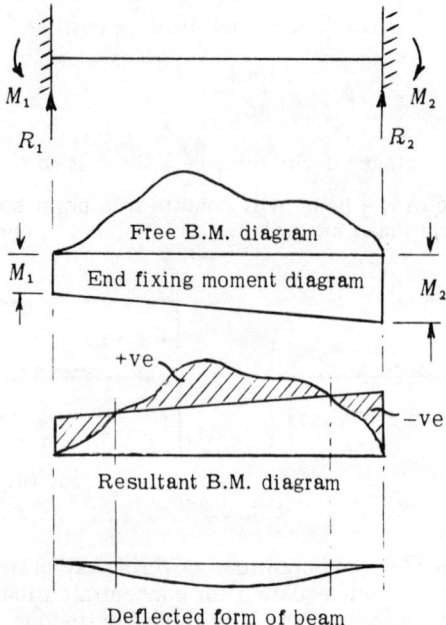

Fig. 5.10

DEFLECTION OF BEAMS

The bending moment at any point is the *vertical* ordinate on the resultant bending moment diagram, which may be replotted on a horizontal base, if required.

The form of the deflected beam follows from the resultant bending moment diagram; the points at which the bending changes from positive to negative are known as the points of *contraflexure*.

5.9 Standard cases

(a) Central concentrated load, Fig. 5.11

Taking the origin at the centre,

B.M. at P $= \dfrac{W}{2}\left(\dfrac{l}{2} - x\right) - M$

$\therefore \dfrac{d^2y}{dx^2} = \dfrac{1}{EI}\left\{\dfrac{W}{2}\left(\dfrac{l}{2} - x\right) - M\right\}$

$\therefore \dfrac{dy}{dx} = \dfrac{1}{EI}\left\{\dfrac{W}{2}\left(\dfrac{lx}{2} - \dfrac{x^2}{2}\right) - Mx\right\} + A$

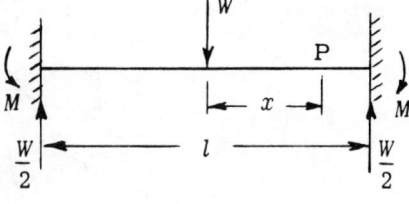

Fig. 5.11

When $x = 0$, $\dfrac{dy}{dx} = 0$, so that $A = 0$

When $x = \dfrac{l}{2}$, $\dfrac{dy}{dx} = 0$, so that $M = \dfrac{Wl}{8}$ \hfill (5.26)

$\therefore \dfrac{dy}{dx} = \dfrac{W}{2EI}\left(\dfrac{lx}{4} - \dfrac{x^2}{2}\right)$

$\therefore y = \dfrac{W}{2EI}\left(\dfrac{lx^2}{8} - \dfrac{x^3}{6}\right) + B$

When $x = \dfrac{l}{2}$, $y = 0$, so that $B = -\dfrac{Wl^3}{192EI}$

$\therefore y = \dfrac{W}{2EI}\left(\dfrac{lx^2}{8} - \dfrac{x^3}{6} - \dfrac{l^3}{96}\right)$

The maximum deflection occurs at the centre, where $x = 0$,

i.e. $\qquad\qquad y = -\dfrac{Wl^3}{192EI}$ \hfill (5.27)

(b) Uniformly distributed load, Fig. 5.12

Taking the origin at the centre,

B.M. at P $= \dfrac{wl}{2}\left(\dfrac{l}{2} - x\right)$

$\qquad\qquad - w\left(\dfrac{l}{2} - x\right)\dfrac{\left(\dfrac{l}{2} - x\right)}{2} - M$

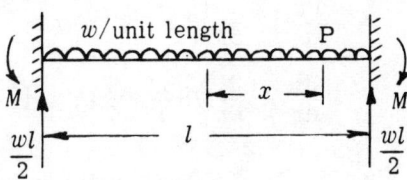

Fig. 5.12

$$\therefore \frac{d^2y}{dx^2} = \frac{1}{EI}\left\{\frac{w}{2}\left(\frac{l^2}{4} - x^2\right) - M\right\}$$

$$\therefore \frac{dy}{dx} = \frac{1}{EI}\left\{\frac{w}{2}\left(\frac{l^2 x}{4} - \frac{x^3}{3}\right) - Mx\right\} + A$$

When $x = 0$, $\frac{dy}{dx} = 0$, so that $A = 0$

When $x = \frac{l}{2}$, $\frac{dy}{dx} = 0$, so that $M = \frac{wl^2}{12}$ \hfill (5.28)

$$\therefore \frac{dy}{dx} = \frac{w}{2EI}\left(\frac{l^2 x}{12} - \frac{x^3}{3}\right)$$

$$\therefore y = \frac{w}{2EI}\left(\frac{l^2 x^2}{24} - \frac{x^4}{12}\right) + B$$

When $x = \frac{l}{2}$, $y = 0$, so that $B = -\frac{wl^4}{384EI}$

$$\therefore y = \frac{w}{2EI}\left(\frac{l^2 x^2}{24} - \frac{x^4}{12} - \frac{l^4}{192}\right)$$

The maximum deflection occurs at the centre, where $x = 0$,

i.e. $$y = -\frac{wl^4}{384EI}$$ \hfill (5.29)

(c) *Single concentrated load not at centre, Fig. 5.13*

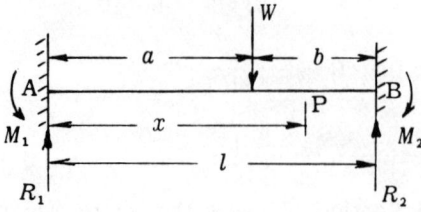

Fig. 5.13

Taking the origin at A and using Macaulay's method,

B.M. at P $= R_1 x - W[x - a] - M_1$

$$\therefore \frac{d^2y}{dx^2} = \frac{1}{EI}\left\{R_1 x - W[x - a] - M_1\right\}$$

$$\therefore \frac{dy}{dx} = \frac{1}{EI}\left\{R_1 \frac{x^2}{2} - \frac{W}{2}[x - a]^2 - M_1 x\right\} + A \hfill (1)$$

When $x = 0$, $\frac{dy}{dx} = 0$, so that $A = 0$, the term $[x - a]$ being negative and hence ignored.

DEFLECTION OF BEAMS

$$\therefore y = \frac{1}{EI}\left\{R_1\frac{x^3}{6} - \frac{W}{6}[x-a]^3 - M_1\frac{x^2}{2}\right\} + B \tag{2}$$

When $x = 0$, $y = 0$, so that $B = 0$, the term $[x-a]$ again being negative.

When $x = l$, $\dfrac{dy}{dx} = 0$ and $y = 0$.

Inserting these conditions in equations (1) and (2) gives

$$R_1 = \frac{Wb^2}{l^3}(l + 2a) \quad \text{and} \quad M_1 = \frac{Wab^2}{l^2}$$

and, by symmetry,

$$R_2 = \frac{Wa^2}{l^3}(l + 2b) \quad \text{and} \quad M_2 = \frac{Wa^2b}{l^2}$$

The deflection under the load is obtained by putting $x = a$, when

$$y = -\frac{Wa^3b^3}{3EIl^3} \tag{5.30}$$

It will be noted that the reactions R_1 and R_2 are not in the ratio b/a as for a simply supported beam; this is due to the unequal fixing moments M_1 and M_2.

5.10 Area-moment method Fig. 5.14 shows part of a beam AB which, when loaded, deflects to A'B'. The points A and B are at distances x_1 and x_2 from an origin O, the slopes at these points are θ_1 and θ_2 and the deflections are y_1 and y_2 respectively.

The sketch below shows the bending moment diagram for AB, the bending moment at a point distant x from O being M.

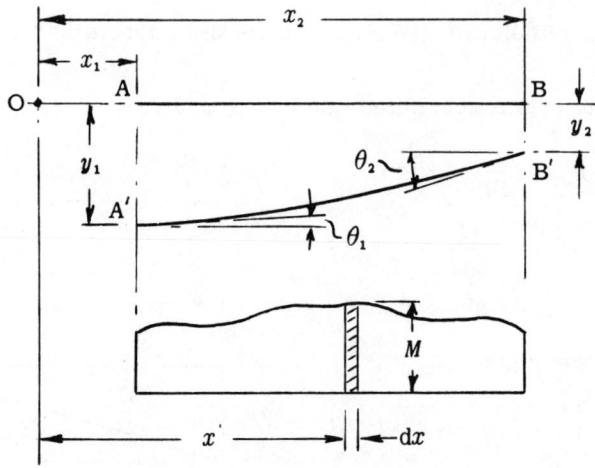

Fig. 5.14

From equation (5.3),

$$\frac{d^2y}{dx^2} = \frac{M}{EI} \qquad (5.31)$$

$$\therefore \frac{dy}{dx} = \frac{1}{EI}\int M\,dx$$

Thus, the change of slope between A and B,

$$\theta_2 - \theta_1 = \frac{1}{EI}\int_{x_1}^{x_2} M\,dx \qquad (5.32)$$

$$= \frac{1}{EI} \times \text{area of B.M. diagram between A and B}$$

Multiplying equation (5.31) by x gives

$$x\frac{d^2y}{dx^2} = \frac{Mx}{EI}$$

Integrating both sides between the limits x_1 and x_2 gives

$$\left[x\frac{dy}{dx} - y\right]_{x_1}^{x_2} = \frac{1}{EI}\int_{x_1}^{x_2} Mx\,dx \qquad (5.33)$$

$$= \frac{1}{EI} \times \text{moment of area of B.M. diagram about O}$$

The origin O is chosen so as to make $x\frac{dy}{dx}$ zero at both limits, so that the left-hand side then gives the difference of deflection between A and B.

5.11 Application of area-moment method to standard cases

Note: Beams with distributed loads require the parabola data given on p. 62.

(a) Cantilever with concentrated end load, Fig. 5.15

$$\theta_b - \theta_a = \frac{1}{EI}(-\tfrac{1}{2}Wl \times l)$$

i.e. $\qquad \theta_b = -\dfrac{Wl^2}{2EI}$

since $\theta_a = 0$

Taking the origin at B,

$$\left[x\frac{dy}{dx} - y\right]_0^l = \frac{1}{EI}(-\tfrac{1}{2}Wl \times l \times \tfrac{2}{3}l)$$

i.e.

$$(0-0) - (0-y_b) = y_b = -\frac{Wl^3}{3EI}$$

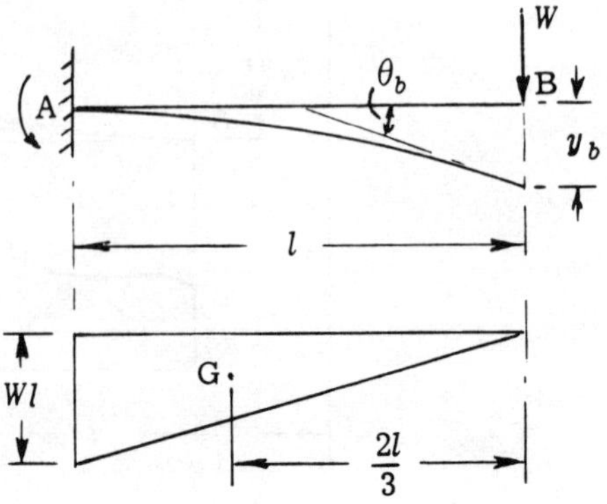

Fig. 5.15

DEFLECTION OF BEAMS

(b) *Cantilever with uniformly distributed load, Fig. 5.16*

$$\theta_b - \theta_a = \frac{1}{EI}(-\tfrac{1}{3}\frac{wl^2}{2} \times l)$$

i.e. $\quad \theta_b = -\dfrac{wl^3}{6EI}$ since $\theta_a = 0$

Taking the origin at B,

$$\left[x\frac{dy}{dx} - y\right]_0^l = \frac{1}{EI}(-\tfrac{1}{3}\frac{wl^2}{2} \times l \times \tfrac{3}{4}l)$$

i.e.

$(0-0) - (0-y_b) = y_b = -\dfrac{wl^4}{8EI}$

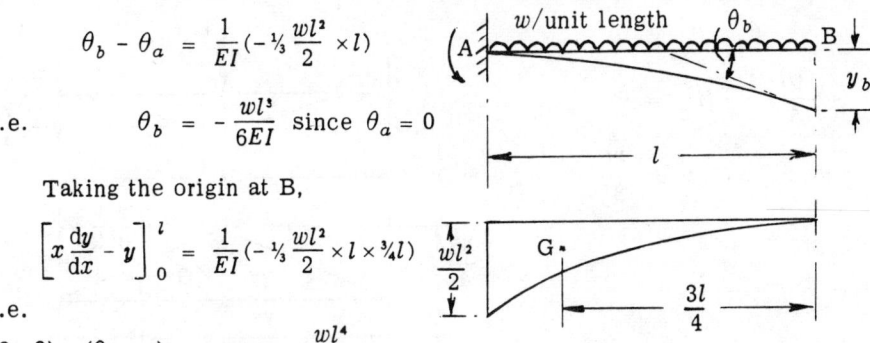

Fig. 5.16

(c) *Simply supported beam with central concentrated load, Fig. 5.17*

$$\theta_b - \theta_a = \frac{1}{EI} \times \tfrac{1}{2} \frac{Wl}{4} \times \frac{l}{2}$$

i.e. $\quad \theta_b = \dfrac{Wl^2}{16EI}$ since $\theta_a = 0$

Taking the origin at B,

$$\left[x\frac{dy}{dx} - y\right]_0^{\frac{l}{2}} = \frac{1}{EI} \times \tfrac{1}{2} \frac{Wl}{4} \times \frac{l}{2} \times \tfrac{2}{3}\frac{l}{2}$$

i.e.

$(0-y_a) - (0-0) = \dfrac{Wl^3}{48EI}$

or $\quad y_a = -\dfrac{Wl^3}{48EI}$

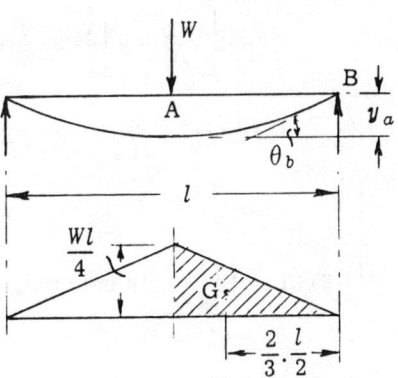

Fig. 5.17

(d) *Simply supported beam with uniformly distributed load, Fig. 5.18*

$$\theta_b - \theta_a = \frac{1}{EI} \times \tfrac{2}{3}\frac{wl^2}{8} \times \frac{l}{2}$$

i.e. $\quad \theta_b = \dfrac{wl^3}{24EI}$ since $\theta_a = 0$

Taking the origin at B,

$$\left[x\frac{dy}{dx} - y\right]_0^{\frac{l}{2}} = \frac{1}{EI} \times \tfrac{2}{3}\frac{wl^2}{8} \times \frac{l}{2} \times \tfrac{5}{8}\frac{l}{2}$$

i.e.

$(0-y_a) - (0-0) = \dfrac{5}{384}\dfrac{wl^4}{EI}$

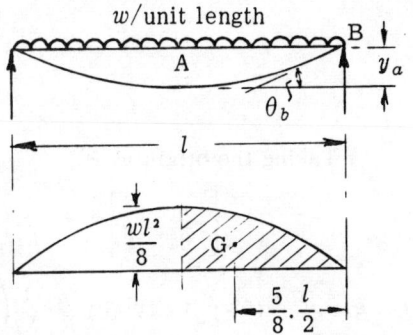

Fig. 5.18

or $\quad y_a = -\dfrac{5}{384}\dfrac{wl^4}{EI}$

DEFLECTION OF BEAMS

(e) Built-in beam with central concentrated load, Fig. 5.19

$$\theta_b - \theta_a = 0$$

$$\therefore \tfrac{1}{2}\frac{Wl}{4} \times \frac{l}{2} - M\frac{l}{2} = 0$$

$$\therefore M = \frac{Wl}{8}$$

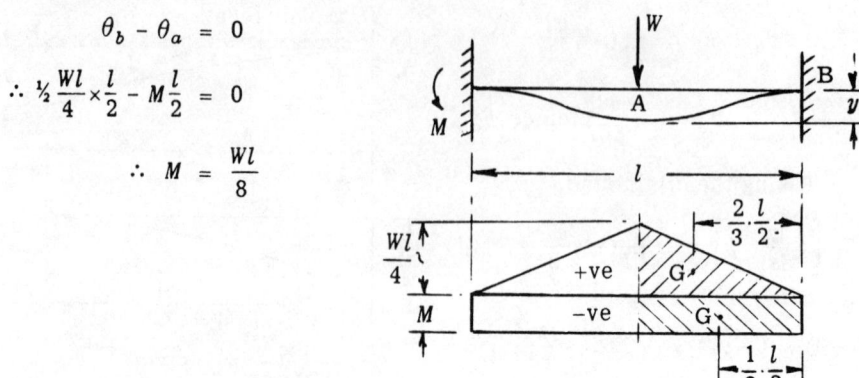

Fig. 5.19

Taking the origin at B,

$$\left[x\frac{dy}{dx} - y\right]_0^{\frac{l}{2}} = \frac{1}{EI}\left(\tfrac{1}{2}\frac{Wl}{4} \times \frac{l}{2} \times \tfrac{2}{3}\frac{l}{2} - M\frac{l}{2} \times \tfrac{1}{2}\frac{l}{2}\right)$$

i.e. $\quad (0 - y_a) - (0 - 0) = \dfrac{1}{EI}\left(\dfrac{Wl^3}{48} - \dfrac{Wl^3}{64}\right)$

$$\therefore y_a = -\frac{Wl^3}{192EI}$$

(f) Built-in beam with uniformly distributed load, Fig. 5.20

$$\theta_b - \theta_a = 0$$

$$\therefore \tfrac{2}{3}\frac{wl^2}{8} \times \frac{l}{2} - M\frac{l}{2} = 0$$

$$\therefore M = \frac{wl^2}{12}$$

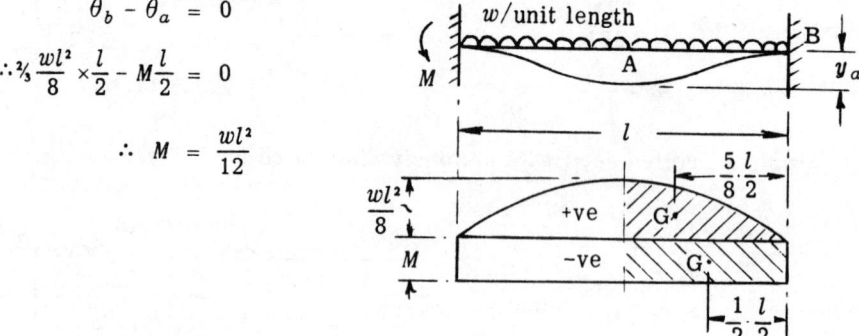

Fig. 5.20

Taking the origin at B,

$$\left[x\frac{dy}{dx} - y\right]_0^{\frac{l}{2}} = \frac{1}{EI}\left(\tfrac{2}{3}\frac{wl^2}{8} \times \frac{l}{2} \times \tfrac{5}{8}\frac{l}{2} - M\frac{l}{2} \times \tfrac{1}{2}\frac{l}{2}\right)$$

i.e. $\quad (0 - y_a) - (0 - 0) = \dfrac{1}{EI}\left(\dfrac{5wl^4}{384} - \dfrac{wl^4}{96}\right)$

$$\therefore y_a = -\frac{wl^4}{384EI}$$

DEFLECTION OF BEAMS

1. *A cantilever of length l carries a uniformly distributed load w over the entire length and is propped at a point l/4 from the free end, the level of the prop being adjusted so that there is no deflection at the free end.*

Derive a formula for the reaction at the prop and for the deflection of the beam at the prop.

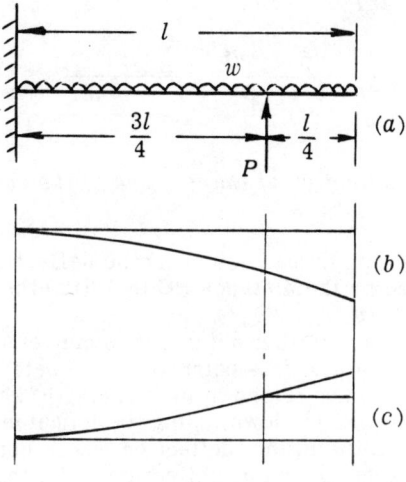

Fig. 5.21

The downward deflection at the free end due to w alone, shown in Fig. 5.21(b), is given by

$$y = -\frac{wl^4}{8EI} \quad \text{from equation (5.11)}$$

The upward deflection at the free end due to the prop force P alone, shown in Fig. 5.21(c), is made up of the deflection at the prop, together with the slope of the prop multiplied by the overhanging length $l/4$, this part of the beam being straight.

Thus

$$y = \frac{P\left(\frac{3l}{4}\right)^3}{3EI} + \frac{P\left(\frac{3l}{4}\right)^2}{2EI} \times \frac{l}{4} \quad \text{from equations (5.9) and (5.8)}$$

$$= \frac{27}{128}\frac{Pl^3}{EI}$$

If there is to be resultant deflection at the free end, then

$$-\frac{wl^4}{8EI} + \frac{27}{128}\frac{Pl^3}{EI} = 0$$

from which

$$P = \frac{16}{27}wl$$

The deflection at the prop is the upward deflection at the prop due to P, less the downward deflection at that point due to w, derived in Art. 5.4(b),

i.e.
$$y = \frac{P\left(\frac{3l}{4}\right)^3}{3EI} - \frac{w}{2EI}\left[\frac{l^2\left(\frac{3l}{4}\right)^2}{2} - \frac{l\left(\frac{3l}{4}\right)^3}{3} + \frac{\left(\frac{3l}{4}\right)^4}{12}\right]$$

$$= \frac{wl^4}{12EI} - \frac{513}{6\,144}\frac{wl^4}{EI} = -\frac{wl^4}{6\,144EI}$$

2. *Calculate the deflection at the free end of the cantilever shown in Fig. 5.22(a).*

The total deflection is made up of (i) the deflection at B (y_1), (ii) the slope at B multiplied by the distance BC (y_2), (iii) the further deflection due to bending of BC (y_3).

In calculating y_1 and y_2 (but *not* y_3), it is convenient to move the load to B, adding a moment Wa at this point to compensate for this movement.

The equivalent system is shown in Fig. 5.22(b) and the slope and deflection at B can then be written down, using the formulae derived in Art. 5.4(a) and (c). y_3 represents the further deflection due to bending of BC due to the load at C relative to the slope and deflection at B; this is given by equation (5.9).

$$\therefore y_1 = -\frac{Wa^3}{3E(2I)} - \frac{Wa.a^2}{2E(2I)}$$

$$= -\frac{5}{12}\frac{Wa^3}{EI}$$

$$y_2 = -\frac{Wa^2}{2E(2I)} - \frac{Wa.a}{2E(2I)} \times a$$

$$= -\frac{3}{4}\frac{Wa^3}{EI}$$

and

$$y_3 = -\frac{Wa^3}{3EI}$$

$$\therefore y = -\frac{Wa^3}{EI}\left(\frac{5}{12} + \frac{3}{4} + \frac{1}{3}\right)$$

$$= -\frac{3}{2}\frac{Wa^3}{EI}$$

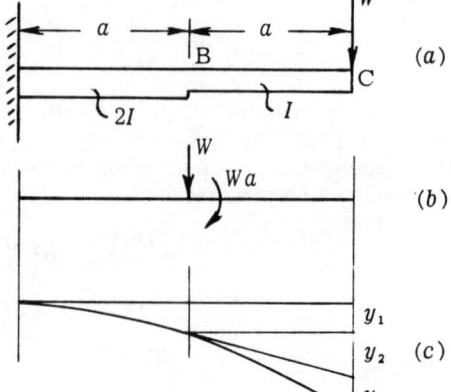

Fig. 5.22

DEFLECTION OF BEAMS

This problem may also be solved by the area-moment method. Since I is not constant, it is necessary to express equation (5.33) as

$$\left[x\frac{dy}{dx} - y\right]_{x_1}^{x_2} = \frac{1}{E} \times \text{moment of area of } \frac{M}{I} \text{ diagram about O}$$

The $\frac{M}{I}$ diagram for the cantilever is shown in Fig. 5.23(b), which may conveniently be broken down into the two parts shown in Fig. 5.23(c).

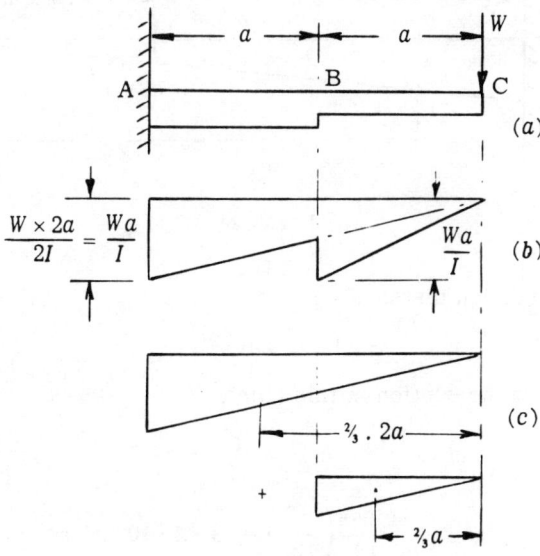

Fig. 5.23

Taking moments about C,

$$\left[x\frac{dy}{dx} - y\right]_0^{2a} = \frac{1}{E}\left(-\tfrac{1}{2}\frac{Wa}{I} \times 2a \times \tfrac{2}{3}.2a - \tfrac{1}{2}\frac{Wa}{2I} \times a \times \tfrac{2}{3}a\right)$$

i.e. $(0-0) - (0-y_c) = -\dfrac{1}{E}\left(\dfrac{4Wa^3}{3I} + \dfrac{Wa^3}{6I}\right)$

$$\therefore y_c = -\underline{\frac{3Wa^3}{2EI}}$$

54 DEFLECTION OF BEAMS

3. *A horizontal cantilever, 1·5 m long, tapers in section from 200 mm deep by 75 mm wide at the fixed end to 75 mm square at the extreme end and carries an end load of 3 kN. Calculate the deflection at the loaded end. E = 14 GN/m².*

The simplest solution will be obtained by taking the origin at the point of convergence of the top and bottom faces of the beam which, by similar triangles, will be found to be 0·9 m beyond the free-end of the cantilever, Fig. 5.24.

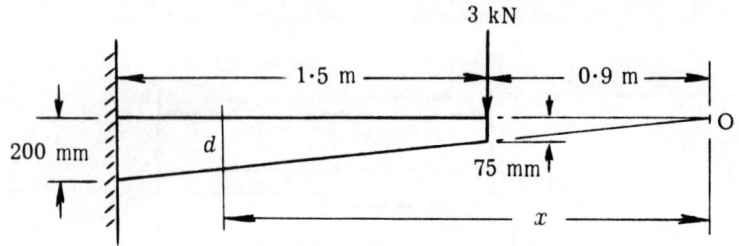

Fig. 5.24

Then at a section distance x from O,

$$M = -3 \times 10^3 (x - 0.9)$$

The depth of the section at this point,

$$d = \frac{x}{2.4} \times 0.2 = \frac{x}{12} \text{ m}$$

$$\therefore I = \frac{0.075}{12}\left(\frac{x}{12}\right)^3 = 3.62 \times 10^{-6} x^3 \text{ m}^4$$

$$\therefore \frac{d^2y}{dx^2} = \frac{M}{EI} = \frac{-3 \times 10^3 (x - 0.9)}{14 \times 10^9 \times 3.62 \times 10^{-6} x^3}$$

$$= -0.0592 (x^{-2} - 0.9 x^{-3})$$

$$\therefore \frac{dy}{dx} = -0.0592 (-x^{-1} + 0.45 x^{-2} + A)$$

When $x = 2.4$ m, $\frac{dy}{dx} = 0$, so that $A = 0.3386$

$$\therefore y = -0.0592 (-\log_e x - 0.45 x^{-1} + 0.3386 x + B)$$

When $x = 2.4$ m, $y = 0$, so that $B = 0.2504$

When $x = 0.9$ m, $y = -0.0592 \left(-\log_e 0.9 - \frac{0.45}{0.9} + 0.3386 \times 0.9 + 0.2504\right)$

$$= -0.00953 \text{ m} \text{ or } \underline{-9.53 \text{ mm}}$$

DEFLECTION OF BEAMS

4. *A uniform horizontal beam 6 m long, with a rectangular cross-section 50 mm wide by 250 mm deep, is simply supported at the ends and also at the middle by a vertical wire which stretches 0·45 μm/N. When the beam is unloaded, there is no tension in the wire.*

Calculate the deflection at the centre of the beam and the tension in the wire when a load of 100 kN is uniformly distributed over one half of the length of the beam. $E = 200 \text{ GN/m}^2$.

When a uniformly distributed load covers the entire span of a simply supported beam, each half contributes equally to the central deflection, so that the central deflection due to a uniformly distributed load over one half of the span, Fig. 5.25,

$$= \tfrac{1}{2} \times -\frac{5}{384} \frac{wl^4}{EI}$$

$$= -\frac{5}{768} \frac{wl^4}{EI} \quad \ldots \quad \text{from equation (5.17)}$$

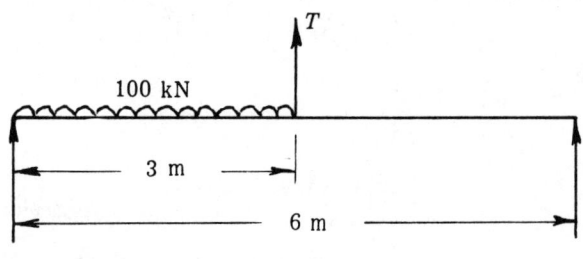

Fig. 5.25

The upward deflection at the centre due to the tension T in the wire is given by

$$y = \frac{Tl^3}{48EI} \quad \ldots \quad \text{from equation (5.15)}$$

Therefore the residual deflection at the centre $= -\dfrac{5}{768}\dfrac{wl^4}{EI} + \dfrac{Tl^3}{48EI}$

Under a tension T, the centre of the beam moves down a distance $0.45 \times 10^{-6} T$, i.e., the stretch of the wire.

Hence $\qquad -\dfrac{5}{768}\dfrac{wl^4}{EI} + \dfrac{Tl^3}{48EI} = -0.45 \times 10^{-6}\, T$

i.e. $\quad -\dfrac{5}{768} \times \dfrac{10^5}{3} \times 6^4 + \dfrac{T \times 6^3}{48} = -0.45 \times 10^{-6} \times 200 \times 10^9 \times \dfrac{50 \times 250^3}{12 \times 10^{12}} T$

from which $\qquad\qquad T = \underline{27 \cdot 1 \times 10^3 \text{ N}}$

Central deflection $= -0.45 \times 10^{-6} \times 27.1 \times 10^3$

$\qquad\qquad\qquad\qquad = \underline{-0.0122 \text{ m}}$

5. *A simply supported beam of span 2l carries a distributed load which varies uniformly from zero at each end to a maximum of w_0 per unit length at the centre. Obtain an expression for the central deflection.*

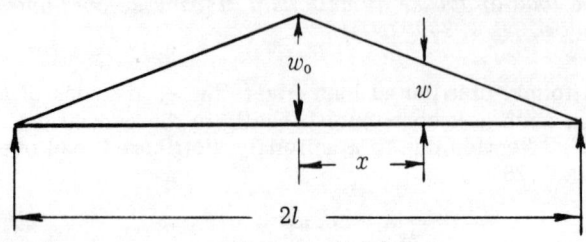

Fig. 5.26

Taking the origin at the centre, Fig. 5.26, the intensity of loading at a distance x from the centre,

$$w = w_0\left(1 - \frac{x}{l}\right)$$

$$\therefore \frac{d^4y}{dx^4} = -\frac{w}{EI}$$

$$= -\frac{w_0}{EI}\left(1 - \frac{x}{l}\right) \quad \cdots \quad \text{from equation (5.7)}$$

$$\therefore \frac{d^3y}{dx^3} = -\frac{w_0}{EI}\left(x - \frac{x^2}{2l} + A\right)$$

When $x = 0$, the S.F. $= 0$, so that $A = 0$

$$\therefore \frac{d^2y}{dx^2} = -\frac{w_0}{EI}\left(\frac{x^2}{2} - \frac{x^3}{6l} + B\right)$$

When $x = l$, the B.M. $= 0$, so that $B = -\frac{l^2}{3}$

$$\therefore \frac{dy}{dx} = -\frac{w_0}{EI}\left(\frac{x^3}{6} - \frac{x^4}{24l} - \frac{l^2 x}{3} + C\right)$$

When $x = 0$, $\frac{dy}{dx} = 0$, so that $C = 0$

$$\therefore y = -\frac{w_0}{EI}\left(\frac{x^4}{24} - \frac{x^5}{120l} - \frac{l^2 x^2}{6} + D\right)$$

When $x = l$, $y = 0$, so that $D = \frac{2l^4}{15}$

When $x = 0$, $\qquad y = -\frac{2}{15}\frac{w_0 l^4}{EI}$

DEFLECTION OF BEAMS

6. *A beam of length 8 m is simply supported at its ends and carries two concentrated loads of 20 kN and 40 kN respectively 2 m and 6 m from the left-hand end together with a distributed load of 15 kN/m on the 4 m between the concentrated loads.*

Calculate the deflection at the centre and also the position and magnitude of the maximum deflection. $E = 200$ GN/m^2 and $I = 0.18 \times 10^{-3}$ m^4.

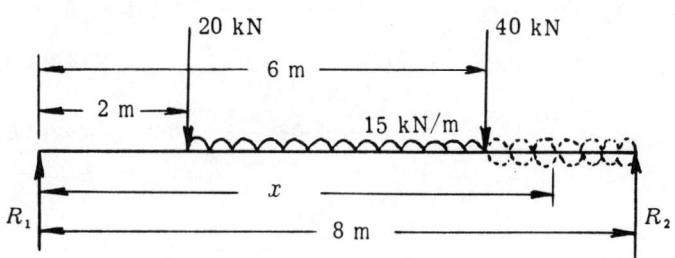

Fig. 5.27

Taking moments about R_2, Fig. 5.27,

$$8R_1 = 20 \times 6 + 40 \times 2 + 15 \times 4 \times 4$$

$$\therefore R_1 = 55 \text{ kN}$$

Taking the origin at R_1, the distributed load must be continued to the far end and a negative load added to compensate, as shown dotted.

Considering a section in the extreme range of the beam from the origin and taking moments of all forces to the left of the section,

$$\frac{d^2y}{dx^2} = \frac{10^3}{EI}\left\{55x - 20[x-2] - 40[x-6] - \frac{15}{2}[x-2]^2 + \frac{15}{2}[x-6]^2\right\}$$

$$\therefore \frac{dy}{dx} = \frac{10^3}{EI}\left\{\frac{55}{2}x^2 - 10[x-2]^2 - 20[x-6]^2 - \frac{5}{2}[x-2]^3 + \frac{5}{2}[x-6]^3 + A\right\}$$

and $y = \frac{10^3}{EI}\left\{\frac{55}{6}x^3 - \frac{10}{3}[x-2]^3 - \frac{20}{3}[x-6]^3 - \frac{5}{8}[x-2]^4 + \frac{5}{8}[x-6]^4 + Ax + B\right\}$

When $x = 0$, $y = 0$, so that $B = 0$ since all terms in square brackets are negative and hence ignored.

When $x = 8$, $y = 0$, so that $A = -390$, all terms in square brackets being positive and hence included.

When $x = 4$, $y = \frac{10^3}{EI}\left\{\frac{55}{6} \times 4^3 - \frac{10}{3} \times 2^3 - \frac{5}{8} \times 2^4 - 390 \times 4\right\}$, other terms being negative

$$= -\frac{10^3 \times 1010}{200 \times 10^9 \times 0.18 \times 10^{-3}}$$

$$= -0.02805 \text{ m} \quad \text{or} \quad \underline{-28.05 \text{ mm}}$$

At the point of maximum deflection $\frac{dy}{dx} = 0$ and it will be assumed that this point lies within the range $2 < x < 6$, so that terms involving $[x-6]$ must be ignored.

Hence $\quad \frac{55}{2}x^2 - 10[x-2]^2 - \frac{5}{2}[x-2]^3 - 390 = 0$

which reduces to $\quad\quad\quad\quad x^3 - 13x^2 - 4x + 164 = 0$

By Newton's method or plotting, $\quad\quad\quad x = 4\cdot06$ m

$$\therefore y_{max} = \frac{10^3}{EI}\left\{\frac{55}{6} \times 4\cdot06^3 - \frac{10}{3} \times 2\cdot06^3 - \frac{5}{8} \times 2\cdot06^4 - 390 \times 4\cdot06\right\}$$

$$= -\frac{10^3 \times 1011}{200 \times 10^9 \times 0\cdot18 \times 10^{-3}}$$

$$= -0\cdot02808 \text{ m}$$

If it is assumed that the maximum deflection occurs either to the left of the 20 kN load or to the right of the 40 kN load and the terms in square brackets are ignored or included to correspond, the solution for x will not be within the assumed range, thus indicating that a false assumption has been made.

In all cases of simply supported beams, the point of maximum deflection will be very close to the centre of the beam.

7. *A horizontal I-beam, built-in at both ends and 8 m long, carries a uniformly distributed load of 12 kN/m and a central concentrated load of 40 kN. If the bending stress is limited to 75 MN/m² and the deflection is not to exceed 2·5 mm, find the depth of section required. E = 200 GN/m².*

Due to the concentrated load alone,

$M = \dfrac{Wl}{8}$ from equation (5.26)

$\quad = \dfrac{40 \times 8}{8} = 40$ kNm

Due to the distributed load alone,

$M = \dfrac{wl^2}{12}$ from equation (5.28)

$\quad = \dfrac{12 \times 8^2}{12} = 64$ kNm

Hence total end fixing moment = 104 kNm

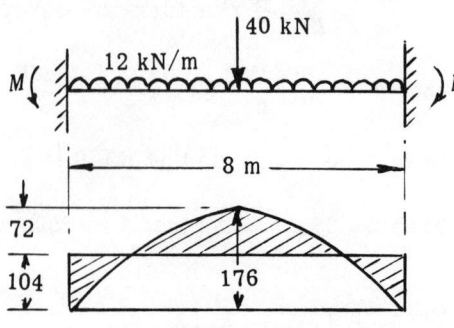

Fig. 5.28

DEFLECTION OF BEAMS   59

Free bending moment at centre due to concentrated load

$$= \frac{Wl}{4} = \frac{40 \times 8}{4} = 80 \text{ kN m}$$

Free bending moment at centre due to distributed load

$$= \frac{wl^2}{8} = \frac{12 \times 8^2}{8} = 96 \text{ kN m}$$

Hence total free bending moment at centre = 176 kN m.

The combined bending moment diagram for the two loads is as shown in Fig. 5.28 and the maximum bending moment to which the beam is subjected is 104 kN m.

Hence maximum stress, $\sigma = \dfrac{M}{I} \cdot y$

i.e.
$$75 \times 10^6 = \frac{104 \times 10^3}{I} \cdot y \qquad (1)$$

Central deflection due to concentrated load

$$= -\frac{Wl^3}{192EI} \qquad \text{from equation (5.27)}$$

$$= -\frac{40 \times 10^3 \times 8^3}{192 \times 200 \times 10^9 \, I} = -\frac{0.5333}{10^6 \, I} \text{ m}$$

Central deflection due to distributed load

$$= -\frac{wl^4}{384EI} \qquad \text{from equation (5.29)}$$

$$= -\frac{12 \times 10^3 \times 8^4}{384 \times 200 \times 10^9 \, I} = -\frac{0.64}{10^6 \, I} \text{ m}$$

$$\therefore \frac{0.5333}{10^6 \, I} + \frac{0.64}{10^6 \, I} = 0.0025$$

from which
$$I = 469.2 \times 10^{-6} \text{ m}^4$$

Substituting in equation (1), $y = 721 \times 469.2 \times 10^{-6} = 0.338$ m

∴ depth of section = 0.676 m   or   <u>676 mm</u>

8. *Find the reactions, end fixing moments, central deflection and maximum deflection for the built-in beam, shown in Fig. 5.29.*
EI = 10^6 N m².

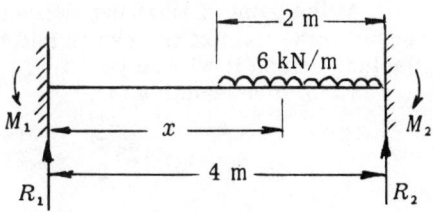

Taking the origin at R_1 and using Macaulay's method,

$$\frac{d^2y}{dx^2} = \frac{10^3}{EI}\left\{R_1 x - \frac{6}{2}[x-2]^2 - M_1\right\}$$

$$\therefore \frac{dy}{dx} = \frac{10^3}{EI}\left\{R_1 \frac{x^2}{2} - \frac{6}{6}[x-2]^3 - M_1 x + A\right\} \quad (1)$$

When $x = 0$, $\frac{dy}{dx} = 0$, so that $A = 0$, the term $[x-2]$ being negative.

$$\therefore y = \frac{10^3}{EI}\left\{R_1 \frac{x^3}{6} - \frac{6}{24}[x-2]^4 - M_1 \frac{x^2}{2} + B\right\} \quad (2)$$

When $x = 0$, $y = 0$, so that $B = 0$, the term $[x-2]$ being negative.

When $x = 4$, $\frac{dy}{dx} = 0$ so that, from equation (1)

$$2R_1 - M_1 = 2 \quad (3)$$

When $x = 4$, $y = 0$ so that, from equation (2)

$$4R_1 - 3M_1 = 1 \cdot 5 \quad (4)$$

Solving equations (3) and (4) gives

$$R_1 = \underline{2 \cdot 25 \text{ kN}} \quad \text{and} \quad M_1 = \underline{2 \cdot 5 \text{ kN m}}$$

Equating vertical forces,

$$R_2 = 6 \times 2 - 2 \cdot 25 = \underline{9 \cdot 75 \text{ kN}}$$

Equating moments about RH end

$$M_2 = 6 \times 2 \times 1 + 2 \cdot 5 - 2 \cdot 25 \times 4 = \underline{5 \cdot 5 \text{ kN m}}$$

The central deflection is obtained by putting $x = 2$ in equation (2),

i.e.
$$y = \frac{10^3}{10^6}\left\{2 \cdot 25 \times \frac{2^3}{6} - 2 \cdot 5 \times \frac{2^2}{2}\right\}$$

$$= -\frac{2}{10^3} \text{ m} \quad \text{or} \quad \underline{-2 \text{ mm}}$$

At the point of maximum deflection, the slope is zero. This point will occur to the right of the centre and hence the term involving $[x-2]$ must be included since it will be positive.

Thus, from equation (1),

$$2 \cdot 25 \frac{x^2}{2} - [x-2]^3 - 2 \cdot 5x = 0$$

or
$$-x^3 + 7 \cdot 125 x^2 - 14 \cdot 5 x + 8 = 0$$

DEFLECTION OF BEAMS

By Newton's method or plotting, $\quad x = 2 \cdot 23$ m

$$\therefore y = \frac{10^3}{10^6}\left\{ 2 \cdot 25 \times \frac{2 \cdot 23^3}{6} - \frac{1}{4}[2 \cdot 23 - 2]^4 - 2 \cdot 5 \times \frac{2 \cdot 23^2}{2} \right\}$$

$$= -\frac{2 \cdot 06}{10^3} \text{ m} \quad \text{or} \quad \underline{-2 \cdot 06 \text{ mm}}$$

9.. A cantilever consists of a steel tube 3 m long, 120 mm outside diameter and 6 mm thick. Calculate the load which, acting 1·8 m from the fixed end, will give a deflection of 2·5 mm at the free end. $E = 200 \text{ GN/m}^2$. (*Ans.*: 450 N)

δ. A cantilever of length l with a concentrated load W at the free end is propped at a distance a from the fixed end to the same level as the fixed end. Find (*a*) the load on the prop, (*b*) the distance of the point of inflexion from the fixed end.

$$\left(Ans.: \; \frac{W}{2a}(3l - a); \; \frac{a}{3} \right)$$

11. A cantilever of length l supports a load W uniformly distributed along its length. The cantilever is propped to the level of the fixed end at a distance $\tfrac{3}{4}l$ from that end. Determine the load on the prop. (*Ans.*: 0·593 W)

12. A cantilever of uniform section and 6 m long is maintained horizontal at one end and supported by a rigid column at a distance of 4 m from the fixed end. The beam carries a load of 80 kN midway between the fixed end and the column and a load of 15 kN at the free end. Determine the force on the column. (*Ans.*: 51·25 kN)

13. A propped cantilever of length l is securely fixed at one end and freely supported at the other. It is subjected to a bending couple M in the vertical plane containing the axis of the beam, applied about an axis 0·75 l from the fixed end. Determine the end fixing moment and the reaction at the support.

(*Ans.*: 0·406 M; 1·406 M/l)

14. A wooden post 6 m high is 50 mm square for the upper 3 m and 100 mm square for the lower 3 m. Find the deflection at the top due to a horizontal pull of 40 N at that point, applied in a direction parallel to one edge of the section.
$E = 10 \text{ GN/m}^2$. (*Ans.*: 99·36 mm)

15. A simply supported beam of T-section has a horizontal flange at the top 100 mm wide and 10 mm thick and a vertical web 10 mm thick and 50 mm deep. The span is 1 m and the central load causes a maximum stress of 120 MN/m². Calculate the central deflection if $E = 200 \text{ GN/m}^2$. (*Ans.*: 1·11 mm)

16. A wooden beam is 240 mm wide and 80 mm deep. It is supported at each end of a span of 4 m and carries concentrated loads of 1 kN each at distances of 1·2 m from each end. Calculate the deflection under the loads and at the centre.
$E = 14 \text{ GN/m}^2$. (*Ans.*: 12·05 mm; 14·75 mm)

17. A beam AB, 8 m long, rests symmetrically on supports C and D, 4 m apart. A load of 40 kN is applied at each of the ends A and B. Calculate the deflection relative to the supports (*a*) at the ends A and B, (*b*) at the centre of CD.
$EI = 10 \text{ MN m}^2$. (*Ans.*: 42·67 mm; 16 mm)

18. A uniform beam of length l is simply supported at its ends and carries a concentrated load W at a distance $l/3$ from one end. Calculate the deflection (*a*) under the load, (*b*) at the centre, (*c*) at the point of maximum deflection.

$$\left(Ans.: \; 0 \cdot 01646 \frac{Wl^3}{EI}; \; 0 \cdot 01775 \frac{Wl^3}{EI}; \; 0 \cdot 01794 \frac{Wl^3}{EI} \right)$$

19. Calculate the deflection at the centre of a simply supported beam of span l carrying a uniformly distributed load w per unit length over the central portion equal to one half of the span.

$$\left(Ans.: \; \frac{19 \, wl^4}{2048 \, EI} \right)$$

20. A simply supported beam of span l carries a uniformly distributed load w per unit length extending for a length $l/3$ from one end. Determine the deflection at mid-span and at the point of maximum deflection. $\left(Ans.: \dfrac{wl^4}{311\,EI};\ \dfrac{wl^4}{305\,EI}\right)$

21. A simply supported beam of span 5 m carries a load which varies uniformly from 20 kN/m at one end to 50 kN/m at the other end. Find the magnitude of the maximum bending moment.

If the depth of the beam is 0·4 m and the maximum bending stress is 100 MN/m², find the central deflection. (*Ans.*: 109·6 kN m; 6·48 mm)

22. A simply supported beam 6 m long carries concentrated loads of 48 kN and 40 kN at points 1 m and 3 m respectively from one end. Determine the position and magnitude of the maximum deflection. $E = 200$ GN/m² and $I = 85 \times 10^{-6}$ m⁴.
(*Ans.*: 2·87 m from end; 16·75 mm)

23. A 300 mm × 125 mm I-beam is built-in at the ends of a span of 6 m and carries a uniformly distributed load of 24 kN/m throughout its length. What is the greatest bending stress in the beam?

By how much per cent is the maximum bending stress increased if the right-hand end becomes free in direction but remains supported at the same level?

I for the beam section = $86·5 \times 10^{-6}$ m⁴ and $E = 200$ GN/m².
(*Ans.*: 124·8 MN/m²; 50%)

24. A horizontal beam, built-in each end, has a clear span of 4·5 m, and carries loads of 50 kN at 1·5 m and 70 kN at 2·5 m from its left-hand end. Calculate the fixing moments and the position of the maximum bending moment.
(*Ans.*: 67·5 kN m; 60·5 kN m; 67·5 kN m at L.H. end)

25. A fixed-ended beam of span 9 m carries a uniformly distributed load of 15 kN/m (including its own weight) and two equal point loads of 200 kN at the third points of the span. Assuming rigid end-fixing, find the fixing moments and the deflection at the centre. $EI = 210$ MN m². (*Ans.*: 518 kN m; 6·84 mm)

26. A 250 mm × 112·5 mm steel beam, $I = 47·6 \times 10^{-6}$ m⁴, is used as a horizontal beam with fixed ends and a clear span of 3 m. Calculate the load which can be applied at one-third span if the bending stress is limited to 120 MN/m².
(*Ans.*: 103 kN)

27. A beam of uniform section, $I = 185 \times 10^{-6}$ m⁴, span 6 m, is fixed horizontally at each end. It carries a point load of 120 kN at 3·6 m from one end. Neglecting the weight of the beam itself, find (*a*) the fixing moments; (*b*) the reactions; (*c*) the position and magnitude of the maximum deflection. $E = 200$ GN/m².
(*Ans.*: 69·12 kN m; 103·68 kN m; 42·25 kN; 77·8 kN; 3·35 mm at 0·275 m from mid-span)

28. A steel joist, 400 mm × 150 mm, $I = 283 \times 10^{-6}$ m⁴, 6 m long, is fixed horizontally at each end and carries loads W and $2W$ at 2 m and 4 m respectively from one end. Draw the bending moment diagram for the beam, stating the values at the principal sections. Find the maximum value of W if the bending stress must not exceed 120 MN/m². (*Ans.*: End fixing moments 1·778W and 2·222W; 76·5 kN)

Properties of the parabola The bending moment diagrams for distributed loads are parabolic and the essential properties of the parabola are as follows:

Area of A = $\tfrac{2}{3}ab$

Area of B = $\tfrac{1}{3}ab$

Distance of centroid of A from O = $\tfrac{5}{8}a$

Distance of centroid of B from Q = $\tfrac{3}{4}a$

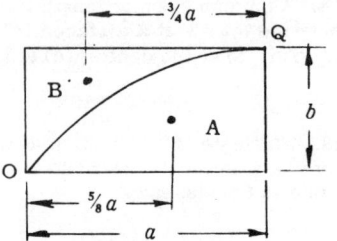

6 Struts

6.1 Introduction A structural member which is subject to a compressive force is called a *strut*. Members which have large cross-sectional area compared with their length will fail by direct compression but those which are slender will fail by buckling before the yield point is reached.

The treatment here is confined to initially straight, homogeneous struts subjected to axial loading applied through the centroids of the ends. Such struts will remain straight until the critical load is reached, when buckling will occur. Any increase in load will cause the strut to collapse and any decrease will cause it to straighten. The value of the critical load depends on the way in which the strut is supported.

6.2 Euler's Theory This assumes that the effect of compressive stress may be neglected and the strut will fail by buckling only.

(a) Strut with both ends pinned, Fig. 6.1

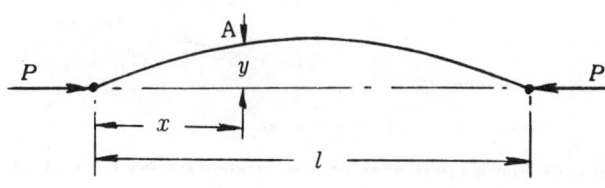

Fig. 6.1

$$\text{B.M. at } A = -Py$$

$$\therefore \frac{d^2y}{dx^2} = \frac{M}{EI} = -\frac{Py}{EI}$$

or

$$\frac{d^2y}{dx^2} + \frac{Py}{EI} = 0$$

i.e.

$$\frac{d^2y}{dx^2} + \mu^2 y = 0 \quad \text{where } \mu = \sqrt{\frac{P}{EI}}$$

The solution is * $\quad y = A\cos\mu x + B\sin\mu x$

When $x = 0$, $y = 0$, so that $A = 0$.

When $x = l$, $y = 0$, so that $B\sin\mu l = 0$,

B cannot be zero, otherwise y is zero for all values of x, i.e., the strut has not buckled. Hence $\sin \mu l = 0$

$$\therefore \mu l = 0, \pi, 2\pi, 3\pi, \text{etc.}$$

μl cannot be zero, so that real values of P are given by

* See Appendix

$$\frac{P}{EI}l^2 = \pi^2, 4\pi^2, 9\pi^2, \text{etc.}$$

$$\text{or} \quad P = \frac{\pi^2 EI}{l^2}, \frac{4\pi^2 EI}{l^2}, \frac{9\pi^2 EI}{l^2}, \text{etc.} \quad (6.1)$$

The lowest value corresponds with the mode of buckling indicated by Fig. 6.1. Higher values correspond with the modes of buckling shown in Fig. 6.2(a) and (b), these cases arising when the strut is prevented from buckling in its fundamental mode.

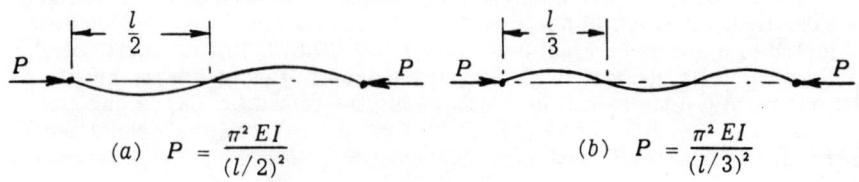

Fig. 6.2

(b) Strut with one end fixed and one end free, Fig. 6.3

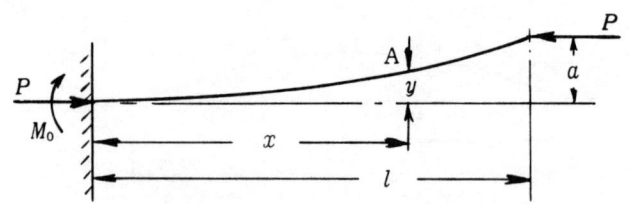

Fig. 6.3

$$\text{B.M. at A} = P(a-y)$$

$$\therefore \frac{d^2y}{dx^2} = \frac{P}{EI}(a-y)$$

or $\quad \dfrac{d^2y}{dx^2} + \mu^2 y = \mu^2 a$

The solution is* $\quad y = A\cos\mu x + B\sin\mu x + a$

When $x = 0$, $y = 0$, so that $A = -a$

When $x = 0$, $\dfrac{dy}{dx} = 0$, so that $B = 0$,

When $x = l$, $y = a$, so that $a = -a\cos\mu l + a$

$$\therefore \cos\mu l = 0$$

$$\therefore \mu l = \frac{\pi}{2}$$

from which $\quad P = \dfrac{\pi^2 EI}{4l^2} \quad (6.2)$

*See Appendix

STRUTS

(c) *Strut with both ends fixed, Fig. 6.4*

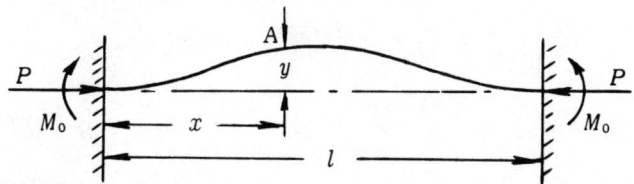

Fig. 6.4

$$\text{B.M. at A} = -Py + M_0$$

$$\therefore \frac{d^2y}{dx^2} = -\frac{Py}{EI} + \frac{M_0}{EI}$$

or

$$\frac{d^2y}{dx^2} + \mu^2 y = \frac{M_0}{EI}$$

The solution is*

$$y = A\cos\mu x + B\sin\mu x + \frac{M_0}{P}$$

When $x = 0$, $y = 0$, so that $A = -\dfrac{M_0}{P}$

When $x = 0$, $\dfrac{dy}{dx} = 0$, so that $B = 0$,

When $x = l$, $y = 0$, so that $0 = \dfrac{M_0}{P}(1 - \cos\mu l)$

$$\therefore \cos\mu l = 1$$

$$\therefore \mu l = 2\pi$$

from which

$$P = \frac{4\pi^2 EI}{l^2} \tag{6.3}$$

(d) *Strut with one end fixed and one end pinned, Fig. 6.5*

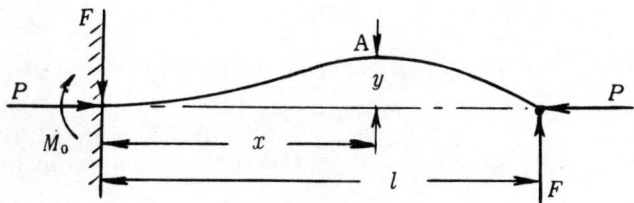

Fig. 6.5

To maintain the pinned end on the axis, a transverse force F is required, the moment Fl balancing the fixing moment M_0.

$$\text{B.M. at A} = -Py + F(l-x)$$

*See Appendix

$$\therefore \frac{d^2y}{dx^2} = -\frac{Py}{EI} + \frac{F}{EI}(l-x)$$

or
$$\frac{d^2y}{dx^2} + \mu^2 y = \frac{F}{EI}(l-x)$$

The solution is*
$$y = A\cos\mu x + B\sin\mu x + \frac{F}{P}(l-x)$$

When $x = 0$, $y = 0$, so that $A = -\dfrac{Fl}{P}$

When $x = 0$, $\dfrac{dy}{dx} = 0$, so that $B = \dfrac{F}{\mu P}$

When $x = l$, $y = 0$, so that $0 = \dfrac{F}{P}\left(-l\cos\mu l + \dfrac{\sin\mu l}{\mu}\right)$

from which $\tan\mu l = \mu l$

The smallest solution is $\mu l = 4\cdot 5$ rad

from which
$$P = \frac{20\cdot 25\,EI}{l^2} \approx \frac{2\pi^2 EI}{l^2} \tag{6.4}$$

6.3 Validity of Euler's Theory As the length of the strut decreases, the critical load increases and, if short enough, the strut will fail by direct compression before the Euler load has been reached. For this condition, the load at failure is given by $\sigma_c a$, where σ_c is the compressive stress at the yield point and a is the cross-sectional area.

The limit of validity of Euler's theory therefore occurs when

$$\sigma_c a = \frac{n\pi^2 EI}{l^2} = \frac{n\pi^2 E(ak^2)}{l^2} = \frac{n\pi^2 Ea}{(l/k)^2}$$

where n has the value 1, ¼, 4 or 2, depending on the end-fixing conditions and k is the least radius of gyration of the cross-section

Thus
$$\frac{l}{k} = \sqrt{\frac{n\pi^2 E}{\sigma_c}} \tag{6.5}$$

The quantity $\dfrac{l}{k}$ is called the *slenderness ratio* of the strut and the value given by equation (6.5) is called the *validity limit*. For a pin-ended mild steel strut, for which $n = 1$, $E = 200$ GN/m² and $\sigma_c = 320$ MN/m², the validity limit is approximately 80 so that Euler's theory is invalid for such struts where $\dfrac{l}{k} < 80$.

Fig. 6.6 shows the hyperbolic relation between P and $\dfrac{l}{k}$ and the dotted line shows the Rankine-Gordon relation (Art. 6.4) which makes allowance for the effect of direct compression. These curves converge at high slenderness ratios.

*See Appendix

STRUTS

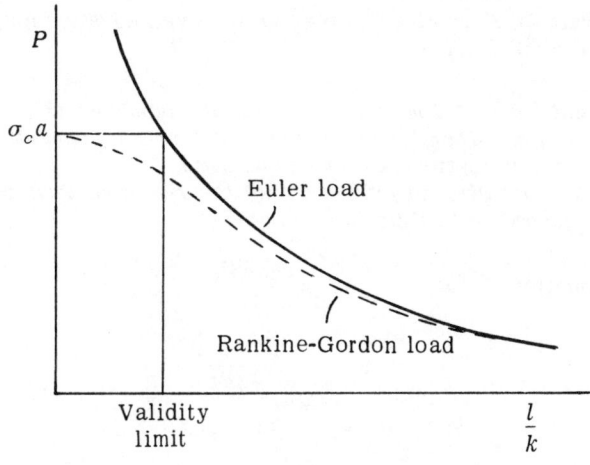

Fig. 6.6

6.4 The Rankine-Gordon relation Euler's theory neglects the effect of direct compression and the critical loads given by it are larger than those obtained by experiment. This error increases as the slenderness ratio approaches the validity limit and for mild steel, for which the validity limit is about 80, Euler's theory should not be used for $\frac{l}{k} < 120$.

For such cases, the Rankine-Gordon relation, based on experimental work, suggests that the strut will fail at a load given by

$$\frac{1}{P} = \frac{1}{P_c} + \frac{1}{P_e}$$

where P_c is the load at which the compressive stress reaches the yield point and P_e is the Euler load.

Thus $$P = \frac{P_c P_e}{P_e + P_c} = \frac{P_c}{1 + \frac{P_c}{P_e}} = \frac{\sigma_c a}{1 + \frac{\sigma_c a}{n\pi^2 EI/l^2}} = \frac{\sigma_c a}{1 + \frac{\sigma_c}{n\pi^2 E}\left(\frac{l}{k}\right)^2}$$

$\frac{\sigma_c}{\pi^2 E}$ is a constant for a given material, so that

$$P = \frac{\sigma_c a}{1 + \frac{c}{n}\left(\frac{l}{k}\right)^2} \qquad (6.6)$$

Values of c could be determined from separate measurements of σ_c and E but experiment shows discrepancies between predicted and measured loads and c is therefore determined directly by experiment. For mild steel, for which $\sigma_c = 320$ MN/m², $c = \frac{1}{7\,500}$ and for cast iron, for which $\sigma_c = 540$ MN/m², $c = \frac{1}{1\,600}$.

Other empirical relations are available for failure loads on struts. In particular, BS 449 – 1969, Specification for the use of Structural Steel in

Buildings (Part 2), Appendix B, gives an involved equation but this is still based on the ideal Euler load.

1. *A straight bar is 1·2 m long and is simply supported at its ends in a horizontal position. When loaded at the centre with a concentrated load of 90 N, the central deflection is found to be 5 mm.*

If placed vertically and loaded along its axis, what load would cause it to buckle, according to Euler's theory?

From equation (5.15),
$$y = -\frac{Wl^3}{48EI}$$

i.e.
$$0 \cdot 005 = \frac{90 \times 1 \cdot 2^3}{48EI}$$

from which
$$EI = 648 \text{ N m}^2$$

When used as a strut,
$$P = \frac{\pi^2 EI}{l^2} \quad \text{. . from equation (6.1)}$$

$$= \frac{\pi^2 \times 648}{1 \cdot 2^2}$$

$$= \underline{4\,440 \text{ N}}$$

2. *A cast iron column 200 mm external diameter is 20 mm thick and 4·5 m long. Assuming that it is rigidly fixed at each end, calculate the safe load by Rankine's formula, using a factor of safety of 4.*

For cast iron, $\sigma_c = 550$ MN/m^2 and $c = \dfrac{1}{1\,600}$.

$$a = \frac{\pi}{4}(D^2 - d^2) = \frac{\pi}{4}\left(\frac{200^2 - 160^2}{10^6}\right) = 0 \cdot 011\,33 \text{ m}^2$$

$$I = \frac{\pi}{64}(D^4 - d^4) = \frac{\pi}{64}\left(\frac{200^4 - 160^4}{10^{12}}\right) = 46 \cdot 4 \times 10^{-6} \text{ m}^4$$

$$k = \sqrt{\frac{I}{a}} = \sqrt{\frac{46 \cdot 4 \times 10^{-6}}{0 \cdot 011\,33}} = 0 \cdot 064 \text{ m}$$

Therefore, from equation (6.6),
$$P = \tfrac{1}{4} \times \frac{\sigma_c a}{1 + \dfrac{c}{4}\left(\dfrac{l}{k}\right)^2}$$

$$= \tfrac{1}{4} \times \frac{550 \times 10^6 \times 0 \cdot 011\,33}{1 + \dfrac{1}{4 \times 1\,600}\left(\dfrac{4 \cdot 5}{0 \cdot 064}\right)^2}$$

$$= \underline{879 \times 10^3 \text{ N}}$$

For this strut, $\dfrac{l}{k} \approx 70$. Taking E for cast iron as approximately 100 GN/m^2, Euler's theory would give a crippling load of about 2·6 times the above value.

STRUTS

3. *A mild steel strut, 8 m long, is fabricated from two flat plates 300 mm wide by 10 mm thick and two channel sections 100 mm wide by 200 mm deep, as shown in Fig. 6.7. The second moments of area of a single channel about its centroid G are:* $I_{XX} = 26 \cdot 20 \times 10^6$ mm^4 *and* $I_{YY} = 1 \cdot 57 \times 10^6$ mm^4. *The cross-sectional area is 4 400 mm^2.*

Assuming pinned ends, determine the critical load (a) by Euler's theory, and (b) by the Rankine-Gordon relation.
$E = 200$ GN/m^2, $c = 1/7500$ and $\sigma_c = 320$ MN/m^2.

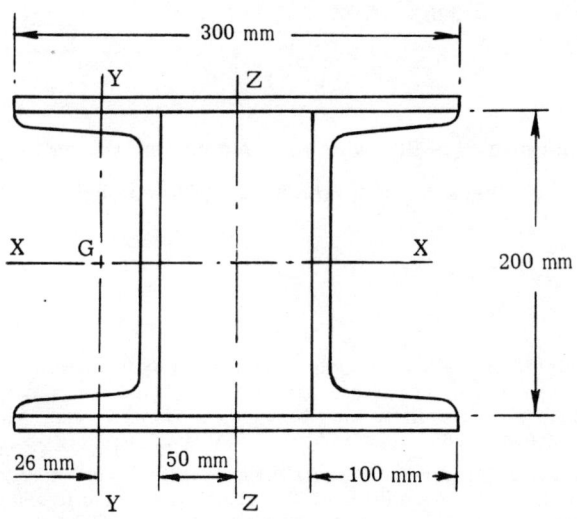

Fig. 6.7

For the complete section

$$a = 2(4\,400 + 300 \times 10)$$

$$= 14\,800 \text{ mm}^2$$

$$I_{XX} = 2 \times 26 \cdot 20 \times 10^6 + 2\left(\frac{300 \times 10^3}{12} + 300 \times 10 \times 105^2\right)$$

$$= 118 \cdot 6 \times 10^6 \text{ mm}^4$$

$$I_{ZZ} = 2(1 \cdot 57 \times 10^6 + 4\,400 \times 76^2) + 2 \times \frac{10 \times 300^3}{12}$$

$$= 99 \cdot 0 \times 10^6 \text{ mm}^4$$

Least radius of gyration of section

$$= \sqrt{\frac{I_{ZZ}}{a}} = \sqrt{\frac{99 \cdot 0 \times 10^6}{14\,800}} = 81 \cdot 7 \text{ mm}$$

(a) $$P = \frac{\pi^2 EI}{l^2} \quad \ldots \quad \text{from equation (6.1)}$$

$$= \frac{\pi^2 \times 200 \times 10^9 \times 99 \times 10^{-6}}{8^2} = \underline{3 \cdot 05 \times 10^6 \text{ N}}$$

(b) $$P = \frac{\sigma_c a}{1 + c\left(\dfrac{l}{k}\right)^2} \quad \ldots \quad \text{from equation (6.6)}$$

$$= \frac{320 \times 10^6 \times 14\,800 \times 10^{-6}}{1 + \dfrac{1}{7\,500}\left(\dfrac{8}{0 \cdot 081\,7}\right)^2} = \underline{2 \cdot 08 \times 10^6 \text{ N}}$$

For this strut, $\dfrac{l}{k} \approx 98$, which is approaching the validity limit for Euler's theory and this leads to an appreciable difference between the results given by the two theories.

4. A straight bar of alloy 1 m long and 12·5 mm by 5 mm in section is mounted in a strut testing machine and loaded axially until it buckles. Using the Euler formula for pin-ended struts, calculate the maximum central deflection before the material yields at a stress of 280 MN/m². $E = 75$ GN/m². (*Ans.*: 0·15 m)

5. Using the Euler formula, calculate the buckling load for a strut 3 m long and hinged at both ends. The strut is of T-section; the flange is 100 mm wide by 10 mm thick and the web is 70 mm deep by 10 mm thick. $E = 200$ GN/m².
(*Ans.*: 209 kN)

6. A steel strip 0·64 m long is arranged as a simply supported beam on a span of 0·5 m and loaded at the centre, when it is found that the central deflection is 0·11 mm/N. Determine the value of EI for the strip and hence calculate the Euler critical load when the strip is tested as a strut with direction-free ends. (*Ans.*: 570 N)

7. Calculate the Euler critical load for a flat steel strip, 75 mm long, 12 mm wide and 0·25 mm thick, used as a strut with built-in ends. $E = 200$ GN/m².
(*Ans.*: 21·9 N)

8. A hollow cast-iron column with fixed ends carries an axial load of 1 MN. The column is 4·5 m long and has an external diameter of 250 mm. Find the thickness of metal required, using the Rankine formula. The constant c for cast iron is 1/1 600 for pinned ends and the working stress is 80 MN/m². (*Ans.*: 28 mm)

9. A pin-ended tubular steel strut 2·25 m long has outer and inner diameters of 38 mm and 33 mm respectively. Determine the crippling loads given by the Euler and Rankine formulae. The yield stress is 325 MN/m², the Rankine constant is 1/7 500 and $E = 200$ GN/m².
For what length of strut does the Euler formula cease to apply?
(*Ans.*: 17·24 kN; 17·20 kN; 0·98 m)

10. In the experimental determination of the buckling loads for 12·5 mm diameter steel struts of various lengths, two of the values obtained were: (i) length 0·5 m, load 9·25 kN, (ii) length 0·2 m, load 25 kN.
 (a) Determine whether either of these results conform with the Euler critical load.
 (b) Assuming that both values conform with the Rankine formula, find the two constants. $E = 200$ GN/m². (*Ans.*: 301·5 MN/m²; 1/8 540)

7 Strain energy

7.1 Introduction The work done in distorting an elastic material is retained in the material as strain-energy and this work can be recovered if the material is subsequently unloaded. It is assumed that straining forces, moments and torques are gradually applied; if suddenly applied, the material will overshoot its equilibrium position and then oscillate about that position. Internal molecular friction, or *hysteresis*, will cause the oscillation to die away, leaving the material, and energy stored, in the same state as if the material had been gradually strained.

7.2 Strain energy due to direct stress Let a tensile load P be gradually applied to an elastic material of cross-sectional area a and length l; the graph of load against extension will then be a straight line, as shown in Fig. 7.1.

An increment of load dP causes an increment of extension dx and the work done is represented by the area da. Thus the total work done by the load P is represented by the area under the graph,

i.e. strain energy, $U = \tfrac{1}{2}Px$

But $x = \dfrac{Pl}{aE}$, so that $U = \tfrac{1}{2}P \cdot \dfrac{Pl}{aE} = \dfrac{P^2 l}{2aE}$ (7.1)

Alternatively, since $\dfrac{P}{a}$ is the stress, σ, $U = \dfrac{\sigma^2 a l}{2E}$

But al is the volume of the material, so that

$$U = \frac{\sigma^2}{2E} \times \text{volume} \qquad (7.2)$$

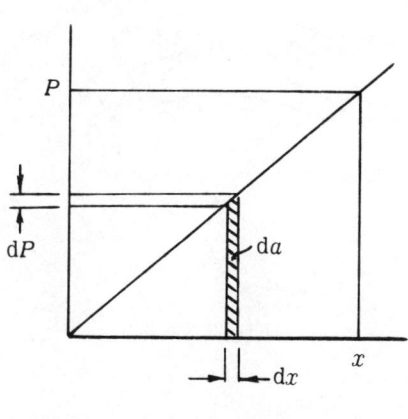

Fig. 7.1

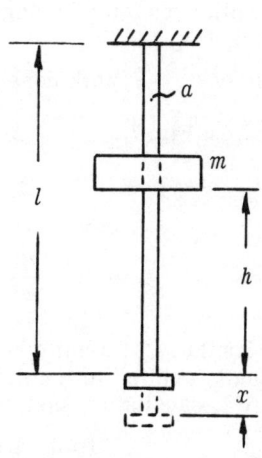

Fig. 7.2

7.3 Impact loading If the load is not gradually applied but is suddenly applied or dropped from a height, the stress, strain and strain energy are all increased. These high values are transient and internal damping of the subsequent oscillation will reduce the values to those obtained under static loading.

Let a mass m be dropped from a height h on to a collar at the lower end of a rod of length l and cross-sectional area a, Fig. 7.2, producing an instantaneous extension x and instantaneous stress σ.

Then loss of potential energy of mass = gain in strain energy of rod

i.e.
$$mg(h + x) = \frac{\sigma^2}{2E} \times \text{volume}$$

But
$$x = \frac{\sigma l}{E}$$

so that
$$mg\left(h + \frac{\sigma l}{E}\right) = \frac{\sigma^2}{2E} \times al$$

from which
$$\sigma = \frac{mg}{a} \pm \sqrt{\left(\frac{mg}{a}\right)^2 - \frac{2mgEh}{al}} \qquad (7.3)$$

The positive sign relates to the stress at the point of maximum extension and the negative sign relates to the stress at the end of the rebound.

When $h = 0$ and the load is suddenly applied,

$$\sigma = \frac{2mg}{a}$$

i.e., the maximum stress is twice that due to a static load.

When h is large compared with the extension x,

$$\sigma = \sqrt{\frac{2mgEh}{al}}$$

7.4 Strain energy due to a uniform moment or torque If a uniform moment M is gradually applied to a uniform beam of length l, causing an angle of bending ϕ, Fig. 7.3,

strain energy = work done = $\tfrac{1}{2}M\phi$

But, from Fig. 7.3, $\phi = \dfrac{l}{R}$

and $\dfrac{1}{R} = \dfrac{M}{EI}$

Hence $U = \tfrac{1}{2}M \dfrac{Ml}{EI} = \dfrac{M^2 l}{2EI} \qquad (7.4)$

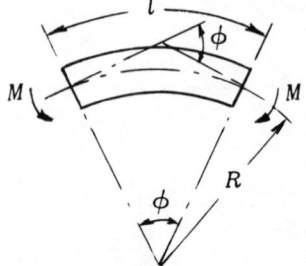

Fig. 7.3

Similarly, if a uniform torque T is gradually applied to a uniform shaft of length l, causing an angle of twist θ,

strain energy = $\tfrac{1}{2}T\theta$

STRAIN ENERGY

But
$$\theta = \frac{Tl}{GJ}$$

$$\therefore U = \tfrac{1}{2} T \cdot \frac{Tl}{GJ} = \frac{T^2 l}{2GJ} \tag{7.5}$$

The torque applied to a shaft is usually constant along its length but it is rare for the bending moment to remain constant along the length of a beam. It is then necessary to sum the energy stored in small elements of the beam, as shown in the following article.

7.5 Strain energy due to a variable bending moment From equation (7.4), the strain energy stored in a short length of beam dx is given by

$$dU = \frac{M^2 \, dx}{2EI}$$

and hence, for the whole beam,

$$U = \int_0^l \frac{M^2 \, dx}{2EI} \tag{7.6}$$

In most cases, the cross-section of the beam is uniform, so that this may be written

$$U = \frac{1}{2EI} \int_0^l M^2 \, dx$$

The appropriate expression for M in terms of x must be substituted before the strain energy can be calculated.

7.6 Castigliano's Theorem Fig. 7.4 shows an elastic body subjected to a number of loads $P_1, P_2, \ldots, P_n$. Let the resultant deflections at the load points *in the directions of the loads* be $y_1, y_2, \ldots, y_n$. If the loads are applied gradually (static loading), the total strain energy,

$$U = \tfrac{1}{2} P_1 y_1 + \tfrac{1}{2} P_2 y_2 + \ldots + \tfrac{1}{2} P_n y_n \tag{7.7}$$

Let P_1 increase to $P_1 + \delta P_1$ and let y_1 increase to $y_1 + \delta y_1$, y_2 to $y_2 + \delta y_2, \ldots, y_n$ to $y_n + \delta y_n$ due to this increase in P_1. Then the increase in strain energy,

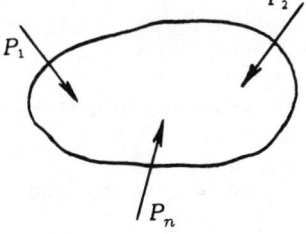

Fig. 7.4

$$\delta U = \tfrac{1}{2} \delta P_1 \delta y_1 + P_1 \delta y_1 + P_2 \delta y_2 + \ldots + P_n \delta y_n$$

Neglecting the first term, which is of the second order of small quantities, this becomes

$$\delta U = P_1 \delta y_1 + P_2 \delta y_2 + \ldots + P_n \delta y_n \tag{7.8}$$

The total strain energy, $U + \delta U$ for the gradually applied loads $P_1 + \delta P_1$, $P_2, \ldots, P_n$, causing deflections $y_1 + \delta y_1, y_2 + \delta y_2, \ldots, y_n + \delta y_n$ is given by

$$U + \delta U = \tfrac{1}{2}(P_1 + \delta P_1)(y_1 + \delta y_1) + \tfrac{1}{2} P_2 (y_2 + \delta y_2) + \ldots + \tfrac{1}{2} P_n (y_n + \delta y_n)$$

and neglecting again the term $½\delta P_1 \delta y_1$, this reduces to

$$U + \delta U = ½P_1 y_1 + ½\delta P_1 y_1 + ½P_1 \delta y_1 + ½P_2 y_2 + ½P_2 \delta y_2 + \ldots + ½P_n y_n + ½P_n \delta y_n \qquad (7.9)$$

Subtracting equation (7.7),

$$\delta U = ½\delta P_1 y_1 + ½P_1 \delta y_1 + ½P_2 \delta y_2 + \ldots + ½P_n \delta y_n \qquad (7.10)$$

Multiplying equation (7.10) by 2 and subtracting equation (7.8),

$$\delta U = \delta P_1 y_1$$

or
$$y_1 = \frac{\delta U}{\delta P_1}$$

In the limit as $\delta P_1 \to 0$, this becomes

$$y_1 = \frac{\partial U}{\partial P_1} \qquad (7.11)$$

Similarly, $\qquad y_2 = \dfrac{\partial U}{\partial P_2}, \ldots, y_n = \dfrac{\partial U}{\partial P_n}$

The partial derivatives arise because the deflection at each point is a function of all the loads.

Equation (7.11) shows that the deflection in the direction of any load due to all the applied loads is the partial derivative of the *total* strain energy of the system with respect to the load considered.

It may also be shown that the rotation θ at any point at which a couple is applied is the partial derivative of the total strain energy with respect to the couple considered.

7.7 Application to deflection of beams and curved bars For simple cases of a body subjected to a single load P, the deflection at the load point, in the direction of the load, may be obtained by equating the strain energy to the work done by the load,

i.e. $\qquad U = \dfrac{1}{2EI} \displaystyle\int_0^l M^2 \, dx = ½Py \qquad (7.12)$

In cases where there are several loads or where the deflection is required in a direction other than in the direction of a load, it is necessary to use Castigliano's Theorem.

$$y = \frac{\partial U}{\partial P} = \frac{\partial}{\partial P}\left(\frac{1}{2EI} \int_0^l M^2 \, dx\right)$$

$$= \frac{1}{2EI} \int_0^l \frac{\partial}{\partial P}(M^2) \, dx$$

$$= \frac{1}{EI} \int_0^l M \frac{\partial M}{\partial P} \, dx \qquad (7.13)$$

STRAIN ENERGY 75

It is legitimate to differentiate M^2 before integrating as the differentiation and integration are with respect to different variables.

If the deflection is required in a direction where these is no applied force, an imaginary force must be applied in that direction. The deflection is then obtained in terms of this force, which is subsequently made zero (see example 4).

7.8 Variable torque If the torque varies along the length of a shaft, the energy stored in a short length dx is given by

$$dU = \frac{T^2\,dx}{2GJ}$$

and hence, for the whole shaft,

$$U = \int_0^l \frac{T^2\,dx}{2GJ} \qquad (7.14)$$

$$= \frac{1}{2GJ}\int_0^l T^2\,dx \quad \text{for a uniform shaft}$$

Castigliano's theorem is equally applicable to torques or couples. Thus, if U is the total strain energy in a body due to various causes, the rotation θ at any point at which a torque T is applied is given by

$$\theta = \frac{\partial U}{\partial T}$$

1. *A metal bar, 40 mm diameter and 1·2 m long, has a collar fitted at the lower end. It is suspended from the upper end and a mass of 2 000 kg is gradually lowered on to the collar, producing an extension in the bar of 0·25 mm. Find the height from which this load could be dropped on to the collar if the maximum tensile stress in the bar is not to exceed 100 MN/m².*

$$E = \frac{Pl}{ax}$$

$$= \frac{2\,000 \times 9 \cdot 81 \times 1 \cdot 2}{\frac{\pi}{4}\times\left(\frac{40}{10^3}\right)^2 \times \frac{0 \cdot 25}{10^3}}$$

$$= 74 \cdot 9 \times 10^9 \text{ N/m}^2$$

When the mass is dropped on to the collar from a height h,

$$mg\left(h + \frac{\sigma l}{E}\right) = \frac{\sigma^2}{2E} \times \text{volume} \qquad . \quad . \quad \text{from Art. 7.3}$$

i.e. $\quad 2\,000 \times 9 \cdot 81\left(h + \dfrac{100 \times 10^6 \times 1 \cdot 2}{74 \cdot 9 \times 10^9}\right) = \dfrac{(100 \times 10^6)^2}{2 \times 74 \cdot 9 \times 10^9} \times 1 \cdot 2 \times \dfrac{\pi}{4}\times\left(\dfrac{40}{10^3}\right)^2$

i.e. $\qquad\qquad h + 0 \cdot 001\,6 = 0 \cdot 005\,13$

$$\therefore h = 0 \cdot 003\,53 \text{ m or } \underline{3 \cdot 53 \text{ mm}}$$

2. *A simply supported beam of length 1·5 m carries a vertical load of 10 kN at the centre. One half of the beam has a diameter of 50 mm and the other half 75 mm. Calculate the central deflection. E = 200 GN/m².*

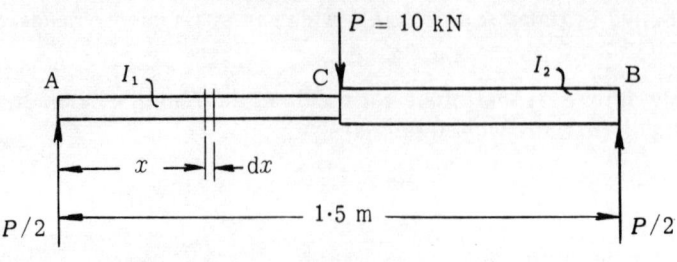

Fig. 7.5

The bending moment at a point distance x from the left-hand support, Fig. 7.5,

$$M = \frac{P}{2}x$$

$$\therefore U_{AC} = \frac{1}{2EI_1}\int_0^{l/2}\left(\frac{P}{2}x\right)^2 dx \quad . \quad . \quad . \quad \text{from equation (7.6)}$$

$$= \frac{P^2}{8EI_1}\left[\frac{x^3}{3}\right]_0^{l/2}$$

$$= \frac{P^2 l^3}{192EI_1}$$

Similarly $U_{BC} = \dfrac{P^2 l^3}{192EI_2}$

$\therefore$ total strain energy,

$$U = \frac{P^2 l^3}{192 E}\left(\frac{1}{I_1} + \frac{1}{I_2}\right) = \tfrac{1}{2}Py \quad . \quad . \quad \text{from equation (7.12)}$$

$$\therefore y = \frac{Pl^3}{96E}\left(\frac{1}{I_1} + \frac{1}{I_2}\right)$$

$$= \frac{10\times 10^3 \times 1\cdot 5^3}{96\times 200\times 10^9}\left(\frac{64\times 10^{12}}{\pi\times 50^4} + \frac{64\times 10^{12}}{\pi\times 75^4}\right)$$

$$= 0\cdot 006\,86 \text{ m} \quad \text{or} \quad \underline{6\cdot 86 \text{ mm}}$$

Note that the sign of M is of no consequence since this is subsequently squared.

3. *A spring is made of steel rod of diameter d, bent to the form shown in Fig. 7.6. Determine the stiffness of the spring if d = 6 mm, r = 40 mm and l = 100 mm. E = 200 GN/m².*

STRAIN ENERGY

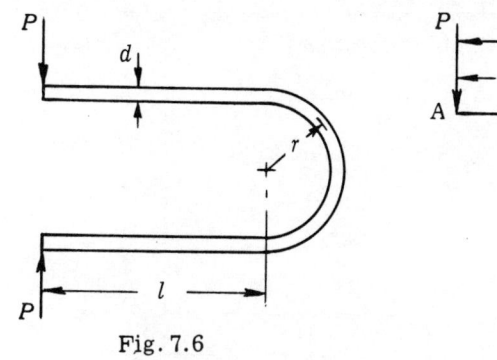
Fig. 7.6

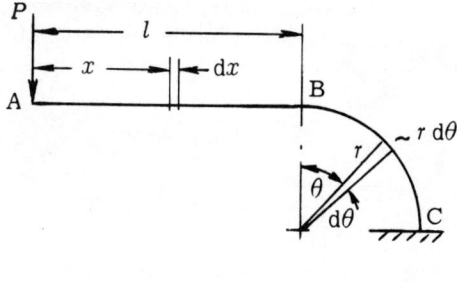
Fig. 7.7

The spring may be imagined to be cut in half and held rigidly at C, Fig. 7.7. The total strain energy is then twice that in the half spring considered.

For AB, $\quad M = Px$

$$\therefore U_{AB} = \frac{1}{2EI} \int_0^l (Px)^2 \, dx$$

$$= \frac{P^2}{2EI} \left[\frac{x^3}{3}\right]_0^l = \frac{P^2 l^3}{6EI}$$

For BC, $\quad M = P(l + r \sin\theta)$

$$\therefore U_{BC} = \frac{1}{2EI} \int_0^{\pi/2} [P(l + r\sin\theta)]^2 r \, d\theta$$

$$= \frac{P^2 r}{2EI} \int_0^{\pi/2} (l^2 + 2lr \sin\theta + r^2 \sin^2\theta) \, d\theta$$

$$= \frac{P^2 r}{2EI} \left[l^2 \theta - 2lr \cos\theta + \frac{r^2}{2}\left(\theta - \frac{\sin 2\theta}{2}\right)\right]_0^{\pi/2}$$

$$= \frac{P^2 r}{2EI} \left(\frac{\pi l^2}{2} + 2lr + \frac{\pi r^2}{4}\right)$$

$$\therefore \text{total strain energy, } U = \frac{P^2}{24EI} [4l^3 + 6\pi l^2 r + 24lr^2 + 3\pi r^3] \times 2$$

$$= \tfrac{1}{2} Py$$

$$\therefore \text{stiffness, } S = \frac{P}{y} = \frac{6EI}{4l^3 + 6\pi l^2 r + 24lr^2 + 3\pi r^3}$$

$$= \frac{6 \times 200 \times 10^9 \times \dfrac{\pi}{64} \times 6^4 \times 10^{-12}}{(4 \times 100^3 + 6\pi \times 100^2 \times 40 + 24 \times 100 \times 40^2 + 3\pi \times 40^3) \times 10^{-9}}$$

$$= 4 \cdot 47 \times 10^3 \text{ N/m}$$

4. *A bar of constant section, second moment of area I, is bent as shown in Fig. 7.8 and fixed at one end. Find the horizontal and vertical deflections at the free end.*

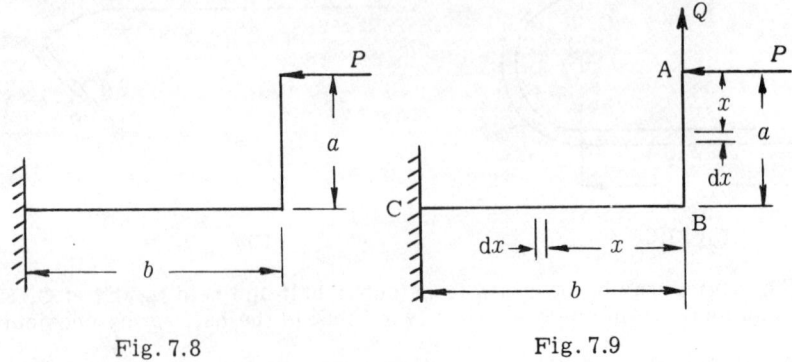

Fig. 7.8   Fig. 7.9

Since there is no force in the vertical direction at the free end, a force Q, of zero magnitude, must be applied there, Fig. 7.9.

Then, for AB, $M = Px$, $\dfrac{\partial M}{\partial P} = x$ and $\dfrac{\partial M}{\partial Q} = 0$

For BC, $M = Pa + Qx$, $\dfrac{\partial M}{\partial P} = a$ and $\dfrac{\partial M}{\partial Q} = x$

For the horizontal deflection,

$$y_h = \frac{\partial U}{\partial P} = \frac{1}{EI} \int M \frac{\partial M}{\partial P} dx$$

$$= \frac{1}{EI} \left[\int_0^a (Px) x \, dx + \int_0^b (Pa + Qx) a \, dx \right]$$

$$= \underline{\frac{Pa^2}{EI} \left(\frac{a}{3} + b \right)}. \quad . \quad . \quad . \quad . \quad \text{since } Q = 0$$

For the vertical deflection,

$$y_v = \frac{\partial U}{\partial Q} = \frac{1}{EI} \int M \frac{\partial M}{\partial Q} dx$$

$$= \frac{1}{EI} \left[\int_0^a (Px) 0 \, dx + \int_0^b (Pa + Qx) x \, dx \right]$$

$$= \underline{\frac{Pab^2}{2EI}}. \quad . \quad . \quad . \quad . \quad . \quad \text{since } Q = 0$$

Note that although Q can be made zero at the integration stage, it cannot be made zero earlier, otherwise no expression for $\dfrac{\partial M}{\partial Q}$ will be obtained.

STRAIN ENERGY

5. *Fig. 7.10 shows a steel rod, 10 mm diameter, with one end firmly fixed. The rod is bent into the form of three-quarters of a circle and the free end is constrained by guides to move in a vertical direction. If the mean radius to which the rod is bent is 150 mm, determine the vertical deflection of the free end when a load of 100 N is gradually applied there.* $E = 200 \text{ GN/m}^2$.

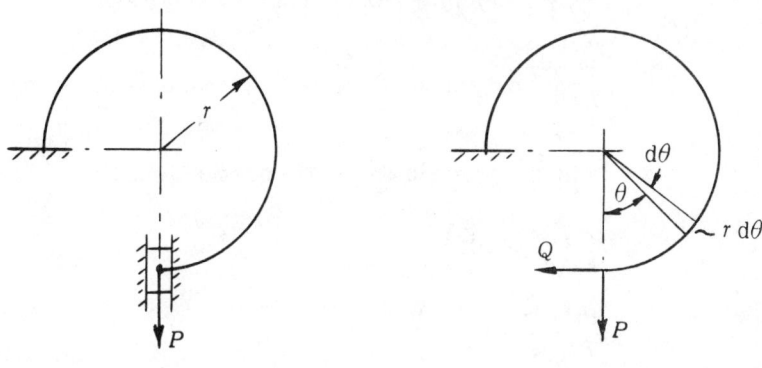

Fig. 7.10   Fig. 7.11

In order to constrain the slider to move vertically, a horizontal force must be applied to the slider by the guide.

Let this force be Q, Fig. 7.11. Then, since there is no horizontal movement in the direction of Q, $\dfrac{\partial U}{\partial Q} = 0$, an equation which will determine Q in terms of P.

At a point subtending an angle θ with the vertical,

$$M = Pr\sin\theta - Qr(1 - \cos\theta)$$

$$\frac{\partial M}{\partial P} = r\sin\theta$$

$$\frac{\partial M}{\partial Q} = -r(1 - \cos\theta)$$

$$\frac{\partial U}{\partial Q} = 0, \quad \therefore \int M \frac{\partial M}{\partial Q} dx = 0$$

i.e. $\displaystyle\int_0^{\frac{3\pi}{2}} \{Pr\sin\theta - Qr(1-\cos\theta)\} \times -r(1-\cos\theta) \times r\, d\theta = 0$

i.e. $\displaystyle P\int_0^{\frac{3\pi}{2}} (\sin\theta - \sin\theta\cos\theta)\, d\theta = Q\int_0^{\frac{3\pi}{2}} (1 - 2\cos\theta + \cos^2\theta)\, d\theta$

i.e. $P\left[-\cos\theta + \tfrac{1}{4}\cos 2\theta\right]_0^{\frac{3\pi}{2}} = Q\left[\theta - 2\sin\theta + \dfrac{\theta}{2} + \tfrac{1}{4}\sin 2\theta\right]_0^{\frac{3\pi}{2}}$

from which $\qquad Q = \dfrac{2P}{9\pi + 8}$

The vertical deflection is given by

$$y = \frac{\partial U}{\partial P} = \frac{1}{EI} \int M \frac{\partial M}{\partial P} dx$$

$$= \frac{1}{EI} \int_0^{\frac{3\pi}{2}} \{Pr \sin\theta - Qr(1-\cos\theta)\} \times r \sin\theta \times r \, d\theta$$

$$= \frac{r^3}{EI} \int_0^{\frac{3\pi}{2}} \{P \sin^2\theta - Q(\sin\theta - \sin\theta\cos\theta)\} d\theta$$

$$= \frac{r^3}{EI} \left[\frac{P}{2}(\theta - \tfrac{1}{2}\sin 2\theta) - Q(-\cos\theta + \tfrac{1}{4}\cos 2\theta) \right]_0^{\frac{3\pi}{2}}$$

$$= \frac{r^3}{EI}\left(\frac{3\pi}{4}P - \frac{Q}{2}\right)$$

$$= \frac{Pr^3}{EI}\left(\frac{3\pi}{4} - \frac{1}{9\pi+8}\right)$$

$$= \frac{100 \times 0 \cdot 15^3}{200 \times 10^9 \times \frac{\pi}{64} \times \frac{10^4}{10^{12}}} \times 2 \cdot 333$$

$$= \underline{0 \cdot 008 \text{ m} \quad \text{or} \quad 8 \cdot 0 \text{ mm}}$$

6. *A cantilever, lying in a horizontal plane, is in the shape of a quadrant of a circle, of radius 150 mm. One end is firmly fixed and a vertical force of 200 N is applied at the other end. Find the vertical deflection of the free end.*

The cantilever is made from circular rod, 10 mm diameter, for which $E = 200 \text{ GN/m}^2$ *and* $G = 80 \text{ GN/m}^2$.

Referring to Fig. 7.12, the force P may first be moved to the point A by adding a moment $P \times a$ to compensate and then to B by adding a moment $P \times b$ to compensate.

The moment $P \times a$ causes bending of the bar at B and the moment $P \times b$ causes twisting of the bar at B.

Hence $\quad M = Pa = Pr \sin\theta$

and $\quad T = Pb = Pr(1 - \cos\theta)$

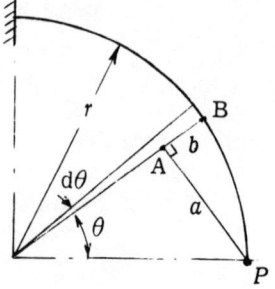

Fig. 7.12

$$\therefore \quad U = \frac{1}{2EI}\int_0^l M^2 \, dx + \frac{1}{2GJ}\int_0^l T^2 \, dx$$

$$= \frac{1}{2EI}\int_0^{\frac{\pi}{2}} (Pr \sin\theta)^2 . r \, d\theta + \frac{1}{2GJ}\int_0^{\frac{\pi}{2}} (Pr[1-\cos\theta])^2 . r \, d\theta$$

STRAIN ENERGY

i.e. $\dfrac{P\delta}{2} = \dfrac{P^2 r^3}{2}\left\{\dfrac{1}{EI}\int_0^{\frac{\pi}{2}} \sin^2\theta\, d\theta + \dfrac{1}{GJ}\int_0^{\frac{\pi}{2}}(1 - 2\cos\theta + \cos^2\theta)\, d\theta\right\}$

$\therefore\ \delta = P r^3\left\{\dfrac{1}{EI}\cdot\dfrac{\pi}{4} + \dfrac{1}{GJ}\left(\dfrac{\pi}{2} - 2 + \dfrac{\pi}{4}\right)\right\}$

$= 200 \times 0.15^3\left[\dfrac{1}{200 \times 10^9 \times \dfrac{\pi}{64} \times 0.01^4}\cdot\dfrac{\pi}{4} + \dfrac{1}{80 \times 10^9 \times \dfrac{\pi}{32} \times 0.01^4}\left(\dfrac{3\pi}{4} - 2\right)\right]$

$= 0.625\,(0.008 + 0.004\,54)$

$= 0.008\,46$ m   or   <u>8·46 mm</u>

7. A ring mass of 25 kg encircles a bar and falls through a distance h before striking a stop fixed to the bottom of the bar which hangs from a rigid support. The bar is of steel 25 mm diameter and 3 m long. What must be the value of h if the maximum extension given to the bar is 0·63 mm? $E = 200$ GN/m². (*Ans.*: 25·9 mm)

8. A mass of 10 kg falls through a height of 150 mm and then commences to stretch a steel bar 18 mm diameter and 0·9 m long. Determine (*a*) the maximum stress induced in the bar, (*b*) the maximum elongation of the bar, (*c*) the energy stored in the bar at the point of maximum entension.
$E = 200$ GN/m². (*Ans.*: 160·3 MN/m²; 0·721 mm; 14·78 J)

9. A simply supported beam carries a central concentrated load W. Determine the total strain energy in the beam due to bending and hence obtain an expression for the deflection under the load. $\left(Ans.:\ \dfrac{W^2 l^3}{96EI};\ \dfrac{W l^3}{48EI}\right)$

10. A simply supported beam of span l carries a uniformly distributed load w per unit length. Determine the total strain energy in the beam due to bending.
$\left(Ans.:\ \dfrac{w^2 l^5}{240EI}\right)$

11. A steel tube 56 mm outside diameter and 50 mm internal diameter is fixed vertically on a rigid base. At a distance of 0·9 m from the base the tube is bent into a quadrant of a circle of radius 0·6 m and a vertical load of 2 kN is applied at the free end. Calculate the vertical deflection at the load. $E = 200$ GN/m².
(*Ans.*: 28·1 mm)

12. A steel ring of rectangular cross-section 7·5 mm wide by 5 mm thick has a mean diameter of 300 mm. A narrow radial saw cut is made and tangential separating forces of 5 N each are applied at the cut in the plane of the ring. Determine the additional separation due to these forces. $E = 200$ GN/m². (*Ans.*: 3·83 mm)

13. A steel spring of uniform section is shown in Fig. 7.13. Derive an expression for the vertical movement of the free end due to the vertical force P.
(*Ans.*: $23.92\, P r^3/EI$)

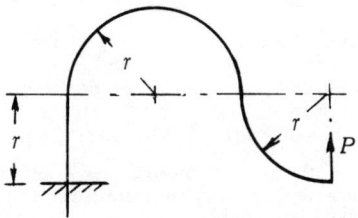

Fig. 7.13

14. The stiff frame shown in Fig. 7.14 is supported on a smooth surface and loaded at the centre of the span. Derive an expression for the deflection at the load point.
(*Ans.*: $0.3175\,Pa^3/EI$)

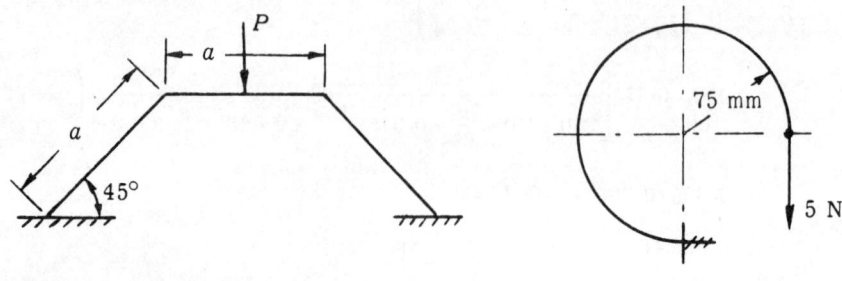

Fig. 7.14   Fig. 7.15

15. The steel spring shown in Fig. 7.15 is rigidly fixed at the lower end and a vertical force of 5 N is applied at the free end. If the section of the spring is 12 mm by 3 mm thick, calculate the vertical and horizontal displacements of the free end. $E = 200 \text{ GN/m}^2$.   (*Ans.*: 3·65 mm; 0·195 mm)

16. The ring shown in Fig. 7.16 is made of flat steel strip 20 mm by 3 mm. It is cut at the point B and a pull P is applied to the ring along a diameter perpendicular to AB. If the maximum tensile stress due to P is 125 MN/m², find the increase in the opening at B. $E = 200 \text{ GN/m}^2$.   (*Ans.*: 14·9 mm)

17. A steel tube, having outside and inside diameters of 60 mm and 45 mm respectively, is bent in the form of a quadrant 2 m radius. One end is rigidly attached to a horizontal base-plate to which a tangent to that end is perpendicular and the free end supports a vertical load of 500 N. Determine the vertical and horizontal deflecof the free end.   (*Ans.*: 36·1 mm; 23·0 mm)

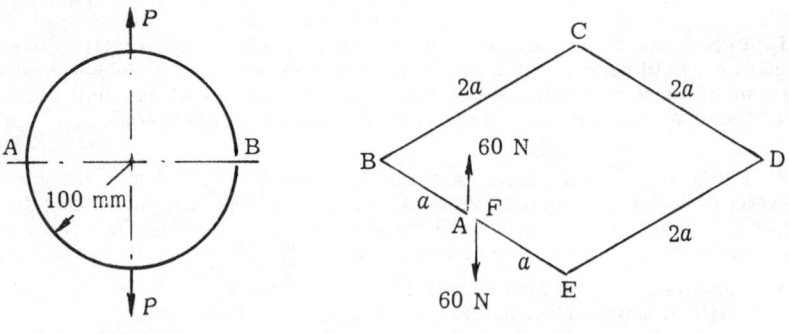

Fig. 7.16   Fig. 7.17

18. A 10 mm diameter steel rod is bent to form a square with sides $2a = 40$ mm long. The ends meet at the mid-point of one side and are separated by equal and opposite forces of 60 N applied in a direction perpendicular to the plane of the square as shown in perspective in Fig. 7.17. Calculate the amount by which they will be out of alignment. $E = 200 \text{ GN/m}^2$; $G = 80 \text{ GN/m}^2$.   (*Ans.*: 9·78 mm)

19. A cantilever forming a circular arc in plan, and subtending $\pi/3$ rad at the centre, has a circular cross-section of 50 mm diameter and the radius of the centre line is 0·75 m. Find the maximum deflection when a load of 400 N is acting at the free end. $E = 200 \text{ GN/m}^2$; $G = 80 \text{ GN/m}^2$.   (*Ans.*: 2·75 mm)

8 Shear stress in beams

8.1 Shear stress distribution When a shearing force is applied to a beam, there will be a shear stress on the transverse section and a complementary shear stress on longitudinal sections. The shear stress is not uniform across the transverse section and cannot be obtained by dividing the shearing force by the cross-sectional area. The equilibrium of longitudinal forces is used to determine the complementary shear stress and this is equal to the transverse stress.

It is assumed that the shear stress is uniform across planes parallel to the neutral axis, an assumption which is reasonable if there is no sudden change in width.

Fig. 8.1 shows a portion of a beam ABCD, of length dx, which is in equilibrium under the action of forces on AC and BD due to bending stresses and on CD due to the shear stress τ.

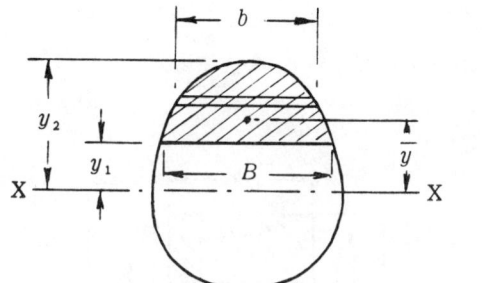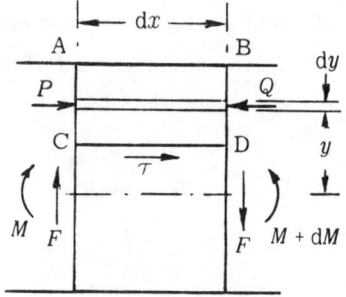

Fig. 8.1

The force P acting on the end of an element of breadth b and thickness dy

$$P = \frac{My}{I} \cdot b \, dy$$

Similarly,
$$Q = \frac{(M + dM)y}{I} \cdot b \, dy$$

Therefore, resultant force on element $= Q - P = \dfrac{dM}{I} y \, b \, dy$

Therefore, for the equilibrium of ABCD,

$$\int_{y_1}^{y_2} \frac{dM}{I} y \, b \, dy = \tau B \, dx$$

from which
$$\tau = \frac{1}{IB} \frac{dM}{dx} \int_{y_1}^{y_2} y \, b \, dy$$

83

But
$$\frac{dM}{dx} = F \quad \ldots \quad \text{from equation (2.2)}$$

$$\therefore \tau = \frac{F}{IB} \int_{y_1}^{y_2} y\, b\, dy \qquad (8.1)$$

If a is the area of the cross-section between the limits y_1 and y_2 and $\bar{y}$ is the distance of the centroid of this area from XX, then

$$\int_{y_1}^{y_2} y\, b\, dy = a\bar{y}$$

so that equation (8.1) can be written in the form

$$\tau = \frac{F}{IB} a\bar{y} \qquad (8.2)$$

Equation (8.2) shows that the shear stress is always zero at edges distant from the neutral axis.

8.2 Application to common cases

(a) *Rectangular section of breadth b and depth d, Fig. 8.2.*

At a layer distance h from XX,

$$a = b\left(\frac{d}{2} - h\right)$$

$$\bar{y} = \tfrac{1}{2}\left(\frac{d}{2} + h\right)$$

$$B = b$$

and $\quad I = \dfrac{bd^3}{12}$

Hence $\quad \tau = \dfrac{6F}{bd^3}\left(\dfrac{d^2}{4} - h^2\right) \qquad (8.3)$

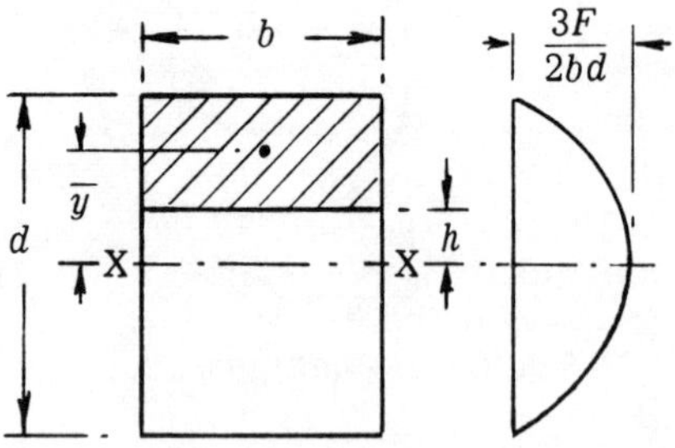

Fig. 8.2

This is a parabolic distribution with a maximum at the neutral axis of $\dfrac{3}{2}\dfrac{F}{bd}$, i.e., $\tfrac{3}{2}$ times the average shear stress.

(b) *Circular section, radius r, Fig. 8.3.*

At a layer distance h from XX,

$$\int_h^r y\, b\, dy = \int_h^r y \times 2\sqrt{r^2 - y^2}\, dy$$

$$= \frac{2}{3}(r^2 - h^2)^{3/2}$$

SHEAR STRESS IN BEAMS

$$B = 2\sqrt{r^2 - h^2}$$

and

$$I = \frac{\pi r^4}{4}$$

Hence

$$\tau = \frac{4F}{3\pi r^4}(r^2 - h^2) \tag{8.4}$$

This is again a parabolic stress distribution with a maximum at the neutral axis of $\frac{4F}{3\pi r^2}$, i.e., $\frac{4}{3}$ times the average shear stress.

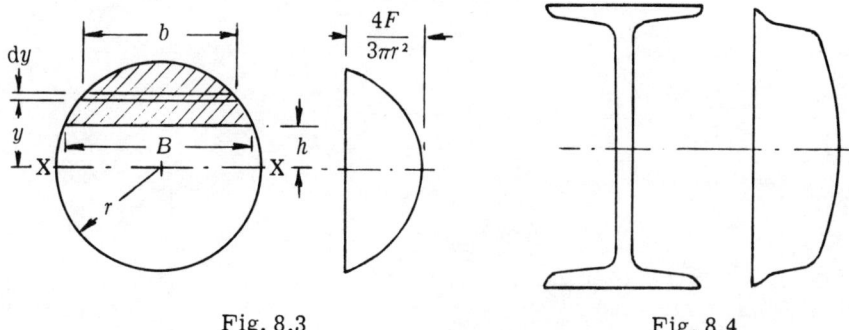

Fig. 8.3   Fig. 8.4

(c) *I-beam, Fig. 8.4.* When there is an abrupt increase in the width of section, as in an I-beam or T-beam, the shear stress decreases accordingly but the simple theory that this stress is uniform across planes parallel to the neutral axis breaks down. However, the stress in the flange is sufficiently small that this is of little consequence and it is customary to assume that the whole of the shear force is carried by the web (See Ex. 3).

1. *A beam has a symmetrical triangular section of breadth b and depth d and is subjected to a transverse shearing force F acting in the direction of the axis of symmetry. Deduce the shearing stress at any depth Z from the vertex of the section and plot a graph to show how the shearing stress varies across the section.*

Find the ratio of the average shearing stress over the section to the maximum shearing stress.

Using the equimomental system (see Art. 3.1), the area of the triangle may be replaced by areas $\frac{bd}{6}$ considered concentrated at the mid-points of the sides, Fig. 8.5.

Then
$$I_{XX} = \frac{bd}{6}\left(\frac{d}{3}\right)^3 + 2 \times \frac{bd}{6}\left(\frac{d}{6}\right)^2$$
$$= \frac{bd^3}{36}$$

Fig. 8.5

At a point distance z below the apex, Fig. 8.6,

$$B = \frac{zb}{d}$$

$$a = \tfrac{1}{2} \cdot \frac{zb}{d} \cdot z = \frac{z^2 b}{2d}$$

$$\bar{y} = \tfrac{2}{3}d - \tfrac{2}{3}z = \tfrac{2}{3}(d - z)$$

$$\therefore \tau = \frac{F}{\frac{bd^3}{36} \cdot \frac{zb}{d}} \times \frac{z^2 b}{2d} \times \tfrac{2}{3}(d - z)$$

$$= \underline{\frac{12Fz}{bd^3}(d - z)}$$

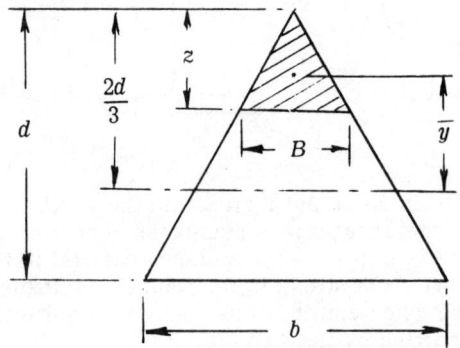 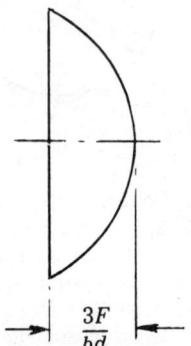

Fig. 8.6

For maximum shear stress, $\quad \dfrac{d\tau}{dz} = 0$

from which $\quad z = \dfrac{d}{2}$

Hence $\quad \tau_{max} = \dfrac{12F}{bd^3} \cdot \dfrac{d}{2} \cdot \dfrac{d}{2} = \dfrac{3F}{bd}$

$$\tau_{mean} = \frac{2F}{bd}$$

$$\frac{\tau_{mean}}{\tau_{max}} = \frac{2}{3}$$

It should be noted that the maximum shear stress does not necessarily occur at the neutral axis.

SHEAR STRESS IN BEAMS

2. *A steel bar of the section shown in Fig. 8.7 is subjected to a shearing force of 200 kN applied in the direction YY. Determine the shearing stress at sections AA and BB.*

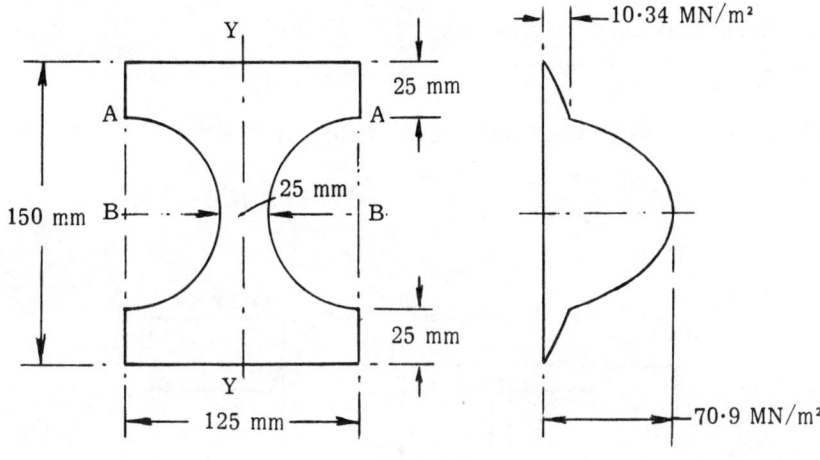

Fig. 8.7

$$I = \left(\frac{125 \times 150^3}{12} - \frac{\pi}{64} \times 100^4\right) \times 10^{-12} = 30 \cdot 25 \times 10^{-6} \text{ m}^4$$

At section AA,

$$a\bar{y} = 125 \times 25 \times 62 \cdot 5 \times 10^{-9} = 195 \cdot 4 \times 10^{-6} \text{ m}^3$$

$$B = 0 \cdot 125 \text{ m}$$

$$\therefore \tau = \frac{200 \times 10^3}{30 \cdot 25 \times 10^{-6} \times 0 \cdot 125} \times 195 \cdot 4 \times 10^{-6} = \underline{10 \cdot 34 \times 10^6 \text{ N/m}^2}$$

At section BB,

$$a\bar{y} = \left(125 \times 75 \times 37 \cdot 5 - \frac{\pi}{8} \times 100^2 \times \frac{4 \times 50}{3\pi}\right) \times 10^{-9} = 268 \cdot 2 \times 10^{-6} \text{ m}^3$$

$$B = 0 \cdot 025 \text{ m}$$

$$\therefore \tau = \frac{200 \times 10^3}{30 \cdot 25 \times 10^{-6} \times 0 \cdot 025} \times 268 \cdot 2 \times 10^{-6} = \underline{70 \cdot 9 \times 10^6 \text{ N/m}^2}$$

3. *A 120 mm × 300 mm I-beam has flanges 15 mm thick and a web 10 mm thick and at a certain section it is subjected to a shearing force of 200 kN. Draw a diagram showing the distribution of shearing stress across the section and determine the percentage of the shearing force which is carried by the web.*

Referring to Fig. 8.8,

$$I = \left(\frac{120 \times 300^3}{12} - \frac{110 \times 270^3}{12}\right) \times 10^{-12} = 89 \cdot 5 \times 10^{-6} \text{ m}^4$$

Immediately above section AA,

$B = 0 \cdot 12$ m

and $\quad a\bar{y} = 0 \cdot 12 \times 0 \cdot 015 \times 0 \cdot 1425 = 256 \cdot 5 \times 10^{-6}$ m^3

$$\therefore \tau = \frac{200 \times 10^3 \times 256 \cdot 5 \times 10^{-6}}{89 \cdot 5 \times 10^{-6} \times 0 \cdot 12} = 4 \cdot 77 \times 10^6 \text{ N/m}^2$$

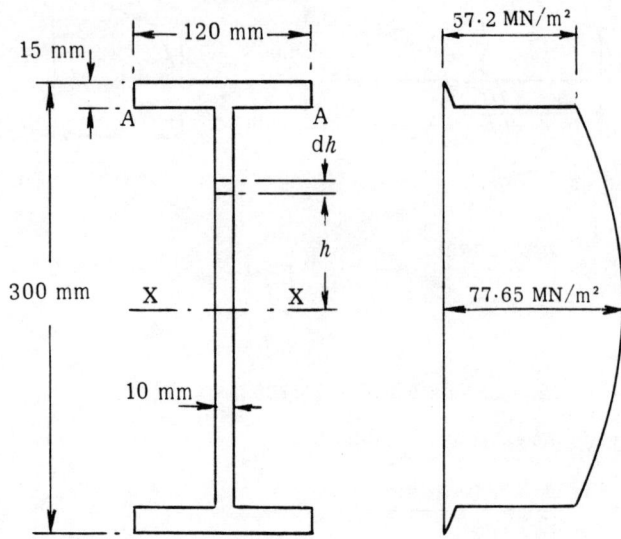

Fig. 8.8

Immediately below section AA, $B = 0 \cdot 01$ m and $a\bar{y}$ remains as before.

$$\therefore \tau = 4 \cdot 77 \times 10^6 \times \frac{0 \cdot 12}{0 \cdot 01} = 57 \cdot 2 \times 10^6 \text{ N/m}^2$$

At a section distance h above XX,

$$a\bar{y} = 0 \cdot 12 \times 0 \cdot 015 \times 0 \cdot 1425 + (0 \cdot 135 - h) \times 0 \cdot 01 \times \left(\frac{0 \cdot 135 + h}{2}\right)$$

$$= (0 \cdot 348 - 5h^2) \times 10^{-3} \text{ m}^3$$

$$\therefore \tau = \frac{200 \times 10^3 \times (0 \cdot 348 - 5h^2) \times 10^{-3}}{89 \cdot 5 \times 10^{-6} \times 0 \cdot 01} = 223 \cdot 5(0 \cdot 348 - 5h^2) \times 10^6 \text{ N/m}^2$$

SHEAR STRESS IN BEAMS

The graph of this equation is shown in Fig. 8.8. The maximum value is given by

$$\tau_{max} = 223 \cdot 5 \times 0 \cdot 348 \times 10^6 = 77 \cdot 65 \times 10^6 \text{ N/m}^2 \text{ when } h = 0$$

S.F. carried by web $= 2 \int_0^{0 \cdot 135} \tau B \, dh$

$$= 2 \int_0^{0 \cdot 135} 223 \cdot 5 (0 \cdot 348 - 5h^2) \times 10^6 \times 0 \cdot 01 \, dh$$

$$= 4 \cdot 47 \times 10^6 \left[0 \cdot 348 h - \tfrac{5}{3} h^3 \right]_0^{0 \cdot 135} = 192 \times 10^3 \text{ N}$$

Therefore percentage of S.F. carried by web $= \dfrac{192}{200} \times 100 = \underline{96\%}$

If the applied S.F. is assumed to be uniformly distributed over the web area alone,

$$\tau = \frac{200 \times 10^3}{0 \cdot 27 \times 0 \cdot 01} = 74 \times 10^6 \text{ N/m}^2$$

which is very near to the actual maximum stress.

4. An I-beam 350 mm × 200 mm has a web thickness of 12·5 mm and a flange thickness of 25 mm. Calculate the ratio of maximum to mean shearing stress in the section and the percentage of the total shear carried by the web. (*Ans.*: 3·57; 92·9 %)

5. A cast-iron beam has a top flange 125 mm × 50 mm and bottom flange 200 mm × 50 mm. The web is 40 mm thick and the overall depth of the section is 250 mm. If the transverse shearing force is 140 kN, calculate the shear stress in the web at the top and bottom junctions with the flanges and also the maximum shear stress.
(*Ans.*: 15 MN/m²; 17·1 MN/m²; 18·5 MN/m²)

6. A cantilever of I-section 200 mm × 100 mm has rectangular flanges 10 mm thick and web 75 mm thick. It carries a uniformly distributed load. Determine the length of the cantilever if the maximum bending stress is three times the maximum shearing stress. What is the ratio of the stresses half-way along the cantilever?
(*Ans.*: 1 m; 1·5)

7. A T-beam has a flange and web each 150 mm × 25 mm and is subjected to a shear force of 200 kN. Sketch the shear stress distribution curve and determine the maximum shear stress. (*Ans.*: 65·3 MN/m²)

8. A beam of circular section, diameter D, is subjected to a shearing force F. Find the maximum shearing stress in terms of F and D.
If the beam is simply supported on a span L and carries a central concentrated load, find the ratio L/D if the maximum shearing stress is half the maximum bending stress. (*Ans.*: $16F/3\pi D^2$; $\tfrac{2}{3}$)

9. A hollow circular section 100 mm external diameter and 75 mm internal diameter is subjected to a shearing force of 160 kN. Sketch the shear stress distribution curve and determine the maximum shear stress. (*Ans.*: 91·7 MN/m²)

10. A beam of square section, side s, is subjected to a shear force F acting parallel with a diagonal. Draw the shear stress distribution curve and find the position and magnitude of the maximum shear stress. (*Ans.*: $s/4\sqrt{2}$ from axis; $9F/8s^2$)

9 Complex stress and strain

9.1 Stresses on an oblique section The material of cross-sectional area a shown in Fig. 9.1 is subjected to a tensile load P, giving a direct stress $\sigma = P/a$. To determine the stresses on the plane XY inclined at an angle θ to the transverse section, consider the equilibrium of the part ABYX. The area of XY is $a \sec \theta$ so that, if the normal and tangential forces on XY are N and T respectively,

$$N = P \cos \theta \quad \text{and} \quad T = P \sin \theta$$

Therefore the direct stress on XY,

$$\sigma_\theta = \frac{N}{a \sec \theta} = \frac{P}{a} \cos^2 \theta = \sigma \cos^2 \theta \qquad (9.1)$$

and the shear stress on XY,

$$\tau_\theta = \frac{T}{a \sec \theta} = \frac{P}{a} \sin \theta \cos \theta = \frac{\sigma}{2} \sin 2\theta \qquad (9.2)$$

The maximum value of σ_θ occurs when $\theta = 0$ and is equal to the applied stress σ. The maximum value of τ_θ occurs when $\theta = 45°$ and is equal to $\sigma/2$. Thus if a material has an ultimate shear stress of less than half the ultimate direct stress (tensile or compressive), then failure under *direct* load will occur due to *shear* stress on the oblique plane.

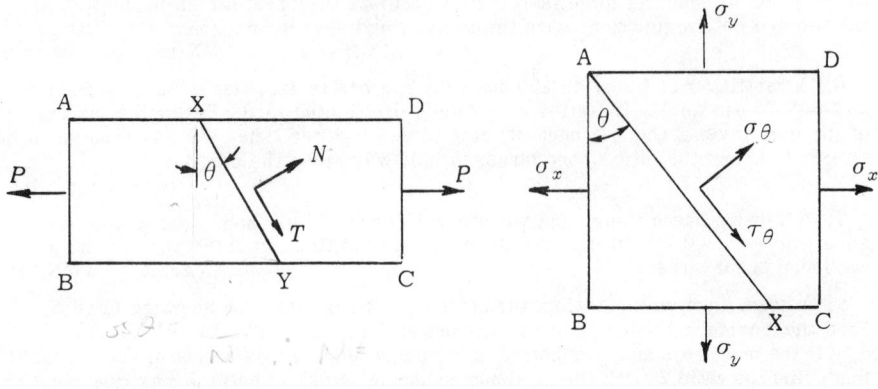

Fig. 9.1   Fig. 9.2

9.2 Material subjected to two perpendicular direct stresses Fig. 9.2 shows an element of material which is subjected to two perpendicular tensile stresses σ_x and σ_y. To obtain the direct and shear stresses, σ_θ and τ_θ respectively, on plane AX inclined at angle θ to AB, consider the equilibrium of the wedge ABX and let the material be of unit thickness.

COMPLEX STRESS AND STRAIN

Equating *forces* normal to **XX**,

$$\sigma_\theta \times AX = \sigma_x \times AB\cos\theta + \sigma_y \times BX\sin\theta$$

$$\therefore \sigma_\theta = \sigma_x \cos^2\theta + \sigma_y \sin^2\theta$$

$$= \frac{\sigma_x + \sigma_y}{2} + \frac{\sigma_x - \sigma_y}{2}\cos 2\theta \qquad (9.3)$$

The maximum and minimum values of σ_θ are σ_x and σ_y (whichever is greater and less respectively) when $\theta = 0$ and $90°$.

Equating *forces* parallel to AX,

$$\tau_\theta \times AX = \sigma_x \times AB\sin\theta - \sigma_y \times BX\cos\theta$$

$$\therefore \tau_\theta = (\sigma_x - \sigma_y)\sin\theta\cos\theta$$

$$= \frac{\sigma_x - \sigma_y}{2}\sin 2\theta \qquad (9.4)$$

The maximum value of τ_θ is $\frac{\sigma_x - \sigma_y}{2}$ when $\theta = 45°$. It is also apparent that if $\sigma_x = \sigma_y$, τ_θ is zero for all values of θ.

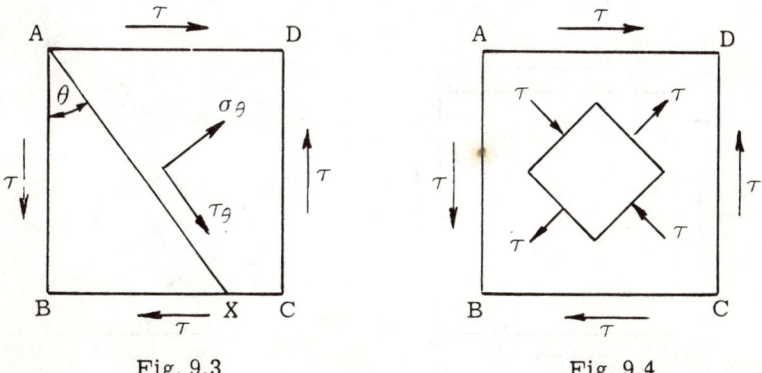

Fig. 9.3   Fig. 9.4

9.3 Material subjected to shear stress Fig. 9.3 shows an element of material of unit thickness which is subjected to applied and complementary shear stresses τ.

Equating *forces* normal to AX for the equilibrium of ABX,

$$\sigma_\theta \times AX = \tau \times AB\sin\theta + \tau \times BX\cos\theta$$

$$\therefore \sigma_\theta = \tau \sin 2\theta \qquad (9.5)$$

The maximum value of σ_θ is τ when $\theta = 45°$.

Equating *forces* parallel to AX,

$$\tau_\theta \times AX = -\tau \times AB\cos\theta + \tau \times BX\sin\theta$$

$$\therefore \tau_\theta = -\tau \cos 2\theta \qquad (9.6)$$

The maximum value of τ_θ is τ when $\theta = 0$ and $90°$ and when $\theta = 45°$, τ_θ is zero.

Equations (9.5) and (9.6) show that the application of shear stress produces tensile and compressive stresses on planes at 45° to the shear planes equal in magnitude to the applied stress, as shown in Fig. 9.4.

9.4 The general case of two-dimensional stress Fig. 9.5 shows an element of material which is subjected to perpendicular tensile stresses and applied and complementary shear stresses.

Combining the stresses obtained in Arts. (9.2) and (9.3),

$$\sigma_\theta = \frac{\sigma_x + \sigma_y}{2} + \frac{\sigma_x - \sigma_y}{2} \cos 2\theta + \tau \sin 2\theta \qquad (9.7)$$

and
$$\tau_\theta = \frac{\sigma_x - \sigma_y}{2} \sin 2\theta - \tau \cos 2\theta \qquad (9.8)$$

σ_θ is a maximum or minimum when $\dfrac{d\sigma_\theta}{d\theta} = 0$,

i.e., when $\quad -(\sigma_x - \sigma_y)\sin 2\theta + 2\tau \cos 2\theta = 0$

or
$$\tan 2\theta = \frac{2\tau}{\sigma_x - \sigma_y} \qquad (9.9)$$

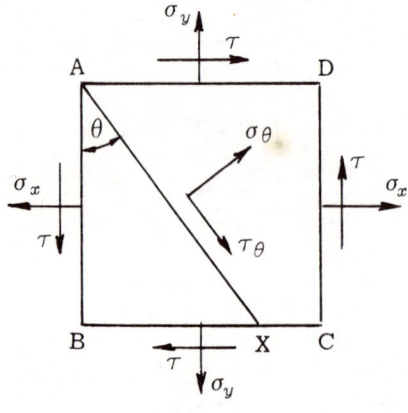
Fig. 9.5

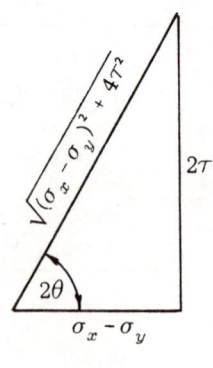
Fig. 9.6

From Fig. 9.6, it will be seen that

$$\cos 2\theta = \frac{\sigma_x - \sigma_y}{\sqrt{(\sigma_x - \sigma_y)^2 + 4\tau^2}} \quad \text{and} \quad \sin 2\theta = \frac{2\tau}{\sqrt{(\sigma_x - \sigma_y)^2 + 4\tau^2}}$$

so that the maximum and minimum values of σ_θ are given by

$$\sigma_\theta = \frac{\sigma_x + \sigma_y}{2} + \frac{\sigma_x - \sigma_y}{2} \cdot \frac{\sigma_x - \sigma_y}{\sqrt{(\sigma_x - \sigma_y)^2 + 4\tau^2}} + \tau \cdot \frac{2\tau}{\sqrt{(\sigma_x - \sigma_y)^2 + 4\tau^2}}$$

$$= \tfrac{1}{2}\left[(\sigma_x + \sigma_y) \pm \sqrt{(\sigma_x - \sigma_y)^2 + 4\tau^2}\right] \qquad (9.10)$$

COMPLEX STRESS AND STRAIN

The planes on which these stresses act are given by equation (9.9),

i.e. $\quad \theta = \frac{1}{2}\tan^{-1}\frac{2\tau}{\sigma_x - \sigma_y} \quad$ and $\quad \frac{1}{2}\tan^{-1}\frac{2\tau}{\sigma_x - \sigma_y} + 90°$

These planes of maximum and minimum direct stress are mutually perpendicular and are known as the *principal planes*; the direct stresses which act on them are the *principal stresses*.

Substituting for $\cos 2\theta$ and $\sin 2\theta$ in equation (9.8) shows that *the shear stress on principal planes is zero*.

Fig. 9.7 shows the principal planes and stresses in the general stress system, the principal stresses being denoted by σ_1 and σ_2. It is often obvious which stress is associated with each principal plane but when in doubt, numerical substitution of values of θ in equation (9.7) will make the determination.

The maximum shear stress in the general case may be determined by differentiating equation (9.8) but from equation (9.4), it is given by $(\sigma_1 - \sigma_2)/2$, acting at $45°$ to the principal planes,

i.e. $\quad \tau_{max} = \dfrac{\frac{1}{2}\left[(\sigma_x+\sigma_y) + \sqrt{(\sigma_x-\sigma_y)^2 + 4\tau^2}\right] - \frac{1}{2}\left[(\sigma_x+\sigma_y) - \sqrt{(\sigma_x-\sigma_y)^2 + 4\tau^2}\right]}{2}$

$\qquad\qquad = \frac{1}{2}\sqrt{(\sigma_x - \sigma_y)^2 + 4\tau^2} \qquad\qquad\qquad\qquad (9.11)$

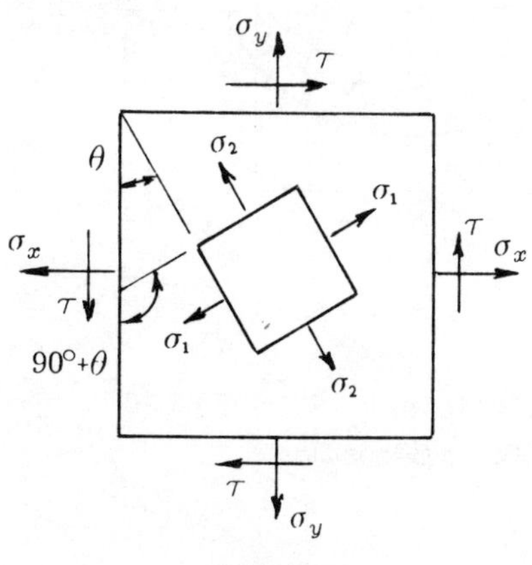

Fig. 9.7 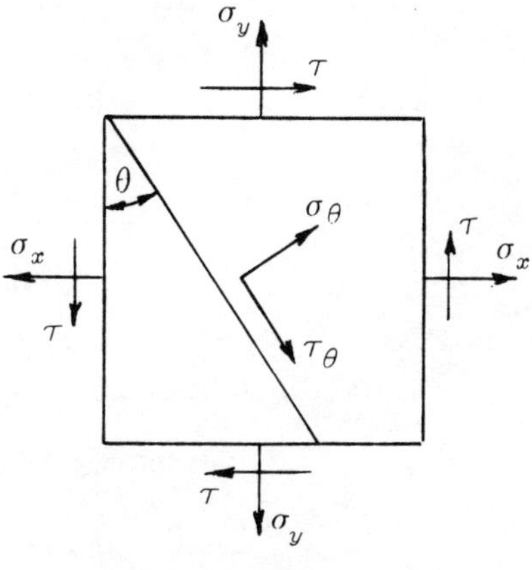   Fig. 9.8

9.5 Mohr's stress circle An alternative graphical method of determination of principal stresses and planes is given by Mohr's stress circle.

For the stress system shown in Fig. 9.8, choose a pole P, Fig. 9.9, and set off PA $= \sigma_x$ and PB $= \sigma_y$ (if σ_x or σ_y is negative, then PA or PB must be set off in the opposite direction). The shear stress on the plane of σ_x results in an anticlockwise couple, which is considered negative while the shear on the plane of σ_y results in a clockwise (positive) couple. Set off AM $= -\tau$ and BN $= +\tau$ and join MN to cut PBA at O. Draw a circle of centre O and radius OM and draw OC such that angle MOC $= 2\theta$. Let angle MOD be β.

It will be seen that $PO = \dfrac{\sigma_x + \sigma_y}{2}$ and $OA = \dfrac{\sigma_x - \sigma_y}{2}$.

Radius $OM = \sqrt{OA^2 + AM^2} = \sqrt{\left(\dfrac{\sigma_x - \sigma_y}{2}\right)^2 + \tau^2} = OC$

$$\begin{aligned} PD &= PO + OC\cos(2\theta - \beta) \\ &= PO + (OM\cos\beta)\cos 2\theta + (OM\sin\beta)\sin 2\theta \\ &= \dfrac{\sigma_x + \sigma_y}{2} + \dfrac{\sigma_x - \sigma_y}{2}\cos 2\theta + \tau\sin 2\theta \end{aligned}$$

From equation (9.7), PD represents σ_θ.

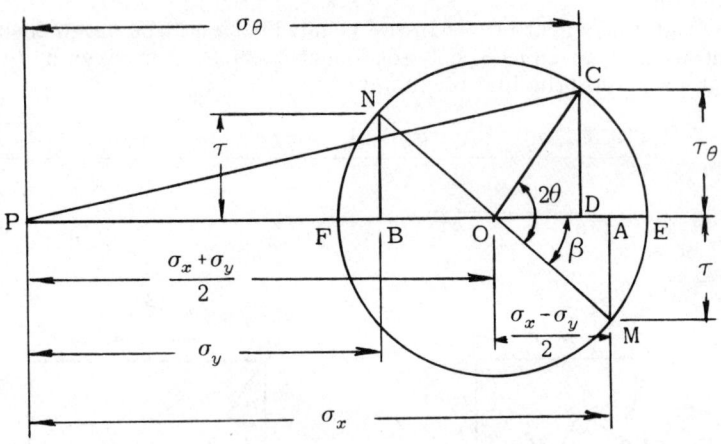

Fig. 9.9

$$\begin{aligned} CD &= OC\sin(2\theta - \beta) \\ &= (OM\cos\beta)\sin 2\theta - (OM\sin\beta)\cos 2\theta \\ &= \dfrac{\sigma_x + \sigma_y}{2}\sin 2\theta - \tau\sin 2\theta \end{aligned}$$

From equation (9.8), CD represents τ_θ.

When τ_θ is zero, PC lies on PE and $2\theta = \beta$ or $\beta + 180°$.

Then
$$\begin{aligned} PE &= PO + OE \\ &= \dfrac{\sigma_x + \sigma_y}{2} + \sqrt{\left(\dfrac{\sigma_x - \sigma_y}{2}\right)^2 + \tau^2} \\ &= \tfrac{1}{2}\left[(\sigma_x + \sigma_y) + \sqrt{(\sigma_x - \sigma_y)^2 + 4\tau^2}\right] \end{aligned}$$

COMPLEX STRESS AND STRAIN

and $\quad$ PF = PO - OF

$$= \frac{\sigma_x + \sigma_y}{2} - \sqrt{\left(\frac{\sigma_x - \sigma_y}{2}\right)^2 + \tau^2}$$

$$= \tfrac{1}{2}[(\sigma_x + \sigma_y) - \sqrt{(\sigma_x - \sigma_y)^2 + 4\tau^2}]$$

By comparison with equation (9.10), PE and PF represent the principal stresses, σ_1 and σ_2 and the principal planes are inclined at $\beta/2$ and $\beta/2 + 90°$ to the plane of σ_x.

If the applied shear stress τ is zero, then AM = BN = 0 and the diagram simplifies to that shown in Fig. 9.10.

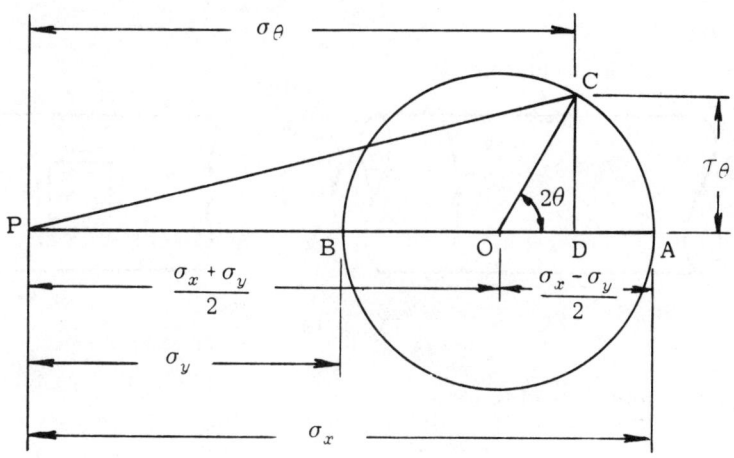

Fig. 9.10

9.6 Simple applications of principal stresses

(a) *Combined thrust and twisting* Fig. 9.11 shows a shaft of radius r which is subjected to a torque T and an axial thrust P. The direct stress due to P is $\dfrac{P}{a}$ and the shear stress at the surface of the shaft due to T is $\dfrac{Tr}{J}$, so that the maximum direct stress,

$$\sigma = \tfrac{1}{2}\left[\frac{P}{a} + \sqrt{\left(\frac{P}{a}\right)^2 + 4\left(\frac{Tr}{J}\right)^2}\right] \quad \cdot \quad \text{from equation (9.10)}$$

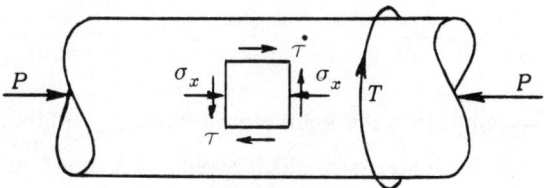

Fig. 9.11

and the maximum shear stress,

$$\tau_{max} = \tfrac{1}{2}\sqrt{\left(\frac{P}{a}\right)^2 + 4\left(\frac{Tr}{J}\right)^2} \quad \ldots \ldots \quad \text{from equation (9.11)}$$

(b) Combined bending and twisting Fig. 9.12 shows a shaft of radius r which is subjected to a torque T and a bending moment M. The direct stress at the surface of the shaft due to M is $\frac{Mr}{I}$ and the shear stress at the surface due to T is $\frac{Tr}{J}$, so that the maximum direct stress,

$$\sigma = \tfrac{1}{2}\left[\frac{Mr}{I} + \sqrt{\left(\frac{Mr}{I}\right)^2 + 4\left(\frac{Tr}{J}\right)^2}\right]$$

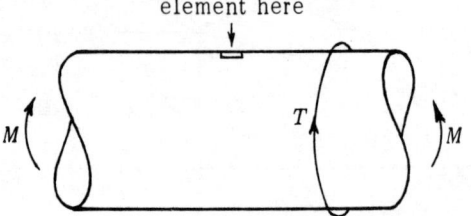 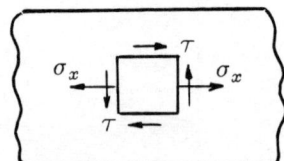

Fig. 9.12

This can be re-arranged to give

$$\frac{\sigma I}{r} = \tfrac{1}{2}[M + \sqrt{M^2 + T^2}] \quad \ldots \ldots \quad \text{since } J = 2I$$

The term $\frac{\sigma I}{r}$ represents the equivalent bending moment, M_e, which, acting alone, would produce the same maximum direct stress as M and T acting together,

i.e. $\qquad M_e = \tfrac{1}{2}[M + \sqrt{M^2 + T^2}] \qquad (9.12)$

The maximum shear stress,

$$\tau_{max} = \tfrac{1}{2}\sqrt{\left(\frac{Mr}{I}\right)^2 + 4\left(\frac{Tr}{J}\right)^2}$$

This can be re-arranged to give

$$\frac{\tau_{max} J}{r} = \sqrt{M^2 + T^2} \quad \ldots \ldots \quad \text{since } I = \frac{J}{2}$$

The term $\frac{\tau_{max} J}{r}$ represents the equivalent torque, T_e, which, acting alone, would produce the same maximum shear stress as M and T acting together,

i.e. $\qquad T_e = \sqrt{M^2 + T^2} \qquad (9.13)$

9.7 Theories of elastic failure In simple cases of direct stress, elastic failure is assumed to have occurred when the stress reaches the elastic limit stress for the material. In cases of complex stress, however, stresses on planes perpendicular to that of the maximum principal stress may affect the point at which failure occurs. This depends on whether the smaller stresses are of the same or opposite sign to that of the greater stress and whether the material is brittle or ductile.

In the case of brittle materials, failure is found to occur when the maximum principal stress reaches the elastic limit, regardless of the value or nature of the stresses on perpendicular planes. For ductile materials, however, the stresses on planes perpendicular to that of the maximum principal stress affect the value of the maximum principal stress at which failure occurs and one theory suggests that failure occurs when the maximum *shear* stress reaches that at the elastic limit in the case of simple direct stress.

If the stress at the elastic limit in the case of direct strees only is denoted by σ_0, the maximum shear stress is $\sigma_0/2$, from equation (9.2).

Denoting the principal stresses in a complex stress situation by σ_x, σ_y and σ_z, the maximum shear stress is half the difference between the greatest and least stresses, from equation (9.4).

If $\sigma_x > \sigma_y > \sigma_z$ and all are of the same sign, then

$$\tau_{max} = \frac{\sigma_x - \sigma_z}{2}$$

Thus, for the maximum shear stress theory of failure,

$$\frac{\sigma_x - \sigma_z}{2} = \frac{\sigma_0}{2}$$

or

$$\sigma_x - \sigma_z = \sigma_0$$

If the principal stresses are not all of the same sign, the maximum shear stress will involve the numerical *sum* of two stresses instead of the difference and each individual case must be investigated to find the combination of principal stresses which gives the greatest shear stress.

In the case of two dimensional stress, the third stress $\sigma_z = 0$. Thus, if σ_x and σ_y are of opposite signs, the maximum shear stress is $\frac{\sigma_x + \sigma_y}{2}$ but if they are of like sign, the maximum shear stress is $\frac{\sigma_x + 0}{2} = \frac{\sigma_x}{2}$.

This shear stress theory finds experimental support for ductile materials and fracture in cases of complex stress occurs on planes inclined to those of the principal stresses, suggesting failure by shear.

More sophisticated theories of failure involve three-dimensional strain energy, which is beyond the scope of this book.

The application of the maximum principal stress theory and the maximum shear stress theory to a shaft subjected to combined bending and twisting is shown in Example 4.

9.8 Principal strains The principal strains are those in the direction of the principal planes. If these strains are ϵ_1 and ϵ_2, then

$$\epsilon_1 = \frac{\sigma_1}{E} - \nu \frac{\sigma_2}{E} \qquad (9.14)$$

and

$$\epsilon_2 = \frac{\sigma_2}{E} - \nu \frac{\sigma_1}{E} \qquad (9.15)$$

If the principal strains are measured, then, by re-arrangement of equations (9.14) and (9.15), the principal stresses are given by

$$\sigma_1 = \frac{E}{1-\nu^2}(\epsilon_1 + \nu\epsilon_2) \qquad (9.16)$$

and

$$\sigma_2 = \frac{E}{1-\nu^2}(\epsilon_2 + \nu\epsilon_1) \qquad (9.17)$$

If the direction of the principal planes is unknown, the magnitude and direction of the principal strains and planes may be determined by measuring the strain in three different directions by means of strain gauge rosettes, Art. 9.9, and then solving for the three unknown quantities ϵ_1, ϵ_2 and θ.

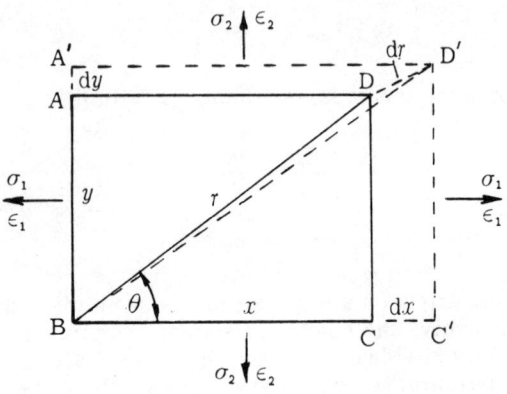

Fig. 9.13

Fig. 9.13 shows an element of material ABCD which distorts to $A'BC'D'$ under the action of principal stresses σ_1 and σ_2. The principal strains are ϵ_1 and ϵ_2 and the strain on BD, inclined at angle θ to BC is ϵ_θ.

Since these strains are very small, the change in length of BD may be taken as DD',

i.e.
$$\epsilon_\theta = \frac{dr}{r}$$

$$r^2 = x^2 + y^2$$

$$\therefore 2r\, dr = 2x\, dx + 2y\, dy$$

$$\therefore \frac{dr}{r} = \frac{dx}{x}\left(\frac{x}{r}\right)^2 + \frac{dy}{y}\left(\frac{y}{r}\right)^2$$

COMPLEX STRESS AND STRAIN   99

i.e. $\epsilon_\theta = \epsilon_1 \cos^2\theta + \epsilon_2 \sin^2\theta$

                    $= \dfrac{\epsilon_1 + \epsilon_2}{2} + \dfrac{\epsilon_1 - \epsilon_2}{2} \cos 2\theta$ (9.18)

This result may be obtained by Mohr's strain circle, Fig. 9.14, constructed in a similar manner to the stress circle.

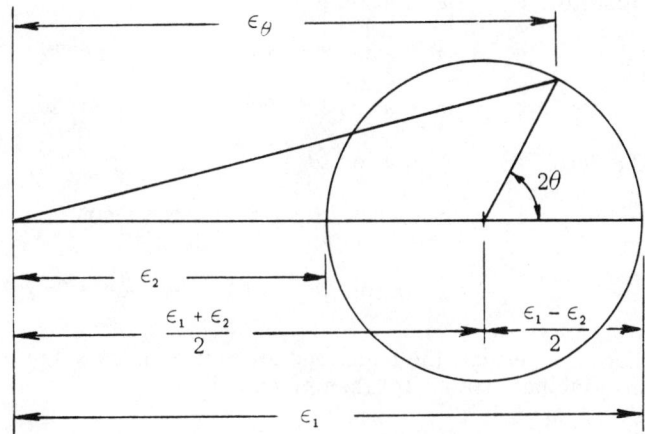

Fig. 9.14

9.9 Electric resistance strain gauges An electric resistance strain gauge consists of a fine wire arranged as shown in Fig. 9.15(a) and fixed to a paper backing. This is then cemented to the surface to be investigated and distortion in the direction of the gauge axis changes the electrical resistance of the wire, which is measured by a Wheatstone bridge circuit. Strain in a lateral direction does not affect the resistance of the wire.

The *gauge factor*, defined as the ratio $\dfrac{\text{fractional change in resistance}}{\text{fractional change in length}}$, is approximately 2, this being calibrated by the manufacturer, and strain bridges are available calibrated directly in strain.

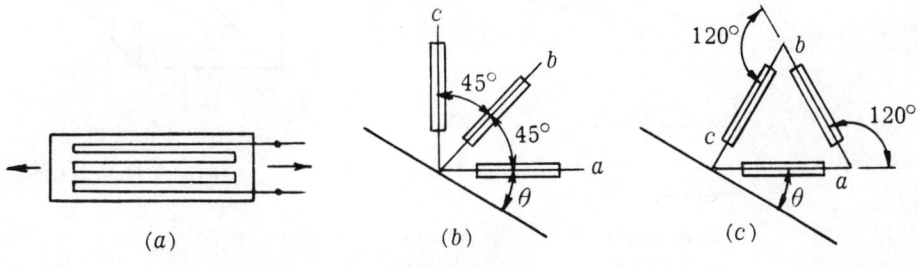

Fig. 9.15

To determine the principal strains and planes, strain rosettes are used. These consist of three strain gauges which are glued to the surface with their axes inclined either at 45° intervals, as in Fig. 9.15(b), or at 120° intervals, as in Fig. 9.15(c).

COMPLEX STRESS AND STRAIN

From equation (9.18), $\epsilon_\theta = \dfrac{\epsilon_1 + \epsilon_2}{2} + \dfrac{\epsilon_1 - \epsilon_2}{2} \cos 2\theta$

which may be written $\epsilon_\theta = m + n \cos 2\theta$ where $m = \dfrac{\epsilon_1 + \epsilon_2}{2}$
and $n = \dfrac{\epsilon_1 - \epsilon_2}{2}$

For the 45° rosette, $\epsilon_a = m + n \cos 2\theta$

$\epsilon_b = m + n \cos 2(\theta + 45°) = m - n \sin 2\theta$

and $\epsilon_c = m + n \cos 2(\theta + 90°) = m - n \cos 2\theta$

For the 120° rosette, $\epsilon_a = m + n \cos 2\theta$

$\epsilon_b = m + n \cos 2(\theta + 120°) = m - \dfrac{n}{2} \cos 2\theta + \dfrac{\sqrt{3}n}{2} \sin 2\theta$

and $\epsilon_c = m + n \cos 2(\theta + 240°) = m - \dfrac{n}{2} \cos 2\theta - \dfrac{\sqrt{3}n}{2} \sin 2\theta$

For either arrangement, the equations are sufficient to solve for m, n and θ and the principal strains are then given by

$$\epsilon_1 = m + n \quad \text{and} \quad \epsilon_2 = m - n$$

The principal stresses are obtained by substitution in equations (9.16) and (9.17).

9.10 Volumetric strain and bulk modulus Fig. 9.16 shows an element of material subjected to principal stresses σ_x, σ_y and σ_z.

Strain on side x,

$$\epsilon_x = \dfrac{\sigma_x - \nu(\sigma_y + \sigma_z)}{E} \quad (9.19)$$

Similarly,

$$\epsilon_y = \dfrac{\sigma_y - \nu(\sigma_x + \sigma_z)}{E} \quad (9.20)$$

and $\epsilon_z = \dfrac{\sigma_z - \nu(\sigma_x + \sigma_y)}{E} \quad (9.21)$

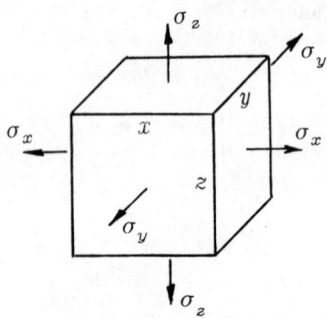

Fig. 9.16

New length of side $x = x(1 + \epsilon_x)$
New length of side $y = y(1 + \epsilon_y)$
New length of side $z = z(1 + \epsilon_z)$

$\therefore$ new volume $= xyz(1 + \epsilon_x)(1 + \epsilon_y)(1 + \epsilon_z)$

$= xyz(1 + \epsilon_x + \epsilon_y + \epsilon_z)$

if products of strains are neglected

COMPLEX STRESS AND STRAIN

$$\text{Volumetric strain} = \frac{\text{change in volume}}{\text{original volume}}$$

i.e.
$$e_v = \frac{xyz(1 + \epsilon_x + \epsilon_y + \epsilon_z) - xyz}{xyz}$$

$$= \epsilon_x + \epsilon_y + \epsilon_z \qquad (9.22)$$

In the special case of equal stresses, such as with fluid pressure,

$$\sigma_x = \sigma_y = \sigma_z = \sigma$$

and
$$\epsilon_x = \epsilon_y = \epsilon_z = \frac{\sigma}{E}(1 - 2\nu)$$

so that
$$\epsilon_v = \frac{3\sigma}{E}(1 - 2\nu) \qquad (9.23)$$

For such a case, the *bulk modulus* (K) is defined as the ratio $\dfrac{\text{stress}}{\text{volumetric strain}}$

i.e.
$$K = \frac{\sigma}{\epsilon_v}$$

$$\therefore \epsilon_v = \frac{\sigma}{K} \qquad (9.24)$$

Hence, from equations (9.23) and (9.24),

$$\frac{3\sigma}{E}(1 - 2\nu) = \frac{\sigma}{K}$$

from which
$$E = 3K(1 - 2\nu) \qquad (9.25)$$

In the case of unequal stresses,

$$\epsilon_v = \epsilon_x + \epsilon_y + \epsilon_z$$

$$= \frac{\sigma_x - \nu(\sigma_y + \sigma_z)}{E} + \frac{\sigma_y - \nu(\sigma_x + \sigma_z)}{E} + \frac{\sigma_z - \nu(\sigma_x + \sigma_y)}{E}$$

$$= \frac{1 - 2\nu}{E}(\sigma_x + \sigma_y + \sigma_z)$$

$$= \frac{\sigma_x + \sigma_y + \sigma_z}{3K} \qquad (9.26)$$

9.11 Relation between E, G and ν Fig. 9.17(*a*) shows an element of material subjected to shear stress τ. From Art. 9.3, it was seen that this results in direct tensile and compressive stresses τ on an element inclined at 45° to the planes of shear and hence

$$\text{strain on the diagonal BD} = \frac{\tau}{E} + \nu\frac{\tau}{E} \qquad (9.27)$$

102　　　　　　　　　　　　　　　　　　　　COMPLEX STRESS AND STRAIN

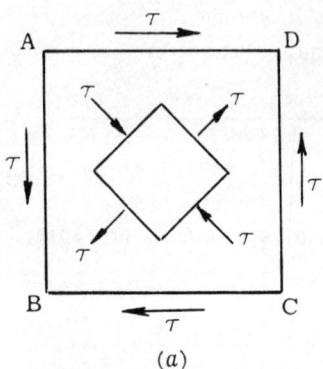
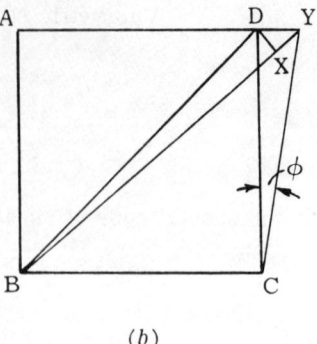

(a)　　　　　　　　　　　　　　(b)

Fig. 9.17

Also, in Fig. 9.17(b),

$$\text{strain on BD} = \frac{XY}{BD} = \frac{DY/\sqrt{2}}{\sqrt{2}\,CD} = \frac{\phi}{2} = \frac{\tau}{2G} \qquad (9.28)$$

Hence, from equations (9.27) and (9.28),

$$\frac{\tau}{E}(1 + \nu) = \frac{\tau}{2G}$$

from which

$$E = 2G(1 + \nu) \qquad (9.29)$$

1. *The principal stresses at a point under two-dimensional stress are 80 MN/m² tension and 50 MN/m² compression. Calculate from first principles the resultant stress on a plane inclined at 30° to the line of action of the tensile stress.*

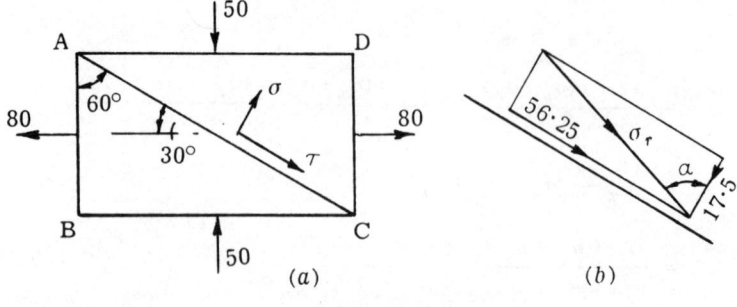

Fig. 9.18

Equating forces normal to AC, Fig. 9.18(a),

$$\sigma \times AC = 80\,AB\cos 60° - 50\,BC\sin 60°$$

$$\therefore \sigma = 80\cos^2 60° - 50\sin^2 60°$$

$$= 80 \times \tfrac{1}{4} - 50 \times \tfrac{3}{4} = -17\cdot 5 \text{ MN/m}^2$$

COMPLEX STRESS AND STRAIN

Equating forces parallel to AC,

$$\tau \times AC = 80\,AB\sin 60° + 50\,BC\cos 60°$$

$$\therefore \tau = 80\cos 60°\sin 60° + 50\sin 60°\cos 60°$$

$$= \frac{80+50}{2}\sin 120°$$

$$= 65 \times \frac{\sqrt{3}}{2} = \underline{56 \cdot 25 \; MN/m^2}$$

From Fig. 9.18(b), the resultant stress,

$$\sigma_r = \sqrt{17 \cdot 5^2 + 56 \cdot 25^2} = \underline{58 \cdot 9 \; MN/m^2}$$

and

$$\alpha = \tan^{-1}\frac{56 \cdot 25}{17 \cdot 5} = \underline{72° 48'}$$

2. *A thin cylinder with closed ends has an internal diameter of 50 mm and a wall thickness of 2·5 mm. It is subjected to an axial pull of 10 kN and an axial torque of 500 Nm while under an internal pressure of 6 MN/m². Determine the principal stresses in the tube and the maximum shear stress.*

The arrangement is shown in Fig. 9.19(a).

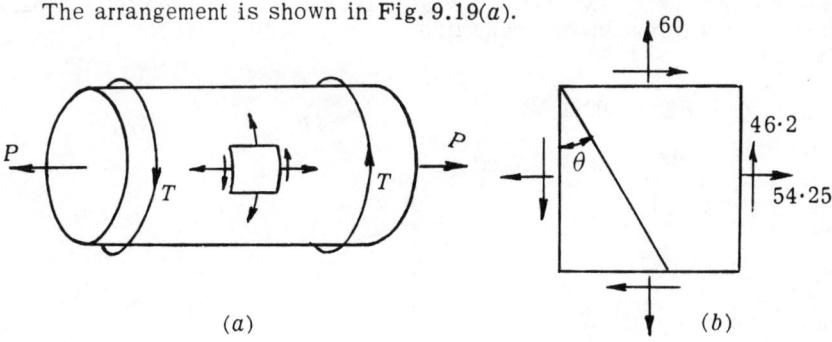

(a)   (b)

Fig. 9.19

Due to internal pressure,

$$\sigma_c = \frac{pd}{2t} \quad \cdot \quad \cdot \quad \cdot \quad \cdot \quad \cdot \quad \text{from equation (1.7)}$$

$$= \frac{6 \times 10^6 \times 50 \times 10^{-3}}{2 \times 2 \cdot 5 \times 10^{-3}} = 60 \times 10^6 \; N/m^2$$

$$\sigma_l = \frac{\sigma_c}{2} = 30 \times 10^6 \; N/m^2 \quad \cdot \quad \cdot \quad \cdot \quad \text{from equation (1.8)}$$

Due to the axial load,

$$\sigma_l = \frac{P}{a} = \frac{10 \times 10^3}{\pi \times 52 \cdot 5 \times 2 \cdot 5 \times 10^{-6}} = 24 \cdot 25 \times 10^6 \; N/m^2$$

$$\therefore \text{total axial stress} = (30 + 24 \cdot 25) \times 10^6 = 54 \cdot 25 \times 10^6 \; N/m^2$$

104 COMPLEX STRESS AND STRAIN

Due to the torque, $\tau = \dfrac{\text{torque}}{\text{mean radius} \times \text{area}}$ since the tube is thin

$$= \dfrac{500}{26 \cdot 25 \times \pi \times 52 \cdot 5 \times 2 \cdot 5 \times 10^{-9}} = 46 \cdot 2 \times 10^{6} \text{ N/m}^2$$

The stresses on an element of the tube are shown in Fig. 9.19(b). From equation (9.10), the principal stresses are given by

$$\sigma = \tfrac{1}{2}\{(54 \cdot 25 + 60) \pm \sqrt{(54 \cdot 25 - 60)^2 + 4 \times 46 \cdot 2^2}\}$$

$$= \underline{103 \cdot 4 \text{ and } 10 \cdot 8 \text{ MN/m}^2}$$

From equation (9.4),

$$\tau_{max} = \dfrac{103 \cdot 4 - 10 \cdot 8}{2} = \underline{46 \cdot 3 \text{ MN/m}^2}$$

From equation (9.9),

$$\theta = \tfrac{1}{2} \tan^{-1} \dfrac{2 \times 46 \cdot 2}{54 \cdot 25 - 60} = \tfrac{1}{2}(180° - 86°26') = \underline{46°47'}$$

Alternatively, using Mohr's stress circle, Fig. 9.20, set off PA = 54·25, PB = 60 and AM = BN = 46·2 MN/m². Draw a circle of centre O to pass through M and N, cutting the base line at E and F.

Then σ_1 = PE = 103·4 MN/m² $\theta = \dfrac{\beta}{2} = 46°47'$

σ_2 = PF = 10·8 MN/m²

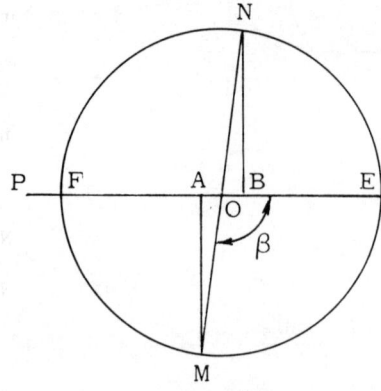

Fig. 9.20

3. On a plane passing through a point in a stressed material, there is a tensile stress of 60 MN/m² and a shearing stress of 40 MN/m². On another plane at 45° to the first plane, there is a tensile stress of 120 MN/m² and an unknown shearing stress.

Find (a) the principal stresses, (b) the maximum shearing stress, (c) the positions of the principal planes, (d) the principal strains. $E = 200 \text{ GN/m}^2$ and $\nu = 0.3$.

COMPLEX STRESS AND STRAIN

In Fig. 9.21, let σ_1 and σ_2 be the principal stresses. Then, from equations (9.3) and (9.4),

$$\sigma_\theta = \frac{\sigma_1 + \sigma_2}{2} + \frac{\sigma_1 - \sigma_2}{2} \cos 2\theta$$

$$= m + n \cos 2\theta$$

where $m = \dfrac{\sigma_1 + \sigma_2}{2}$ and $n = \dfrac{\sigma_1 - \sigma_2}{2}$

and $\tau_\theta = n \sin 2\theta$

$\therefore \quad 60 = m + n \cos 2\theta \qquad (1)$

$\qquad 40 = n \sin 2\theta \qquad (2)$

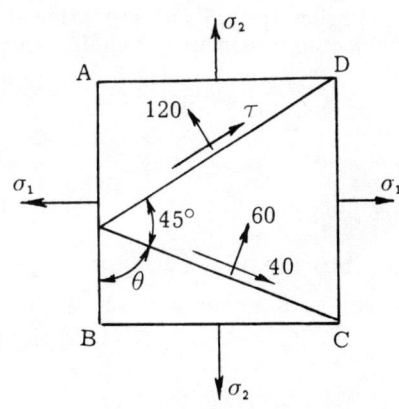

Fig. 9.21

and $120 = m + n \cos 2(\theta + 45°) = m - n \sin 2\theta \qquad (3)$

From equations (2) and (3), $\quad m = 160$

so that from equation (1), $\quad n \cos 2\theta = -100 \qquad (4)$

and from equations (2) and (4), $\quad n = \sqrt{(-100)^2 + 40^2} = 107\cdot 6$

and $\qquad \tan 2\theta = \dfrac{40}{-100}$

from which $\qquad \theta = \tfrac{1}{2}(180° - 21°48') = \underline{79°6'}$

$\qquad \sigma_1 = m + n = 160 + 107\cdot 6 = \underline{267\cdot 6 \text{ MN/m}^2}$

and $\qquad \sigma_2 = m - n = 160 - 107\cdot 6 = \underline{52\cdot 4 \text{ MN/m}^2}$

$\qquad \tau_{max} = \dfrac{\sigma_1 - \sigma_2}{2} = \dfrac{267\cdot 6 - 52\cdot 4}{2} = \underline{107\cdot 6 \text{ MN/m}^2}$

$\epsilon_1 = \dfrac{\sigma_1}{E} - \nu \dfrac{\sigma_2}{E} = \dfrac{(267\cdot 6 - 0\cdot 3 \times 52\cdot 4) \times 10^6}{200 \times 10^9} = \underline{1\cdot 26 \times 10^{-3}}$

and $\epsilon_2 = \dfrac{\sigma_2}{E} - \nu \dfrac{\sigma_1}{E} = \dfrac{(52\cdot 4 - 0\cdot 3 \times 267\cdot 6) \times 10^6}{200 \times 10^9} = \underline{-0\cdot 139 \times 10^{-3}}$

4. *A shaft having a diameter of 100 mm is subjected to a bending moment of 6·5 kN m in addition to the torque which it transmits. Find the maximum allowable torque if (a) the material is brittle and failure is assumed to occur when the maximum direct stress reaches 120 MN/m²; (b) the material is ductile and failure is assumed to occur when the maximum shear stress reaches that at the elastic limit stress of 300 MN/m² in pure tension.*

From Art. 9.6(*b*), the maximum and minimum principal stresses in a shaft subjected to combined bending and twisting are given by

$$\sigma = \frac{r}{2I}\left[M \pm \sqrt{M^2 + T^2}\right] = \frac{0.05}{2 \times \frac{\pi}{64} \times 0.1^4}\left[6.5 \times 10^3 \pm \sqrt{(6.5 \times 10^3)^2 + T^2}\right]$$

$$= 5\,093\left[6.5 \times 10^3 \pm \sqrt{42.25 \times 10^6 + T^2}\right]$$

(*a*) Maximum principal stress, $\sigma_x = 5\,093\left[6.5 \times 10^3 + \sqrt{42.25 \times 10^6 + T^2}\right]$
which is to be equal to 120 MN/m².

Hence $\qquad T = \underline{15.77\ \text{kN m}}$

(*b*) Maximum principal stress, $\sigma_x = 5\,093\left[6.5 \times 10^3 + \sqrt{42.25 \times 10^6 + T^2}\right]$
and minimum principal stress, $\sigma_y = 5\,093\left[6.5 \times 10^3 - \sqrt{42.25 \times 10^6 + T^2}\right]$

Since $\sqrt{42.25 \times 10^6 + T^2} > 6.5 \times 10^3$, σ_y will be negative and hence

maximum shear stress $= \dfrac{\sigma_x - \sigma_y}{2}$

From the maximum shear stress theory of failure,

$$\sigma_x - \sigma_y = \sigma_0 \quad . \quad . \quad . \quad . \quad . \quad \text{from Art. 9.7}$$

$$\therefore\ 5\,093\left[6.5 \times 10^3 + \sqrt{42.25 \times 10^6 + T^2}\right] - 5\,093\left[6.5 \times 10^3 - \sqrt{42.25 \times 10^6 + T^2}\right]$$

$$= 300 \times 10^6$$

from which $\qquad T = \underline{28.7\ \text{kN m}}$

5. *In a two-dimensional strain system, the following readings were taken with a 45° strain rosette:*

$$\epsilon_{0°} = 0.4 \times 10^{-3},\quad \epsilon_{45°} = 0.4 \times 10^{-3},\quad \epsilon_{90°} = 0.1 \times 10^{-3}.$$

Determine the magnitude and directions of the principal strains. If $E = 200\ \text{GN/m}^2$ and $\nu = 0.3$, find the principal stresses.

From Art. (9.9), $\quad \epsilon_1 = m + n\cos 2\theta = 0.4 \times 10^{-3}$ $\qquad$ (1)

$\qquad\qquad\qquad \epsilon_2 = m - n\sin 2\theta = 0.4 \times 10^{-3}$ $\qquad$ (2)

and $\qquad\qquad\qquad \epsilon_3 = m - n\cos 2\theta = 0.1 \times 10^{-3}$ $\qquad$ (3)

Hence, from equations (1) and (3), $m = 0.25 \times 10^{-3}$

Equations (1) and (2) then become $n\sin 2\theta = -0.15 \times 10^{-3}$

and $\qquad\qquad\qquad n\cos 2\theta = 0.15 \times 10^{-3}$

Therefore $\quad n = \sqrt{(-0.15 \times 10^{-3})^2 + (0.15 \times 10^{-3})^2} = 0.212 \times 10^{-3}$

and $\qquad \tan 2\theta = -1$

$\qquad\qquad \therefore\ \theta = \tfrac{1}{2}(180° - 45°) = \underline{67°\,30'}$

COMPLEX STRESS AND STRAIN

$$\epsilon_1 = m + n = 0.462 \times 10^{-3}$$

$$\epsilon_2 = m - n = 0.038 \times 10^{-3}$$

$$\sigma_1 = \frac{E}{1-\nu^2}(\epsilon_1 + \nu\epsilon_2) \quad \ldots \quad \text{from equation (9.16)}$$

$$= \frac{200 \times 10^9}{1 - 0.3^2}(0.462 + 0.3 \times 0.038) \times 10^{-3} = \underline{104 \times 10^6 \text{ N/m}^2}$$

Similarly, $\quad \sigma_2 = \dfrac{E}{1-\nu^2}(\epsilon_2 + \nu\epsilon_1)$

$$= \frac{200 \times 10^9}{1 - 0.3^2}(0.038 + 0.3 \times 0.462) \times 10^{-3} = \underline{38.8 \times 10^6 \text{ N/m}^2}$$

6. *A bar of metal 20 mm diameter is tested in tension and extends $3 \cdot 1 \times 10^{-9}$ m/N on a gauge length of 200 mm. Another bar of the same metal and same diameter is tested in torsion and twists $0 \cdot 00912°/N\,m$ on a gauge length of 200 mm. Calculate the values of E, G and ν.*

A bar of this material and 20 mm diameter is subjected to an axial compressive load of 60 kN together with a lateral pressure of 80 MN/m² applied to the circumference. Determine the change in length on a 200 mm gauge length and the change in diameter.

$$E = \frac{Pl}{ax} = \frac{1 \times 0.2}{\frac{\pi}{4} \times 0.02^2 \times 3.1 \times 10^{-9}} = \underline{205 \times 10^9 \text{ N/m}^2}$$

$$G = \frac{Tl}{J\theta} = \frac{1 \times 0.2}{\frac{\pi}{32} \times 0.02^4 \times \left(0.00912 \times \frac{\pi}{180}\right)} = \underline{80 \times 10^9 \text{ N/m}^2}$$

$$E = 2G(1 + \nu) \quad \ldots \quad \text{from equation (9.29)}$$

$$\therefore \nu = \frac{E}{2G} - 1 = \frac{205}{2 \times 80} - 1 = \underline{0.28}$$

$$\text{Axial stress} = \frac{60 \times 10^3}{\frac{\pi}{4} \times 20^2 \times 10^{-6}} = 191 \times 10^6 \text{ N/m}^2$$

$$\therefore \text{ axial strain} = \frac{191 \times 10^6 - 2 \times 0.28 \times 80 \times 10^6}{205 \times 10^9} \quad \text{from equation (9.19)}$$

$$= 0.712 \times 10^{-3}$$

$$\therefore \text{ decrease in length} = 200 \times 0.712 \times 10^{-3} = \underline{0.1424 \text{ mm}}$$

$$\text{Lateral strain} = \frac{80 \times 10^6 - 0.28 \times 80 \times 10^6 - 0.28 \times 191 \times 10^6}{205 \times 10^9}$$

$$= 0.02 \times 10^{-3}$$

$$\therefore \text{ decrease in diameter} = 20 \times 0.02 \times 10^{-3} = \underline{0.0004 \text{ mm}}$$

7. *Determine the change in volume of a steel bar 80 mm square in section and 1·2 m long when subjected to an axial compressive load of 20 kN. $E = 200 \text{ GN}/m^2$ and $G = 80 \text{ GN}/m^2$.*

$$\sigma_x = \frac{20 \times 10^3}{80^2 \times 10^{-6}} = 3 \cdot 125 \times 10^6 \text{ N}/m^2$$

$$\sigma_y = \sigma_z = 0$$

As in example 6, $\quad \nu = \dfrac{E}{2G} - 1 = \dfrac{200}{2 \times 80} - 1 = 0 \cdot 25$

$$\therefore \epsilon_x = \frac{3 \cdot 125 \times 10^6}{200 \times 10^9} = \frac{3 \cdot 125}{200} \times 10^{-3}$$

$$\epsilon_y = \epsilon_z = -0 \cdot 25 \times \frac{3 \cdot 125}{200} \times 10^{-3}$$

$$\therefore \epsilon_v = \epsilon_x + \epsilon_y + \epsilon_z \quad . \quad . \quad \text{from equation (9.22)}$$

$$= \frac{3 \cdot 125}{200} \times 10^{-3} (1 - 2 \times 0 \cdot 25)$$

$$= \frac{3 \cdot 125}{400} \times 10^{-3}$$

Alternatively, $\quad E = 2G(1 + \nu) \quad . \quad . \quad$ from equation (9.29)

and $\quad E = 3K(1 - 2\nu) \quad . \quad . \quad$ from equation (9.23)

Eliminating ν between these equations gives

$$K = \frac{EG}{3(3G - E)}$$

$$= \frac{200 \times 80 \times 10^{18}}{3(3 \times 80 - 200) \times 10^9}$$

$$= \frac{400}{3} \times 10^9 \text{ N}/m^2$$

$$\therefore \epsilon_v = \frac{\sigma_x + \sigma_y + \sigma_z}{3K} \quad . \quad . \quad \text{from equation (9.26)}$$

$$= \frac{3 \cdot 125 \times 10^6}{400 \times 10^9}$$

$$= \frac{3 \cdot 125}{400} \times 10^{-3}$$

$$\therefore \text{change in volume} = \frac{3 \cdot 125}{400} \times 10^{-3} \times 80^2 \times 1\,200$$

$$= \underline{60 \cdot 3 \text{ mm}^3}$$

COMPLEX STRESS AND STRAIN

8. *A compressed air cylinder, 1·8 m long and 0·6 m internal diameter has a wall thickness of 10 mm. Find the increase in volume when it is subjected to an internal pressure of 4 MN/m². E = 200 GN/m² and ν = 0·3.*

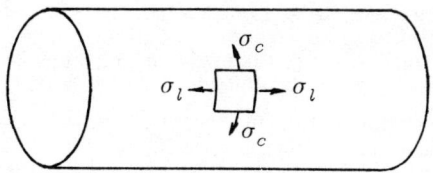

Fig. 9.22

Referring to Fig. 9.22,

$$\sigma_c = \frac{pd}{2t} \quad \ldots \quad \text{from equation (10.1)}$$

$$= \frac{4 \times 10^6 \times 0.6}{2 \times 0.01}$$

$$= 120 \times 10^6 \text{ N/m}^2$$

$$\sigma_l = \frac{\sigma_c}{2} \quad \ldots \quad \text{from equation (10.2)}$$

$$= 60 \times 10^6 \text{ N/m}^2$$

The volumetric strain is the sum of the strain in the axial direction and the strains on two perpendicular diameters. The diametral strain is the same as the circumferential strain, since the diameter and circumference each expand in the same proportion.

Neglecting the effect of radial stress, which is relatively insignificant,

$$\epsilon_d = \epsilon_c = \frac{\sigma_c}{E} - \nu \frac{\sigma_l}{E}$$

$$= \frac{(120 - 0.3 \times 60) \times 10^6}{200 \times 10^9} = 0.51 \times 10^{-3}$$

$$\epsilon_l = \frac{\sigma_l}{E} - \nu \frac{\sigma_c}{E}$$

$$= \frac{(60 - 0.3 \times 120) \times 10^6}{200 \times 10^9} = 0.12 \times 10^{-3}$$

$$\therefore \epsilon_v = \epsilon_l + 2\epsilon_d$$

$$= (0.12 + 2 \times 0.51) \times 10^{-3} = 1.14 \times 10^{-3}$$

$$\therefore \text{increase in volume} = \frac{\pi}{4} \times 0.6^2 \times 1.8 \times 1.14 \times 10^{-3} = \underline{0.000\,58 \text{ m}^3}$$

9. A thin cylinder, 300 mm diameter and 3 mm thick is subjected to an internal pressure of 3·5 MN/m² and an axial tensile force of 100 kN. Determine the normal and shear stresses on a plane inclined at 30° to the axis of the cylinder.
(*Ans.*: 161 MN/m²; 22·6 MN/m²)

10. At a point in a stressed material, the major principal stress is 140 MN/m² tension and the maximum shear stress is 80 MN/m². Determine (*a*) the minor principal stress, (*b*) the direct stress on the plane of maximum shear stress, (*c*) the stresses on a plane inclined at 30° to the plane on which the major principal stress acts. (*Ans.*: 20 MN/m²; 60 MN/m²; 100 MN/m²; 69·3 MN/m²)

11. The principal stresses at a point in a material are 50 MN/m² tension and 30 MN/m² tension. Determine, for a plane inclined at 40° to the plane on which the former stress acts, (*a*) the normal and shear stresses, (*b*) the magnitude and inclination of the resultant stress.
(*Ans.*: 41·73 MN/m²; 9·85 MN/m²; 42·8 MN/m² at 13°18' to the normal stress)

12. The principal stresses at a point in a material are 60 MN/m² and 20 MN/m², both tensile. Determine the normal and shear stresses on a plane inclined at an angle of $\tan^{-1} 0.25$ to the plane on which the maximum principal stress acts.
(*Ans.*: 57·7 MN/m²; 9·4 MN/m²)

13. Direct stresses of 80 MN/m² tension and 60 MN/m² compression are applied at a point in a material, on planes at right angles to one another. If the maximum direct stress in the material is limited to 100 MN/m² tension, what shear stress may be applied to the given planes and what will then be the maximum shearing stress at the point? (*Ans.*: 56·56 MN/m²; 90 MN/m²)

14. On a plane in a stressed material, there is a tensile stress of 100 MN/m² and a shearing stress of 55 MN/m². On a plane making an angle of 30° anti-clockwise to this plane, there is a tensile stress of 20 MN/m² and an unknown shearing stress. Find the position of the principal planes and the magnitude of the principal stresses.
(*Ans.*: 69°54' and 159°54'; +120·3 and −49·7 MN/m²)

15. A thin cylinder 75 mm diameter and 5 mm thick is subjected to an internal pressure of 5·5 MN/m² and also to a torque of 1·6 kN m. Determine the maximum and minimum principal stresses and the maximum shearing stress in the tube.
(*Ans.*: 64·3 and −2·5 MN/m²; 33·4 MN/m²)

16. A thin cylinder has a diameter of 50 mm and a thickness of 2·5 mm. The tube is subjected to an internal pressure of 6 MN/m², an axial tensile force of 10 kN and a torque of 500 N m. Determine the maximum and minimum principal stresses in the tube and also the maximum shear stress.
(*Ans.*: 103·3 and 10·98 MN/m²; 57·13 MN/m²)

17. A hollow shaft 200 mm outside diameter and 125 mm inside diameter is subjected to a bending moment of 43 kN m and a torque of 65 kN m. Calculate the maximum shearing stress in the shaft. (*Ans.*: 58·6 MN/m²)

18. A shaft 80 mm diameter is simply supported in bearings 0·6 m apart. A flywheel of mass 500 kg is mounted midway between the bearings and the shaft transmits 30 kW at 360 rev/min. Calculate the principal stresses and maximum shear stress on the shaft at the ends of a vertical and horizontal diameter in the plane of the flywheel. (*Ans.*: 18·1, 10·78, 7·9, 7·9 MN/m²)

19. A solid circular shaft is subjected to a bending moment *M* and a torque *T*. At a point on the circumference, the maximum principal stress is numerically four times the minimum principal stress. Determine the ratio *M/T* and the angle between the plane of the maximum principal stress and the plane of the bending stress.
(*Ans.*: 3/4; 26°30')

20. A hollow shaft 100 mm external diameter and 50 mm internal diameter is subjected to an axial thrust of 50 kN while transmitting 600 kW at 500 rev/min. What bending moment may be applied to the shaft if the greater principal stress is not to exceed 100 MN/m²? What will then be the value of the smaller principal stress?
(*Ans.*: 4·85 kN m; 38·85 MN/m²)

COMPLEX STRESS AND STRAIN

21. A solid shaft 200 mm diameter transmits 2 MW at 250 rev/min and is subjected to a bending moment of 50 kN m. Calculate the maximum permissible end thrust on the shaft if the maximum shearing stress is limited to 80 MN/m^2. (*Ans.*: 2 MN)

22. A circular shaft 0·1 m diameter is subjected to combined bending and twisting moments, the bending moment being three times the twisting moment. If the direct tension yield-point of the material is 350 MN/m^2 and the factor of safety on yield is to be 4, calculate the allowable twisting moment by the following theories of elastic failure: (*a*) maximum principal stress theory; (*b*) maximum shearing stress theory.
(*Ans.*: 2 790 N m; 2 715 N m)

23. Three exactly similar specimens of mild steel tube are 40 mm external diameter and 32 mm internal diameter. One of these is tested in tension and reaches the limit of proportionality at an axial tensile load of 90 kN. The second is tested in simple torsion. The third is also tested in torsion, but with a uniform bending moment of 350 N m applied throughout the test. Assuming maximum shear stress to be the criterion of elastic failure, estimate the torque at which the two torsion specimens should fail. (*Ans.*: 737 N m; 648·6 N m)

24. The principal stresses at a point in a material are 160 and 40 MN/m^2, both tensile. If $E = 200$ GN/m^2 and $\nu = 0·28$, find the strain in a direction inclined at 30° to that of the greater principal stress. In what direction is the strain zero?
(*Ans.*: $0·552 \times 10^{-3}$; 79° 50' to direction of 160 MN/m^2 stress)

25. A strain rosette fixed to a point in a stressed material gave the following readings:

Gauge No.	Direction relative to Gauge No. 1	Strain
1	0°	$+423 \times 10^{-6}$
2	45°	$+542 \times 10^{-6}$
3	90°	$+82 \times 10^{-6}$

If $E = 200$ GN/m^2 and $\nu = 0·3$, find the magnitude of the principal stresses and their directions relative to Gauge No. 1.
(*Ans.*: 124 MN/m^2; 20·8 MN/m^2; 150° 23'; 60° 23')

26. The following strains were recorded with a 120° strain rosette:

$$e_0 = +716 \times 10^{-6} \qquad e_{120°} = +539 \times 10^{-6} \qquad e_{240°} = +155 \times 10^{-6}$$

Find (*a*) the principal strains, (*b*) the principal stresses.
$E = 200$ GN/m^2 and $\nu = 0·3$.
(*Ans.*: $+801 \times 10^{-6}$; $+139 \times 10^{-6}$; $+185·2$ MN/m^2; $+83·4$ MN/m^2)

27. A uniform bar 9 m long carries an axial tensile load of 200 kN. Determine the increase in volume of the bar if $G = 80$ GN/m^2 and $\nu = 0·25$. (*Ans.*: 4 500 mm^2)

28. An axial compressive load of 500 kN is applied to a metal bar 50 mm square in section. The contraction on a 200 mm gauge length is 0·55 mm and the increase in thickness is 0·045 mm. Find the values of E and ν.
If a uniform lateral pressure of 80 MN/m^2 is applied to the four sides of the bar in addition to the axial load, find the contraction on the 200 mm gauge length and the change in thickness. (*Ans.*: 72·7 GN/m^2; 0·327; 0·406 mm; 0·007 9 mm)

29. A solid steel sphere 400 mm diameter is subjected to a uniform hydraulic pressure of 3·5 MN/m^2. Determine the decrease in volume if $E = 200$ GN/m^2 and $\nu = 0·3$.
(*Ans.*: 703·5 mm^3)

30. In tests on a steel bar 25 mm diameter, a tensile load of 50 kN produced an extension of 0·099 4 mm on a gauge length of 200 mm and a torque of 200 N m produced an angle of twist of 0·925° on a gauge length of 250 mm. Calculate the value of Poisson's ratio for the steel. (*Ans.*: 0·27)

10 Cylinders

10.1 Stresses in thin cylindrical shells A thin cylinder is one in which the thickness is sufficiently small in relation to the diameter for the stress across the thickness to be reasonably uniform. When such a cylinder is subjected to internal pressure, tensile stresses σ_c and σ_l are set up in the circumferential and longitudinal directions respectively. If the internal diameter is d, the length is l, the thickness of the plate is t and the internal pressure is p, then the force tending to separate the top and bottom halves, Fig.10.1(a), is $p \times d \times l$. This is resisted by the stress σ_c acting on an area $2 \times t \times l$ (neglecting the strength of the ends) so that, for equilibrium,

$$p d l = 2\sigma_c t l$$

$$\therefore \sigma_c = \frac{pd}{2t} \qquad (10.1)$$

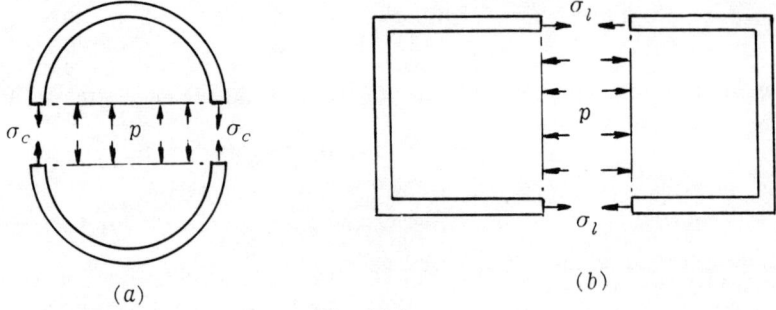

Fig.10.1

The force tending to separate the right and left-hand ends, Fig. 10.1(b), is $p \times \frac{\pi}{4} d^2$. This is resisted by the stress σ_l acting on an area πdt, so that, for equilibrium,

$$p \times \frac{\pi}{4} d^2 = \sigma_l \times \pi d t$$

$$\therefore \sigma_l = \frac{pd}{4t} \qquad (10.2)$$

The circumferential stress acts on a longitudinal section and the longitudinal stress acts on a circumferential section. If the shell is made up of plates riveted together, then if the efficiency of the longitudinal joints, i.e., the ratio strength of joint/strength of undrilled plate, is η_l, the circumferential stress at the joint is given by

$$\sigma_c = \frac{pd}{2t\eta_l} \qquad (10.3)$$

CYLINDERS

Similarly, if the efficiency of the circumferential joints is η_c, the longitudinal stress at the joint is given by

$$\sigma_l = \frac{pd}{4t\eta_c} \qquad (10.4)$$

The efficiency of the circumferential joints therefore need only be half that of the longitudinal joints for equal joint stress.

10.2 Lamé's Theory When a cylinder is too thick for the simple assumption that the radial stress is negligible and the circumferential stress is therefore uniform across the section, i.e., when t is greater than about $0.1d$, Lamé's theory is used to determine the stress distribution.

If the cylinder is long in comparison with its diameter, the longitudinal stress and strain are assumed to be uniform across the thickness of the cylinder wall.

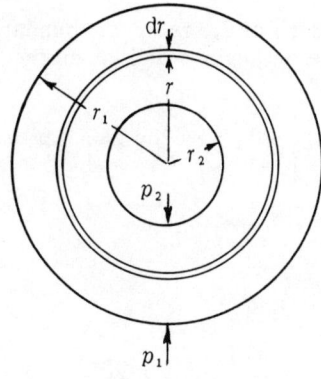   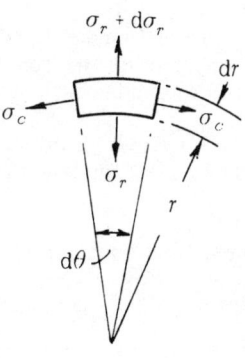

Fig. 10.2   Fig. 10.3

Fig. 10.2 shows a cylinder of external and internal radii r_1 and r_2 subjected to external and internal pressures p_1 and p_2 respectively, and Fig. 10.3 shows an element of the cross-section at radius r, subtending an angle $d\theta$ at the centre. The radial and circumferential stresses on the element are σ_r and σ_c respectively and so, equating radial forces on unit axial length,

$$(\sigma_r + d\sigma_r)(r + dr)d\theta = \sigma_r r\, d\theta + 2\sigma_c\, dr\, \frac{d\theta}{2}$$

which reduces to $\qquad \sigma_r\, dr + r\, d\sigma_r = \sigma_c\, dr$

or

$$\sigma_r + r\frac{d\sigma_r}{dr} = \sigma_c \qquad (10.5)$$

Let the longitudinal stress be σ_l and the longitudinal strain ϵ_l.

Then

$$\epsilon_l = \frac{\sigma_l}{E} - \nu\left(\frac{\sigma_r + \sigma_c}{E}\right) \qquad (10.6)$$

Since ϵ_l and σ_l are assumed to be constant across the section,

$$\sigma_r + \sigma_c = \text{constant} = 2a \qquad (10.7)$$

Substituting $\sigma_c = 2a - \sigma_r$ in equation (10.5) and multiplying throughout by r,

$$2\sigma_r r + r^2 \frac{d\sigma_r}{dr} - 2ar = 0$$

or

$$\frac{d}{dr}(\sigma_r r^2 - ar^2) = 0$$

Integrating,

$$\sigma_r r^2 - ar^2 = \text{constant} = b$$

$$\therefore \sigma_r = a + \frac{b}{r^2} \qquad (10.8)$$

and, substituting in equation (10.7),

$$\sigma_c = a - \frac{b}{r^2} \qquad (10.9)$$

These are Lamé's equations and substitution of the relevant boundary conditions enables the constants a and b to be determined and hence the radial and circumferential stresses at any point.

10.3 Thick cylinder subject to internal pressure only. Let the gauge pressure be p. Then the boundary conditions are $\sigma_r = -p$ when $r = r_2$ and $\sigma_r = 0$ when $r = r_1$.

Thus, from equation (10.8),

$$-p = a + \frac{b}{r_2^2}$$

and

$$0 = a + \frac{b}{r_1^2}$$

from which

$$a = \frac{r_2^2}{r_1^2 - r_2^2}p \quad \text{and} \quad b = \frac{-r_1^2 r_2^2}{r_1^2 - r_2^2}p$$

Therefore the radial stress

$$\sigma_r = \frac{pr_2^2}{r_1^2 - r_2^2}\left(1 - \frac{r_1^2}{r^2}\right) \qquad (10.10)$$

and the circumferential stress

$$\sigma_c = \frac{pr_2^2}{r_1^2 - r_2^2}\left(1 + \frac{r_1^2}{r^2}\right) \qquad (10.11)$$

The distribution of these stresses across the section is shown in Fig. 10.4.

The maximum value of σ_r

$$= -p$$

and the maximum value of σ_c

$$= p\frac{r_1^2 + r_2^2}{r_1^2 - r_2^2} \qquad (10.12)$$

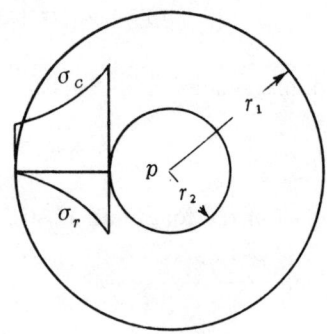

Fig. 10.4

CYLINDERS

10.4 Longitudinal and shear stress The uniform longitudinal stress σ_l is determined by the longitudinal equilibrium of the end of the cylinder, Fig. 10.5. Thus, for the simple case of internal pressure only,

$$\sigma_l \times \pi(r_1^2 - r_2^2) = p \times \pi r_2^2$$

$$\therefore \sigma_l = \frac{pr_2^2}{r_1^2 - r_2^2} \quad (10.13)$$

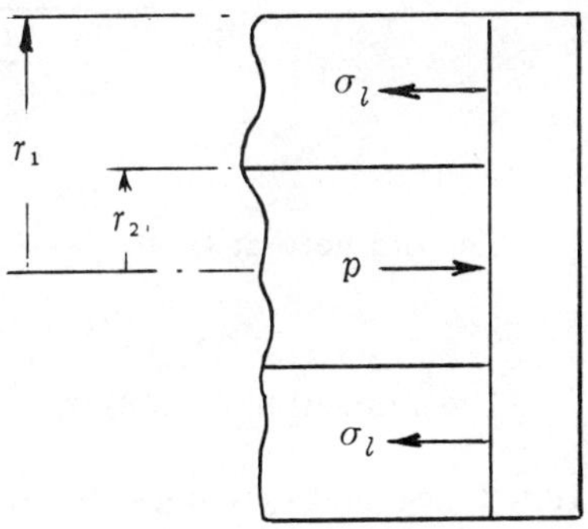

Fig. 10.5

σ_r, σ_c and σ_l are principal stresses, so that the maximum shear stress τ is equal to half the difference between the maximum and minimum principal stresses (see Art. 9.2). Thus for internal pressure only, σ_r is compressive (negative) and σ_c and σ_l are tensile (positive), so that τ is either $\tfrac{1}{2}(\sigma_c - \sigma_r)$ or $\tfrac{1}{2}(\sigma_l - \sigma_r)$, whichever is greater. Since, however, $\sigma_c > \sigma_l$,

$$\tau = \frac{\sigma_c - \sigma_r}{2}$$

$$= \frac{pr_1^2 r_2^2}{(r_1^2 - r_2^2)r^2} \quad \text{from equations (10.10) and (10.11)}$$

Thus the maximum shear stress occurs at the inside surface, where $r = r_2$

i.e. $$\tau_{max} = \frac{pr_1^2}{r_1^2 - r_2^2} \quad (10.14)$$

10.5 The Lamé line By plotting σ_r against $\frac{1}{r^2}$ and σ_c against $-\frac{1}{r^2}$, equations (10.8) and (10.9) are represented graphically by a *single* straight line, having the same intercept a and slope b. This is shown in Fig. 10.6 for the case of internal pressure only where $\sigma_r = -p$ at $r = r_2$ (point A) and $\sigma_r = 0$ at $r = r_1$ (point B).

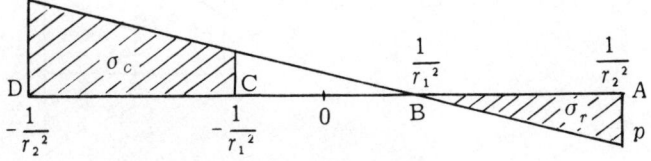

Fig. 10.6

Extending the line to negative values of $1/r^2$ gives values of σ_c between points C and D. The maximum value of σ_c occurs at point D and this is given by

$$\frac{\sigma_c}{p} = \frac{\dfrac{1}{r_1^2} + \dfrac{1}{r_2^2}}{\dfrac{1}{r_2^2} - \dfrac{1}{r_1^2}} = \frac{r_1^2 + r_2^2}{r_1^2 - r_2^2} \quad \text{as given by equation (10.12).}$$

The ordinate at O $= \dfrac{p \times \dfrac{1}{r_1^2}}{\dfrac{1}{r_2^2} - \dfrac{1}{r_1^2}} = \dfrac{p r_2^2}{r_1^2 - r_2^2}$

From equation (10.13), this is the longitudinal stress, σ_l.

10.6 Compound cylinders In order to reduce the high circumferential stress at the inside surface of a cylinder under internal pressure, the cylinder can be pre-stressed by shrinking on to it an outer cylinder. The inner diameter of the outer cylinder is made slightly smaller than the outer diameter of the inner cylinder and assembly is achieved by either heating the outer cylinder or cooling the inner one. This puts the inner cylinder in an initial state of compression so that, when the internal pressure is applied, the resulting tensile stress at the inner surface is less than would have been the case in a homogeneous cylinder of the same total thickness.

If the radial pressure between the two cylinders due to shrinking is p_0, the initial stresses in the two parts of the cylinder can be determined and these are then combined with those due to the initial pressure, treating the compound cylinder as homogeneous, to give the resultant stresses.

The combination of stresses is shown in Fig. 10.7.

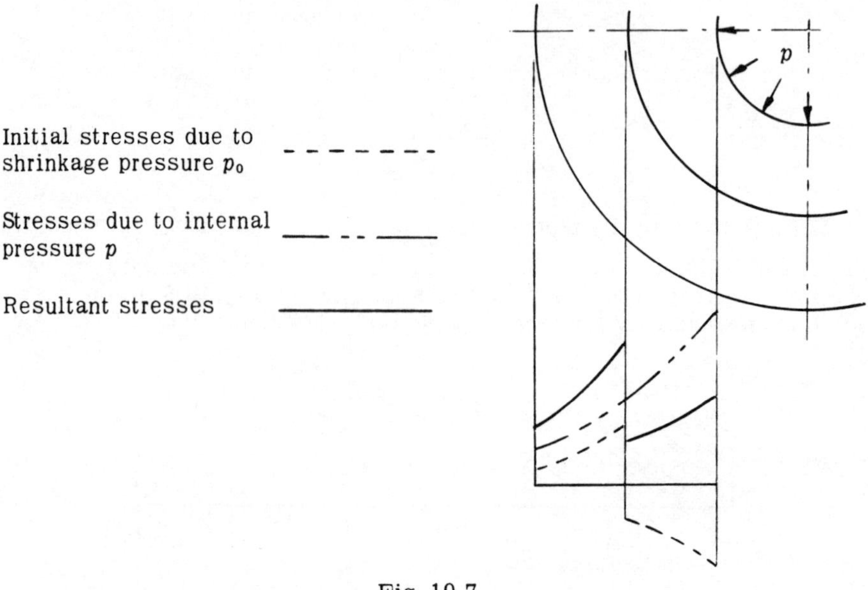

Initial stresses due to shrinkage pressure p_0

Stresses due to internal pressure p

Resultant stresses

Fig. 10.7

CYLINDERS 117

By this process, the maximum tensile stress in a cylinder of given thickness can be reduced or alternatively, a thinner cylinder can be used for a given maximum stress. The optimum design is that which will give the same maximum stress at the inner surface of each cylinder.

The radial stress distribution in a compound cylinder may be deduced in a similar manner but these stresses are usually relatively unimportant.

1. *A cylindrical compressed air drum is 2 m in diameter with plates 12·5 mm thick. The efficiencies of the longitudinal and circumferential joints are respectively 85% and 45%. If the tensile stress in the plating is to be limited to 100 MN/m², find the maximum safe air pressure.*

For a circumferential stress of 100 MN/m²,

$$p = \frac{2t\eta_l \sigma_c}{d} \quad \ldots \quad \text{from equation (10.3)}$$

$$= \frac{2 \times 12 \cdot 5 \times 10^{-3} \times 0 \cdot 85 \times 100 \times 10^6}{2} = 106 \cdot 3 \times 10^3 \text{ N/m}^2$$

For a longitudinal stress of 100 MN/m²,

$$p = \frac{4t\eta_c \sigma_l}{d} \quad \ldots \quad \text{from equation (10.4)}$$

$$= \frac{4 \times 12 \cdot 5 \times 10^{-3} \times 0 \cdot 45 \times 100 \times 10^6}{2} = 112 \cdot 5 \times 10^3 \text{ N/m}^2$$

Thus the maximum safe pressure is 106·3 kN/m², which will produce a circumferential stress at the joint of 100 MN/m² and a longitudinal stress of 94·4 MN/m².

2. *A steel pipe 100 mm external diameter and 50 mm internal diameter is subjected to an internal pressure of 14 MN/m² and an external pressure of 5·5 MN/m². Sketch curves showing the distribution of radial and circumferential stresses across the section.*

From Lamé's equations,

$$\sigma_r = a + \frac{b}{r^2}$$

and $$\sigma_c = a - \frac{b}{r^2}$$

When $r = 25$ mm, $\sigma_r = -14$ MN/m²

$$\therefore \quad -14 = a + \frac{b}{25^2} \qquad (1)$$

When $r = 50$ mm, $\sigma_r = -5\cdot 5$ MN/m²

$$\therefore \quad -5\cdot 5 = a + \frac{b}{50^2} \qquad (2)$$

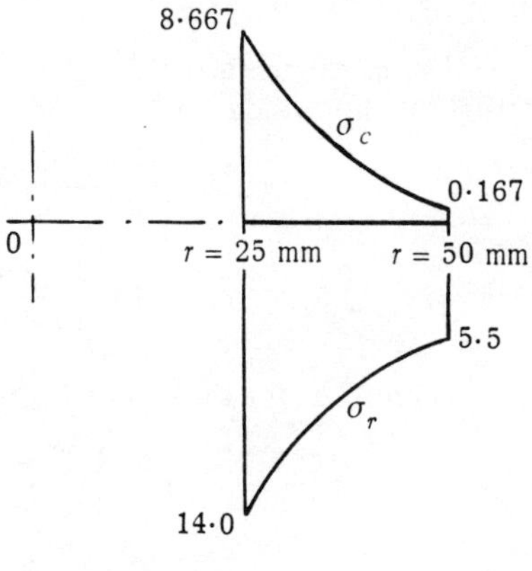

Fig. 10.8

Therefore, from equations (1) and (2),

$$a = -\frac{8}{3} \quad \text{and} \quad b = -\frac{8 \cdot 5 \times 50^2}{3}$$

$$\therefore \sigma_c = -\frac{8}{3} + \frac{8 \cdot 5 \times 50^2}{3r^2}$$

When $r = 25$ mm, $\sigma_c = -\frac{8}{3} + \frac{8 \cdot 5 \times 50^2}{3 \times 25^2} = 8 \cdot 667$ MN/m²

When $r = 50$ mm, $\sigma_c = -\frac{8}{3} + \frac{8 \cdot 5 \times 50^2}{3 \times 50^2} = 0 \cdot 167$ MN/m²

The distribution of σ_r and σ_c across the section is shown in Fig. 10.8.

Alternatively, using the Lamé line, Fig. 10.9, set off $\sigma_r = -14$ MN/m² at $1/25^2$ and $\sigma_r = -5 \cdot 5$ MN/m² at $1/50^2$. The circumferential stresses of $8 \cdot 667$ and $0 \cdot 167$ MN/m² are then given at $-1/25^2$ and $-1/50^2$ respectively.

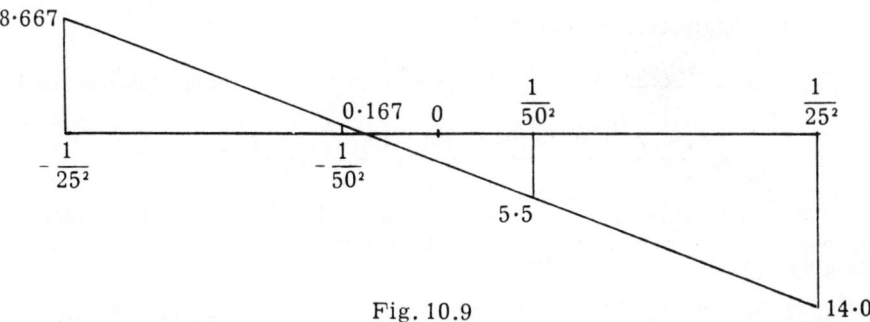

Fig. 10.9

3. *A cast iron pipe 150 mm internal diameter and 200 mm external diameter is tested under pressure and breaks at an internal pressure of 48 MN/m². Find the safe internal pressure for a pipe of the same material and of the same internal diameter with walls 40 mm thick, using a factor of safety of 4.*

The maximum stress due to internal pressure only is the circumferential stress at the inside surface and hence the stress at failure is given by

$$\sigma_c = p \frac{r_1^2 + r_2^2}{r_1^2 - r_2^2} \quad . \quad . \quad . \quad . \quad \text{from equation (10.12)}$$

$$= 48 \times \frac{100^2 + 75^2}{100^2 - 75^2} = 171 \cdot 5 \text{ MN/m}^2$$

Therefore, for the second pipe,

$$\frac{171 \cdot 5}{4} = p \times \frac{115^2 + 75^2}{115^2 - 75^2}$$

from which $p = \underline{17 \cdot 28 \text{ MN/m}^2}$

CYLINDERS

Alternatively, these results can be obtained from the Lamé lines for the two cylinders, shown in Fig. 10.10(a) and (b) respectively.

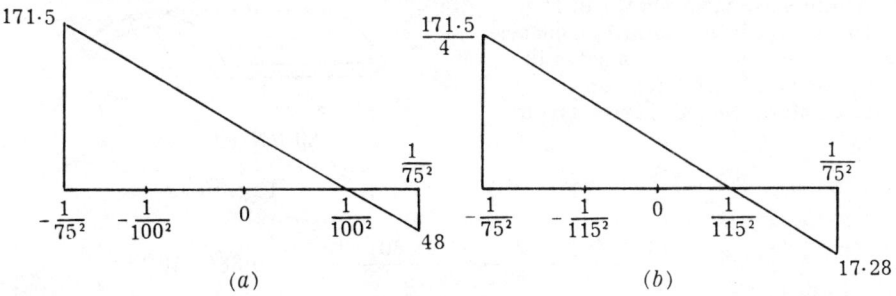

Fig. 10.10

3. *The cylinder of a hydraulic ram supported at the open end is 250 mm internal diameter and is required to sustain an internal pressure of 20 MN/m². Calculate the necessary thickness of the wall if the maximum shearing stress is limited to 50 MN/m².*

Allowing for the effect of the longitudinal stress caused by the pressure on the end of the cylinder, calculate the increase in diameter due to the application of the 20 MN/m² pressure. E = 200 GN/m² and ν = 0·28.

Fig. 10.11 shows the arrangement of the cylinder.

From equation (10.14),

$$\tau_{max} = \frac{pr_1^2}{r_1^2 - r^2}$$

i.e. $$50 = \frac{20 r_1^2}{r_1^2 - 125^2}$$

from which $r_1 = \dfrac{125}{\sqrt{0\cdot 6}} = 161\cdot 4$ mm

$\therefore t = 161\cdot 4 - 125 = \underline{36\cdot 4 \text{ mm}}$

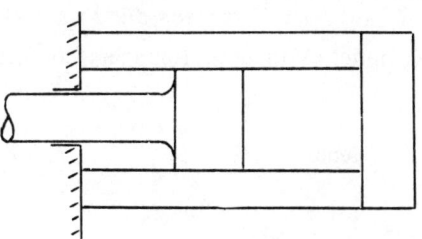

Fig. 10.9

From equation (10.13),

$$\sigma_l = \frac{pr_2^2}{r_1^2 - r_2^2}$$

$$= \frac{20 \times 125^2}{161\cdot 4^2 - 125^2} = 30 \text{ MN/m}^2$$

and from equation (10.12),

$$\sigma_c = p\frac{r_1^2 + r_2^2}{r_1^2 - r_2^2}$$

$$= 20 \times \frac{161\cdot 4^2 + 125^2}{161\cdot 4^2 - 125^2} = 80 \text{ MN/m}^2$$

The state of stress on an element of the cylinder at the inside surface is shown in Fig. 10.12. As in example 8, Chapter 9, the strain on the diameter is the same as the strain on the circumference and this is given by

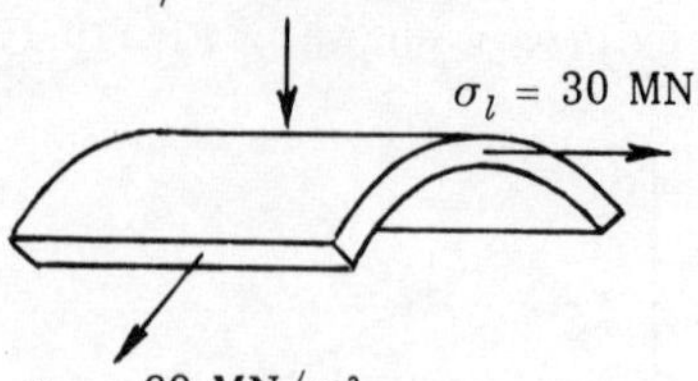

$$\epsilon_c = \frac{\sigma_c}{E} + \nu\frac{\sigma_r}{E} - \nu\frac{\sigma_l}{E}$$

$\sigma_r = 20 \text{ MN/m}^2$

$\sigma_l = 30 \text{ MN/m}^2$

$\sigma_c = 80 \text{ MN/m}^2$

Fig. 10.12

$$= \frac{(80 + 0\cdot28 \times 20 - 0\cdot28 \times 30) \times 10^6}{200 \times 10^9} = 0\cdot386 \times 10^{-3}$$

Therefore increase in diameter $= 0\cdot386 \times 10^{-3} \times 250 = \underline{0\cdot0965 \text{ mm}}$

5. *One steel cylinder is shrunk on to another, the compound cylinder having an inside diameter of 100 mm, an outside diameter of 200 mm and a diameter of 150 mm at the surfaces in contact. If shrinkage produces a radial pressure p_0 at the surfaces in contact, after which the compound cylinder is subjected to an internal pressure of 100 MN/m², find the value of p_0 so that the maximum circumferential stresses in the two cylinders shall be the same and determine the value of this stress.*

(a) *Initial stresses due to shrinkage*

(i) Inner cylinder :— Boundary conditions are $\sigma_r = -p_0$ when $r = 75$ mm

and $\sigma_r = 0$ when $r = 50$ mm

Hence $\quad -p_0 = a + \dfrac{b}{75^2}$

and $\quad 0 = a + \dfrac{b}{50^2}$

from which $\quad a = -1\cdot8p_0 \quad$ and $\quad b = 4500p_0$

Therefore, at inner surface, $\quad \sigma_c = -1\cdot8p_0 - \dfrac{4500p_0}{50^2} = -3\cdot6p_0 \text{ MN/m}^2$

(ii) Outer cylinder :— Boundary conditions are $\sigma_r = -p_0$ when $r = 75$ mm

and $\sigma_r = 0$ when $r = 100$ mm

Hence $\quad -p_0 = a + \dfrac{b}{75^2}$

and $\quad 0 = a + \dfrac{b}{100^2}$

from which $\quad a = \dfrac{9}{7}p_0 \quad$ and $\quad b = -\dfrac{90\,000}{7}p_0$

Therefore, at inner surface, $\quad \sigma_c = \dfrac{9}{7}p_0 + \dfrac{90\,000p_0}{7 \times 75^2} = 3\cdot57p_0 \text{ MN/m}^2$

(b) *Stresses due to internal pressure*

Boundary conditions are $\sigma_r = -100$ MN/m² at $r = 50$ mm

and $\sigma_r = 0$ at $r = 100$ mm

Hence
$$-100 = a + \frac{b}{50^2}$$

and
$$0 = a + \frac{b}{100^2}$$

from which
$$a = \frac{100}{3} \quad \text{and} \quad b = -\frac{10^6}{3}$$

Therefore, at $r = 50$ mm, $\sigma_c = \frac{100}{3} + \frac{10^6}{3 \times 50^2} = 166 \cdot 7$ MN/m²

and at $r = 75$ mm, $\sigma_c = \frac{100}{3} + \frac{10^6}{3 \times 75^2} = 92 \cdot 6$ MN/m²

Thus resultant stress at inner surface of inner cylinder
$$= -3 \cdot 6 p_0 + 166 \cdot 7 \text{ MN/m}^2$$

and resultant stress at inner surface of outer cylinder
$$= 3 \cdot 57 p_0 + 92 \cdot 6 \text{ MN/m}^2$$

Therefore, for equal stresses,
$$-3 \cdot 6 p_0 + 166 \cdot 7 = 3 \cdot 57 p_0 + 92 \cdot 6$$

from which
$$p_0 = \underline{10 \cdot 3 \text{ MN/m}^2}$$

Substituting in either of these equations gives the value of the maximum stress in each cylinder as $\underline{129 \cdot 4 \text{ MN/m}^2}$.

6. A cylindrical vessel 0·75 m diameter is to withstand an internal pressure of 2·8 MN/m², the maximum allowable stress being 85 MN/m². If the efficiency of the riveted joints is 75%, determine (a) the required thickness of the cylinder, (b) the axial stress in the cylinder with this pressure at a point where the full section is undiminished. (*Ans.*: 16·46 mm; 31·9 MN/m²)

7. A cylindrical vessel 1·8 m diameter is subjected to an internal pressure of 1·25 MN/m². The plates are 12 mm thick and have an ultimate tensile strength of 450 MN/m². If the efficiencies of the longitudinal and circumferential joints are 75% and 50% respectively, calculate the factor of safety. (*Ans.*: 3·6)

8. A thick cylinder has outer and inner diameters of 400 mm and 250 mm respectively. If the maximum permissible tensile stress is 140 MN/m², calculate the maximum safe internal pressure. What is the circumferential stress at the outer circumference due to this internal pressure? (*Ans.*: 61·35 MN/m²; 78·65 MN/m²)

9. A thick cylinder having external diameter of 200 mm and an internal diameter of 100 mm is subjected to an internal pressure of 56 MN/m² and an external pressure of 7 MN/m². Find the maximum direct stress in the cylinder and the change of external diameter. $E = 200$ GN/m² and $\nu = 0 \cdot 3$. (*Ans.*: 74·67 MN/m²; 0·0278 mm)

10. A steel tube is 18 mm internal diameter, 3 mm thick and 300 mm long. Calculate the safe internal pressure if the maximum stress is not to exceed 150 MN/m² and also the increase in internal volume under this pressure. $E = 200$ GN/m² and $\nu = 1/3 \cdot 5$. (*Ans.*: 42 MN/m²; 120·6 mm³)

11. A steel cylinder, 1 m inside diameter, is to withstand an internal pressure of 8 MN/m². Calculate the thickness if the maximum shearing stress is not to exceed 35 MN/m².

Calculate the increase in volume due to this internal pressure if the cylinder is 6 m long. $E = 200$ GN/m² and $\nu = \frac{1}{3}$. (*Ans.*: 69·3 mm; 0·002 835 m³)

12. A thick cylinder, 200 mm internal diameter, is to withstand an internal pressure of 50 MN/m². Find the necessary thickness if the maximum shearing stress is not to exceed 100 MN/m². What will then be the greatest and least hoop stresses in the material? (*Ans.*: 41·4 mm; 150 and 100 MN/m²)

13. A thick cylinder has a length of 0·25 m and internal and external diameters of 0·1 m and 0·1414 m respectively. Determine (*a*) the circumferential and longitudinal stresses at the inner surface when the cylinder is subjected to an internal pressure of 10 MN/m², (*b*) the increase in internal volume of the cylinder. $E = 200$ GN/m² and $\nu = 0\cdot3$. (*Ans.*: 30 MN/m²; 10 MN/m²; 628·4 × 10⁻³ m³)

14. A steel cylinder 200 mm external diameter and 150 mm internal diameter has another cylinder 250 mm external diameter shrunk on to it. If the maximum tensile stress induced in the outer cylinder is 80 MN/m², find the radial compressive stress between the cylinders.

Determine the circumferential stresses at inner and outer diameters of both cylinders and show, by means of a diagram, how these stresses vary with the radius.
(*Ans.*: 17·6; 80; 62·5; 80·5; 62·8 MN/m²)

15. A steel cylinder of outside diameter 300 mm and inside diameter 250 mm is shrunk on to one having diameters 250 mm and 200 mm, the interference fit being such that under an internal pressure p the inner tensile stress in both cylinders is 84 MN/m². Find the value of p. (*Ans.*: 36·9 MN/m²)

16. A compound steel cylinder has a bore of 80 mm and an outside diameter of 160 mm, the diameter of the common surface being 120 mm. Find the radial pressure at the common surface which must be provided by shrinkage, if the resultant maximum circumferential tension in the inner cylinder under a superimposed internal pressure of 60 MN/m² is to be half the value of the maximum circumferential tension which would be produced in the inner cylinder if that cylinder alone were subjected to an internal pressure of 60 MN/m². (*Ans.*: 6·11 MN/m²)

11 Dynamics

11.1 Introduction Dynamics involves a study of the motion of systems and the forces associated with the motion, which may be linear, angular or a combination of linear and angular.

11.2 Linear motion The *velocity*, v, of a body is the rate of change of its displacement x from some reference position with respect to time, t,

i.e.
$$v = \frac{dx}{dt} \tag{11.1}$$

The *acceleration*, f, of a body is the rate of change of its velocity v with respect to time, t,

i.e.
$$f = \frac{dv}{dt} = \frac{d^2x}{dt^2} = v\frac{dv}{dx} \tag{11.2}$$

If a body moves with uniform acceleration f such that the velocity changes from u to v in time t while traversing a distance s, then it may be shown that

$$v = u + ft \tag{11.3}$$

$$s = ut + \tfrac{1}{2}ft^2 \tag{11.4}$$

and
$$v^2 = u^2 + 2fs \tag{11.5}$$

If experimental data relating velocity, distance and time are available, equations (11.1) and (11.2) may be interpreted graphically, as shown in Fig. 11.1. Solutions by this method are not restricted to cases of uniform acceleration.

11.3 Angular motion For a body moving with angular velocity ω and angular acceleration α, turning through an angle θ from some datum position,

$$\omega = \frac{d\theta}{dt} \tag{11.6}$$

and
$$\alpha = \frac{d\omega}{dt} = \frac{d^2\theta}{dt^2} = \omega\frac{d\omega}{d\theta} \tag{11.7}$$

If the angular acceleration is uniform and the angular velocity changes from ω_1 to ω_2 in time t while rotating through an angle θ, then

$$\omega_2 = \omega_1 + \alpha t \tag{11.8}$$

$$\theta = \omega_1 t + \tfrac{1}{2}\alpha t^2 \tag{11.9}$$

and
$$\omega_2^2 = \omega_1^2 + 2\alpha\theta \tag{11.10}$$

Similar graphs to those in Fig. 11.1 may be used for experimental data involving angular motion.

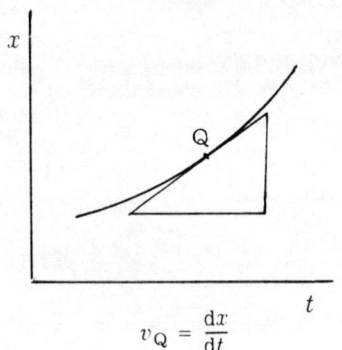

$$v_Q = \frac{dx}{dt}$$

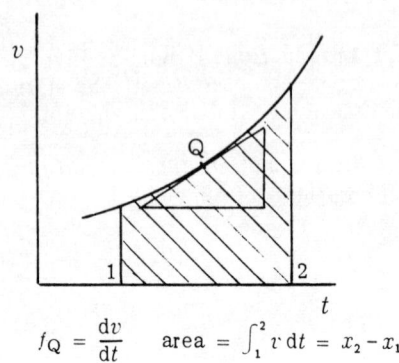

$$f_Q = \frac{dv}{dt} \qquad \text{area} = \int_1^2 v\,dt = x_2 - x_1$$

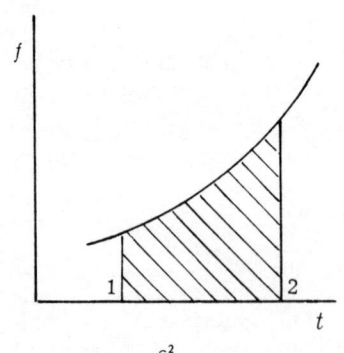

$$\text{area} = \int_1^2 f\,dt = v_2 - v_1$$

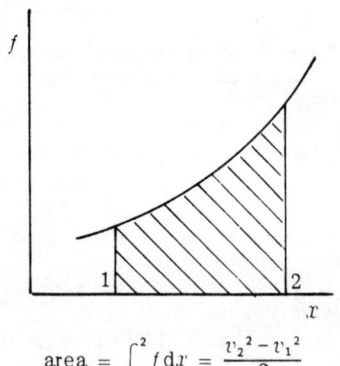

$$\text{area} = \int_1^2 f\,dx = \frac{v_2^2 - v_1^2}{2}$$

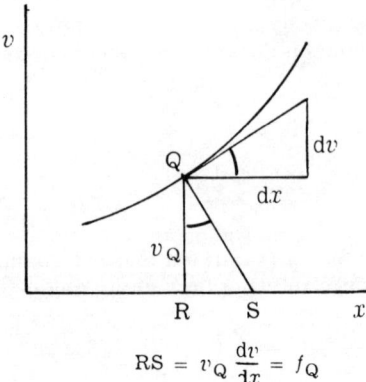

$$RS = v_Q \frac{dv}{dx} = f_Q$$

Fig. 11.1

DYNAMICS

11.4 Mass, force, weight, momentum The *mass*, m, of a body is a measure of the amount of matter in the body, this being independent of geographical location.

Force, P, is that necessary to change the state of rest or uniform motion of a body. Unit force is that required to give unit acceleration to unit mass.

The *weight*, W, of a body is the force of attraction between the body and the earth, which varies slightly in different parts of the world due to the variation in the value of g.

The *momentum* of a body of mass m moving with velocity v is the product mv. The total momentum of a system in a given direction remains constant unless acted upon by an external force.

11.5 Newton's Laws of Motion (i) Every body continues in its state of rest or uniform motion in a straight line unless acted upon by an external force.

(ii) The rate of change of momentum of a body is proportional to the external force and takes place in the direction of the force.

(iii) To every action (force) there is an equal and opposite reaction (force).

From the second law, $\quad P \propto$ rate of change of momentum

$\propto$ mass × rate of change of velocity

i.e. $\qquad P = kmf \quad$ where k is a constant.

The units are chosen to make the constant of proportionality unity. In the *Système International d'Unités*, the unit of mass is the kilogramme and the unit of length is the metre. The unit of force is the newton, which is defined as the force necessary to give a mass of 1 kg an acceleration of 1 m/s²,

i.e. $\qquad P \text{ (N)} = m \text{ (kg)} \times f \text{ (m/s}^2) \qquad (11.11)$

Since a body falling freely under the earth's gravitational pull has an acceleration g, the weight

$$W = mg \qquad (11.12)$$

g has the standard value of 9·806 65 m/s² which, for normal purposes, is taken as 9·81 m/s².

11.6 Impulse The impulse of a constant force P acting for a time t is the product Pt. If this causes a mass m to change its velocity from u to v, then

$$P = mf = m\frac{v - u}{t}$$

or $\qquad Pt = m(v - u) \qquad (11.13)$

i.e. $\qquad$ impulse = change of momentum

11.7 Circular motion Consider a particle of mass m moving in a circular path of radius r with constant velocity v. Fig. 11.2(a). Let the particle move from P to Q in time dt. Then the magnitude of the velocity is unchanged but the direction has changed and from the relative velocity diagram, Fig. 11.2(b), this change is represented by pq.

126   DYNAMICS

Thus there is an inward or *centripetal* acceleration,

$$f = \frac{pq}{dt} = v\frac{d\theta}{dt} = v\omega$$

But $v = \omega r$, so that $f = \omega^2 r$ or $\dfrac{v^2}{r}$ (11.14)

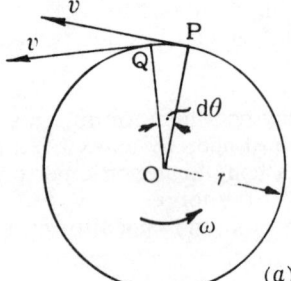
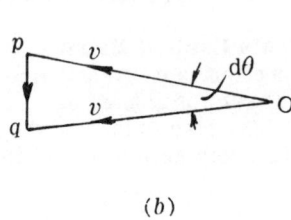

(a)   (b)

Fig. 11.2

The centripetal force necessary to constrain the mass to move in a circular path is given by

$$P = mf = m\omega^2 r \quad \text{or} \quad m\frac{v^2}{r} \qquad (11.15)$$

By Newton's third law, there will be an equal and opposite outward or *centrifugal* reaction to this force, which acts at the centre of rotation.

In problems involving equilibrium of forces, the moving body may be brought to rest by applying a radially outward force, mv^2/r, to the body to convert the dynamic case to the equivalent static case.

11.8 Work, energy and power If a constant force P moves through a distance x, the work done $= P.x$.

If the force is variable, work done $= \Sigma P \delta x$ but if the force is a function of x, this can be expressed in the form

$$\text{work done} = \int_0^x f(x)\,dx \qquad (11.16)$$

The unit of work is the joule (J), which is the work done by a force of 1 N moving through a distance of 1 m.

Work represents a transfer of energy and thus the energy of a body is a measure of its capacity to do work. Energy may exist in various forms, such as mechanical, thermal and electrical energy but it cannot be created or destroyed; a loss in one form is always accompanied by an equal gain in another form.

Mechanical energy may be of the form of potential, kinetic or strain energy. Potential energy is the energy a body possesses by virtue of its position relative to the earth. If a body of mass m is at a height h above some datum level, the potential energy relative to that datum is given by

$$\text{potential energy} = mgh \qquad (11.17)$$

DYNAMICS

Kinetic energy is the energy a body possesses due to its velocity. If a a force P is applied to a body of mass m such that it attains a velocity v in time t whilst moving a distance s, then

$$\text{kinetic energy} = \text{work done} = P \times s$$

$$= mf \times \frac{v^2}{2f} = \tfrac{1}{2}mv^2 \qquad (11.18)$$

Strain energy is the energy stored in a body when it is deformed. If a force P is applied to a body of stiffness S and causes a deformation x,

$$\text{strain energy} = \text{work done} = \tfrac{1}{2}Px = \tfrac{1}{2}Sx^2 \qquad (11.19)$$

Power is the work transfer rate. If a constant force P moves with velocity v, then the power $= Pv$. The unit of power is the watt (W), which is $1\,\text{N\,m/s}$ or $1\,\text{J/s}$.

11.9 Moment of inertia The moment of inertia of a particle about an axis is the product of its mass and the square of its distance from that axis. For a distributed mass, Fig. 11.3, the moment of inertia of a particle of mass dm about an axis through O is $dm \cdot l^2$, so that the total moment of inertia of the body about this axis is $\int dm \cdot l^2$; this is denoted by I.

The *radius of gyration* of a body is the radius at which the mass may be considered concentrated to give the same moment of inertia. This radius is denoted by k, so that $I = mk^2$, where m is the mass of the body.

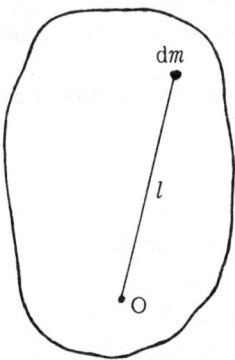
Fig. 11.3

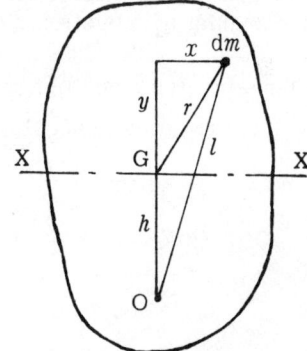
Fig. 11.4

11.10 Theorem of parallel axes Let G be the centre of gravity of the body, Fig. 11.4, and let the distance between axes through G and O, perpendicular to the plane of the body, be h.

Then
$$I_O = \int dm \cdot l^2 = \int dm \{x^2 + (h + y)^2\}$$

$$= \int dm \{x^2 + h^2 + 2yh + y^2\}$$

$$= \int dm \{r^2 + h^2 + 2hy\}$$

$$= \int dm \cdot r^2 + h^2 \int dm + 2h \int dm \cdot y$$

But $\int dm \cdot r^2 = I_G$, $\int dm = m$ and $\int dm \cdot y$ is the total first moment of the mass about the axis XX, which is zero since this axis passes through the centre of gravity.

Thus $\qquad I_O = I_G + mh^2 \qquad (11.20)$

11.11 Values of I and k for common cases

(a) *Thin uniform disc, radius r*

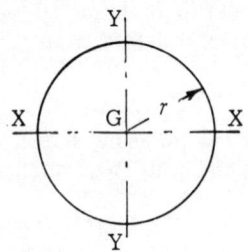

$I_G = \dfrac{mr^2}{2} \qquad k_G = \dfrac{r}{\sqrt{2}}$

$I_{XX} = I_{YY} = \dfrac{mr^2}{4} \qquad k_{XX} = k_{YY} = \dfrac{r}{2}$

(b) *Thin uniform rod, length l*

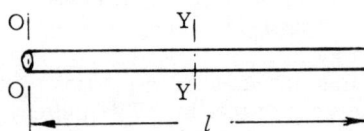

$I_{YY} = \dfrac{ml^2}{12} \qquad k_{YY} = \dfrac{l}{\sqrt{12}}$

$I_{OO} = \dfrac{ml^2}{3} \qquad k_{OO} = \dfrac{l}{\sqrt{3}}$

11.12 Torque and angular acceleration Fig. 11.5 shows a body rotating about O with angular acceleration α due to an applied external torque T.

Linear acceleration of particle of mass $dm = \alpha l$

∴ force required to accelerate particle $= dm \cdot \alpha l$

∴ torque required about O $= dm \cdot \alpha l^2$

∴ total torque for whole body $= \int dm \cdot \alpha l^2$

But $\int dm \cdot l^2 = I_O$

∴ $T = I_O \alpha \qquad (11.21)$

11.13 Angular momentum and angular impulse The angular momentum of a body about an axis is the moment of its momentum about that axis. Fig. 11.6 shows a body rotating about O with angular velocity ω.

Linear velocity of particle of mass $dm = \omega l$

∴ momentum of particle $= dm \cdot \omega l$

∴ moment of momentum about O $= dm \cdot \omega l^2$

∴ total moment of momentum about O $= \int dm \cdot \omega l^2$

$= I_O \omega \qquad (11.22)$

DYNAMICS

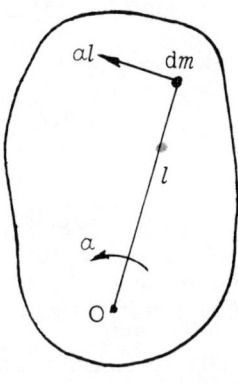

Fig. 11.5   Fig. 11.6

If G is the centre of gravity, then $I_O = I_G + mh^2$

∴ moment of momentum about O $= I_G \omega + mh^2 \omega$

$= I_G \omega + mvh$ (11.23)

The angular momentum about O is therefore the angular momentum about a parallel axis through G, together with the moment of the linear momentum about O.

The angular momentum of a system about any axis remains constant, unless acted upon by an external torque about that axis.

The angular impulse of a constant torque T acting for a time t is the product Tt. If this causes a body of moment of inertia I to change its angular velocity from ω_1 to ω_2, then

$$T = I\alpha = I\frac{\omega_2 - \omega_1}{t}$$

or $Tt = I(\omega_2 - \omega_1)$ (11.24)

i.e.   angular impulse = change of angular momentum

11.14 Angular work, power and kinetic energy If a constant torque T moves through an angle θ, the work done $= T\theta$.

If the torque is variable, work done $= \Sigma T \delta\theta$ but if the torque is a function of θ, this may be expressed in the form

$$\text{work done} = \int_0^\theta f(\theta)\,d\theta \qquad (11.25)$$

If a constant torque T moves with an angular velocity ω, then

$$\text{power} = T\omega$$

Referring to Fig. 11.6

kinetic energy of particle of mass $dm = \tfrac{1}{2} dm (\omega l)^2$

∴ total kinetic energy of body $= \dfrac{\omega^2}{2} \int dm \cdot l^2$

$$= \tfrac{1}{2} I_O \omega^2 \qquad (11.26)$$

Alternatively, since $I_O = I_G + mh^2$, kinetic energy of body

kinetic energy of body $= \tfrac{1}{2} I_G \omega^2 + \tfrac{1}{2} m h^2 \omega^2$

$$= \tfrac{1}{2} I_G \omega^2 + \tfrac{1}{2} m v^2 \qquad (11.27)$$

The total kinetic energy is therefore the kinetic energy due to rotation about G together with the kinetic energy due to the linear velocity of G.

11.15 Equivalent mass of a rotating body It is sometimes convenient to treat combined linear and angular motion as an equivalent linear problem, the angular effects being allowed for by an equivalent mass added to the actual mass of the body.

Consider the body of mass m shown in Fig. 11.7. Let a tangential force P be applied at radius r causing an angular acceleration a.

Then $Pr = I_O a = mk^2 \dfrac{f}{r}$

$$\therefore P = m\left(\dfrac{k^2}{r^2}\right) f \qquad (11.28)$$

The quantity $m\left(\dfrac{k^2}{r^2}\right)$ is the equivalent mass of the rotating body referred to the line of action of P.

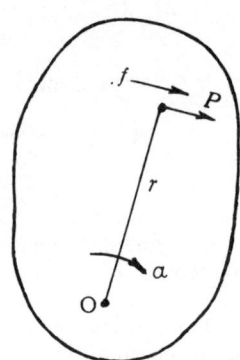

Fig. 11.7

11.16 Acceleration of geared systems Fig. 11.8 shows two gear wheels A and B with moments of inertia I_a and I_b respectively, having a speed ratio $\dfrac{N_b}{N_a} = n$.

If a torque T is applied to shaft A to give the shafts angular accelerations a_a and a_b, then torque required on B to accelerate B

$$= I_b a_b = I_b a_a n$$

since $\dfrac{a_b}{a_a} = n$.

∴ torque required on A to accelerate B

$$= n \times I_b a_a n = n^2 I_b a_a$$

Torque required on A to accelerate A

$$= I_a a_a$$

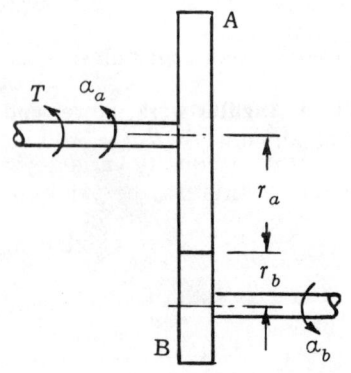

Fig. 11.8

DYNAMICS

∴ torque required on A to accelerate A and B $= I_a \alpha_a + n^2 I_b \alpha_a$

i.e.
$$T = (I_a + n^2 I_b)\alpha_a \qquad (11.29)$$

The quantity $I_a + n^2 I_b$ is the moment of inertia of the system referred to shaft A and this principle may be extended to any number of gears meshing together.

The tangential force P between the teeth is given by

$$P r_b = I_b \alpha_b$$

or
$$P = \frac{I_b \alpha_b}{r_b} \qquad (11.30)$$

11.17 Maximum acceleration of vehicles The maximum acceleration of a vehicle is limited by the friction force between the wheels and the road. This will depend on the normal reactions at the wheels and the coefficient of friction between wheels and road.

Fig. 11.9 shows a vehicle of mass m, moving up a gradient inclined at angle θ to the horizontal with acceleration f. The normal reactions at the front and rear wheels are R_f and R_r respectively.

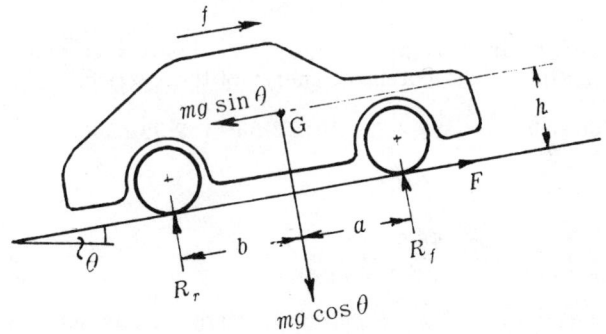

Fig. 11.9

The tractive force $F = \mu R_f$, μR_r or $\mu(R_f + R_r)$, depending on whether the vehicle has front, rear or four-wheel drive. In the latter case, however, the tractive forces at each pair of wheels must be in the ratio of the normal reactions if slipping is to occur simultaneously at all four wheels.

By resolution of forces, $\qquad R_f + R_r = mg \cos \theta \qquad (11.31)$

and $\qquad\qquad\qquad\qquad F = mg \sin \theta + mf \qquad (11.32)$

Taking moments about G, $\qquad Fh = R_r b - R_f a \qquad (11.33)$

For maximum retardation of vehicles, the direction of the force F becomes reversed.

1. *A pile of mass ¾ t can just support a stationary mass of 40 t without subsidence. The mass is removed and the pile is driven to a greater depth by blows of a 2 t hammer dropping on to the top of the pile from a height of 1·22 m. The hammer does not rebound from the top of the pile.*

Calculate (a) the penetration per blow, assuming that the ground resistance is constant, (b) the energy lost per blow, and (c) the efficiency of the operation.

(a) Ground resistance, $R = (40 + ¾) \times 10^3 \times 9·81$

$$= 400 \times 10^3 \text{ N}$$

Velocity of impact $= \sqrt{2gh}$

$$= \sqrt{2 \times 9·81 \times 1·22}$$

$$= 4·89 \text{ m/s}$$

Momentum of hammer before impact = momentum of hammer and pile after impact

i.e. $2 \times 4·89 = (2 + ¾) \times V$ where V is the common velocity

$$\therefore V = 3·56 \text{ m/s}$$

The work done in moving a distance x against a resistance R is equal to the loss of kinetic and potential energy of the system.

i.e. $400 \times 10^3 x = ½ \times 2·75 \times 10^3 \times 3·56^2 + 2·75 \times 10^3 \times 9·81 x$

from which $x = 0·046\,7$ m

or $\underline{46·7 \text{ mm}}$

(b) Energy lost = loss of potential energy of hammer − work done against ground resistance

$$= 2 \times 10^3 \times 9·81(1·22 + 0·046\,7) - 400 \times 10^3 \times 0·046\,7$$

$$= \underline{6·15 \times 10^6 \text{ J}}$$

(c) Efficiency $= \dfrac{\text{work done against ground resistance}}{\text{work done in raising hammer above top of pile}}$

$$= \dfrac{400 \times 10^3 \times 0·046\,7}{2 \times 10^3 \times 9·81 \times 1·22}$$

$$= 0·78$$

or $\underline{78\%}$

DYNAMICS

2. *A shaft is being turned in a lathe which is driven by a motor developing 2·25 kW at 1400 rev/min, the speed reduction between the motor and lathe spindle being 10 to 1. The friction torque at the lathe spindle is 17·5 Nm. The moment of inertia of the rotating parts of the motor is 0·08 kg m² and that of the lathe spindle and work-piece is 1·2 kg m².*

If the tool is suddenly given an excessively heavy cut which stops the shaft in one revolution, calculate the force on the tool if it is cutting at a radius of 140 mm.

Equivalent moment of inertia of spindle S, Fig. 11.10

$$= I_s + I_m \times 10^2 \quad . \quad . \quad . \quad \text{from equation (11.29)}$$

$$= 1 \cdot 2 + 0 \cdot 08 \times 100$$

$$= 9 \cdot 2 \text{ kg m}^2$$

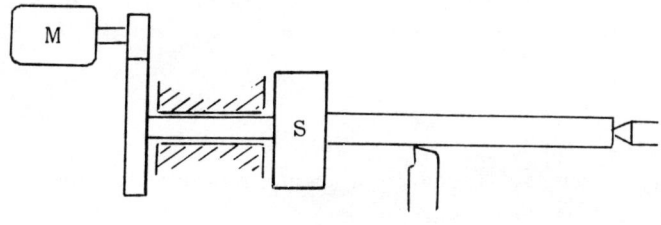

Fig. 11.10

The deceleration of the spindle is given by

$$\omega^2 = 2\alpha\theta \quad . \quad . \quad . \quad . \quad \text{from equation (11.10)}$$

i.e.
$$\alpha = \frac{\left(140 \times \frac{2\pi}{60}\right)^2}{2 \times 2\pi} = 17 \cdot 1 \text{ rad/s}^2$$

$$\text{Torque applied to spindle by motor} = \frac{2 \cdot 25 \times 10^3}{140 \times \frac{2\pi}{60}}$$

$$= 153 \cdot 5 \text{ Nm}$$

The net decelerating torque on the shaft

= torque due to tool force, P + friction torque − driving torque

$$= P \times 0 \cdot 14 + 17 \cdot 5 - 153 \cdot 5$$

$$\therefore \ 0 \cdot 14 P - 136 = I\alpha = 9 \cdot 2 \times 17 \cdot 1$$

from which $\qquad\qquad P = \underline{2090 \text{ N}}$

3. *A drum A of mass 200 kg, external diameter 380 mm and radius of gyration 150 mm, rotates on frictionless bearings at 250 rev/min. A stationary drum B of mass 50 kg, external diameter 200 mm and radius of gyration 80 mm, mounted on a frictionless axis parallel to that of A, is pressed into contact with A with a force of 90 N. The coefficient of friction is 0·25.*
Determine (a) the time of slipping and the final speeds of A and B,
(b) the time of slipping if a torque is applied to A to maintain a constant speed of 250 rev/min.

Let X be the tangential impulse acting at the circumference of the drums, Fig. 11.11, acting so as to decelerate A and accelerate B.

If N is the final speed of A, the final speed of B is $N \times \dfrac{0·19}{0·10}$

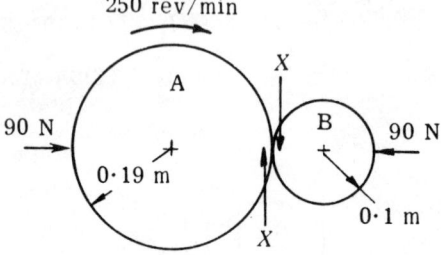

Fig. 11.11

Angular impulse = change of angular momentum   from Art. 11.13

i.e. $X \times 0·19 = 200 \times 0·15^2 (250 - N) \times \dfrac{2\pi}{60}$ (1)

and $X \times 0·10 = 50 \times 0·08^2 \left(N \times \dfrac{0·19}{0·10}\right) \times \dfrac{2\pi}{60}$ (2)

$\therefore\ 1·9 = 4 \times \left(\dfrac{0·15}{0·08}\right)^2 \times \dfrac{250 - N}{1·9N}$

from which $N = \underline{199 \text{ rev/min}}$

Speed of B $= 1·9N = \underline{378 \text{ rev/min}}$

Substituting for N in equation (1),

$$X = \dfrac{200 \times 0·15^2 \times 51 \times 2\pi}{0·19 \times 60} = 126·6 \text{ N s}$$

Friction force between cylinders $= 0·25 \times 90 = 22·5$ N

Therefore time of slipping $= \dfrac{126·6}{22·5} = \underline{5·62 \text{ s}}$

When speed of A remains constant at 250 rev/min, equation (2) gives

$$90 \times 0·25 \times t \times 0·10 = 50 \times 0·08^2 \times 250 \times \dfrac{0·19}{0·10} \times \dfrac{2\pi}{60}$$

from which $t = \underline{7·07 \text{ s}}$

DYNAMICS

4. *A valve of mass 0·25 kg closes horizontally under the action of a spring. In the closed position, the spring is compressed 12 mm and the maximum opening of the valve is 6 mm. If the spring stiffness is 4 kN/m, find the time required for the valve to close and the velocity with which it strikes the seat.*

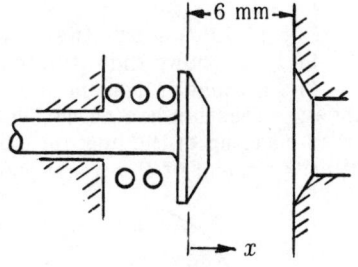

In the open position, Fig. 11.12,

compression of spring = 12 + 6 = 18 mm

Therefore, when the valve has moved a distance x m,

compression of spring = $0.018 - x$ m

Fig. 11.12

The equation of motion of the valve is therefore

$$P = mf = m\frac{d^2x}{dt^2}$$

i.e. $\quad 4 \times 10^3 (0.018 - x) = 0.25 \dfrac{d^2x}{dt^2}$

$\therefore \dfrac{d^2x}{dt^2} + 16 \times 10^3 \, x = 16 \times 10^3 \times 0.018$

The solution is* $\quad x = A \cos 126.4t + B \sin 126.4t + 0.018$

when $t = 0$, $x = 0$. $\quad \therefore A = -0.018$

when $t = 0$, $\dfrac{dx}{dt} = 0$, $\quad \therefore B = 0$

$\therefore x = 0.018(1 - \cos 126.4t)$

When $x = 6$ mm $\quad 0.006 = 0.018(1 - \cos 126.4t)$

$\therefore \cos 126.4t = \tfrac{2}{3}$

$\therefore t = \underline{0.006\,65 \text{ s}}$

$v = \dfrac{dx}{dt} = 0.018 \times 126.4 \sin 126.4t$

When $t = 0.066\,5$ s, $\quad v = 0.018 \times 126.4 \sin 126.4 \times 0.006\,65$

$= 2.275 \sin 48° 12'$

$= \underline{1.7 \text{ m/s}}$

*See Appendix

5. *A solid uniform cylinder rolls without slipping down a plane AO inclined at 45° to the horizontal and then rolls up a plane OB inclined at θ to the horizontal. The centre of gravity of the cylinder descends a vertical distance of 0·6 m while the cylinder is rolling down AO.*

Calculate (a) the minimum value of θ which will cause the cylinder to be brought to rest when it strikes OB, (b) the vertical distance through which the cylinder will rise as it rolls up OB if θ = 45°.

Fig. 11.13(a) shows the situation immediately before impact at P and Fig. 11.13(b) shows the situation immediately after impact.

The moment of momentum of the cylinder about P remains unchanged since the reaction has no moment about that point. This moment of momentum is made up of the angular momentum about the centre of gravity, together with the moment of the linear momentum about P (Art. 11.13).

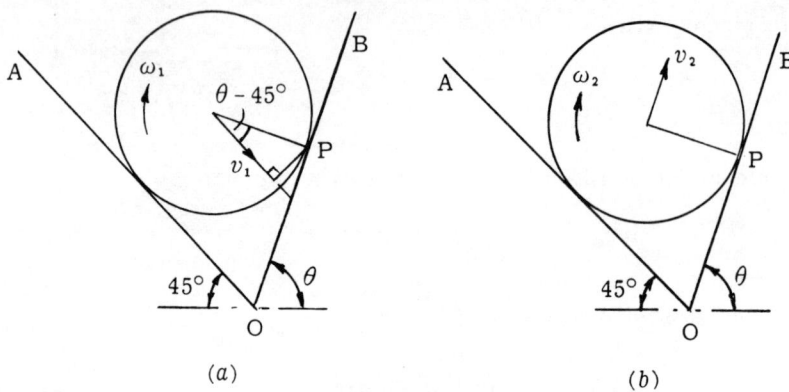

(a)   (b)

Fig. 11.13

Taking clockwise moment of momentum as positive,

initial moment of momentum about P $= I\omega_1 - mv_1 \times r \sin(\theta - 45°)$,

the moment of mv_1 about P being anticlockwise.

Final moment of momentum about P $= I\omega_2 + mv_2 r$

both terms being clockwise.

The moment of inertia of a solid uniform cylinder about its polar axis is $\frac{1}{2}mr^2$ and $\omega = v/r$, so that, equating initial and final moments of momentum about P,

$$m\frac{r^2}{2} \cdot \frac{v_1}{r} - mv_1 r \sin(\theta - 45°) = m\frac{r^2}{2} \cdot \frac{v_2}{r} + mv_2 r$$

from which $\qquad v_2 = \dfrac{v_1}{3}[1 - 2\sin(\theta - 45°)]$

When $v_2 = 0$, $\qquad \sin(\theta - 45°) = 0·5$

$\qquad \therefore \theta - 45° = 30°$

or $\qquad$ or $\theta = \underline{75°}$

DYNAMICS   137

When $\theta = 45°$, $\qquad v_2 = \dfrac{v_1}{3}$

When rolling down OA, loss of potential energy = gain in kinetic energy

i.e.
$$mgh_1 = \tfrac{1}{2}I\omega_1^2 + \tfrac{1}{2}mv_1^2$$
$$= \tfrac{1}{2}m\dfrac{r^2}{2}\left(\dfrac{v_1}{r}\right)^2 + \tfrac{1}{2}mv_1^2$$
$$\therefore \ m \times 9\cdot 81 \times 0\cdot 6 = \tfrac{3}{4}mv_1^2$$

from which $\qquad v_1 = 2\cdot 82$ m/s

$$\therefore \ v_2 = \dfrac{2\cdot 82}{3}\ \text{m/s}$$

When rolling up OB, loss of kinetic energy = gain of potential energy

i.e. $\qquad \tfrac{1}{2}I\omega_2^2 + \tfrac{1}{2}mv_2^2 = mgh_2$

i.e. $\qquad \tfrac{3}{4}mv_2^2 = mgh_2$

$$\therefore \ h_2 = \dfrac{3}{4 \times 9\cdot 81}\left(\dfrac{2\cdot 82}{3}\right)^2$$
$$= 0\cdot 065\ \text{m}$$

6. *One end of a thin uniform rod 0·45 m long is hinged to a rigid support. The rod, which has a mass of 2 kg and initially hangs downwards, is raised through 60° from the vertical and then released. As the rod approaches the end of its travel, it is restrained by a horizontal compression spring situated 0·4 m below the hinge. The stiffness of the spring is 35 kN/m and it is arranged so that the rod just reaches the vertical position at the point of maximum compression.*

Calculate the greatest reaction at the hinge. Where should the spring be placed for this value to be zero?

Let P be the force exerted by the spring at the maximum compression x. When released from the highest position, Fig. 11.14(a),

loss of potential energy of rod = gain of strain energy of spring

i.e. $\qquad mgh = \tfrac{1}{2}Sx^2 = \tfrac{1}{2}\dfrac{P^2}{S}\quad$ since $P = Sx$

$$\therefore \ 2 \times 9\cdot 81 \times \dfrac{0\cdot 45}{2}(1 - \cos 60°) = \tfrac{1}{2} \times \dfrac{P^2}{35 \times 10^3}$$

from which $\qquad P = 393$ N

This is the maximum compressive force in the spring and at this instant, Fig. 11.14(b) the angular acceleration of the rod is given by taking moments about O,

i.e. $\qquad P \times 0\cdot 4 = I_O \alpha = \dfrac{ml^2}{3}\alpha$

$$\therefore \quad 393 \times 0.4 = 2 \times \frac{0.45^2}{3} \alpha$$

$$\therefore \quad \alpha = 1\,165 \text{ rad/s}^2$$

The linear motion of the rod is given by

$$P + Q = mf = m \times \frac{l}{2}\alpha$$

i.e.
$$393 + Q = 2 \times 0.225 \times 1\,165 = 524$$

$$\therefore \quad Q = \underline{131 \text{ N}}$$

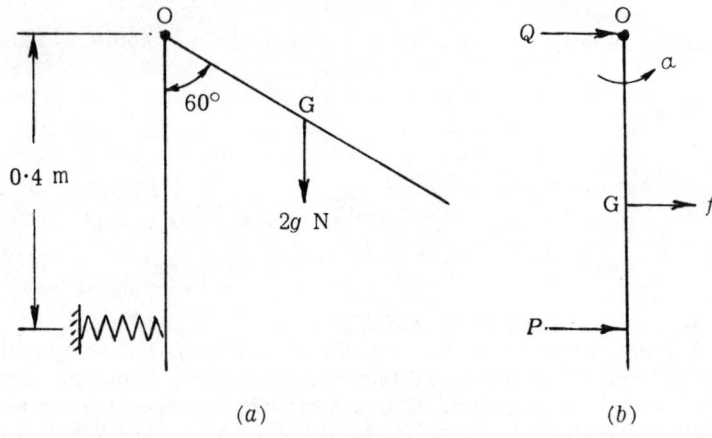

(a) (b)

Fig. 11.14

If the spring is at a distance h below O, then

$$Ph = I_O \alpha$$

$$\therefore \quad \alpha = \frac{Ph}{I_O} = \frac{3Ph}{ml^2}$$

$$Q = mf - P = m\frac{l}{2}\alpha - P$$

$$= \frac{3Ph}{2l} - P$$

Therefore, when $Q = 0$, $\quad \dfrac{3h}{2l} = 1$

$$\therefore \quad h = \tfrac{2}{3}l = \tfrac{2}{3} \times 0.45 = \underline{0.3 \text{ m}}$$

NOTE : The point of impact which produces no reaction at the hinge is called the *centre of percussion*.

DYNAMICS

7. *The loaded cage of a goods hoist has a mass of 1 200 kg. The rope passes over a drum at the top of the shaft and then to a balance mass of 500 kg. The cage and balance mass move in guides and the friction force at each guide is 500 N. The drum has a diameter of 1·5 m, a mass of 600 kg and a radius of gyration of 0·6 m. The maximum acceleration attained is 1·5 m/s², which occurs at a speed of 2·5 m/s. The maximum speed is 4·5 m/s and retardation is at a uniform rate from that speed to zero in the last 4·5 m of travel.*

Determine (a) the power required to drive the drum at the condition of maximum acceleration, (b) the rope tensions during retardation.

Suffices 1 and 2 refer to the cage and balance mass respectively.

Let the rope tensions be P_1 and P_2, Fig. 11.15, and let the friction force be F.

Then during the acceleration period,

$$P_1 = m_1 g + m_1 f + F$$
$$= 1\,200(9\cdot81 + 1\cdot5) + 500$$
$$= 14\,070 \text{ N}$$

and
$$P_2 = m_2 g - m_2 f - F$$
$$= 500(9\cdot81 - 1\cdot5) - 500$$
$$= 3\,660 \text{ N}$$

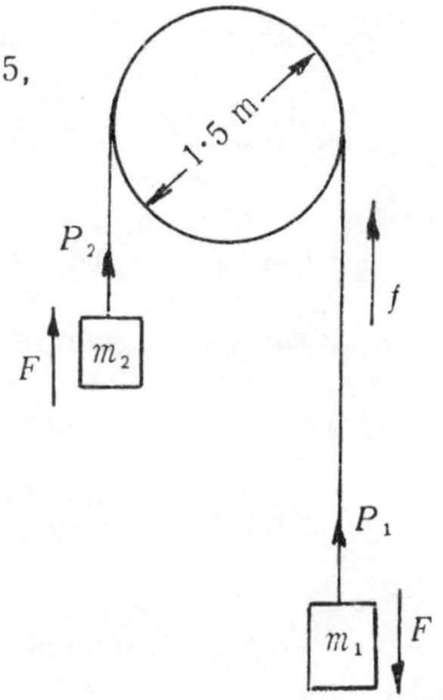

Therefore, torque on drum,

$$T = I\alpha + (P_1 - P_2)r$$
$$= 600 \times 0\cdot6^2 \times \frac{1\cdot5}{0\cdot75} + (14\,070 - 3\,660) \times 0\cdot75 = 8\,239 \text{ N m}$$

$$\text{Power} = T\omega = 8\,239 \times \frac{2\cdot5}{0\cdot75} = \underline{27\,460 \text{ W}}$$

Fig. 11.15

During retardation, the deceleration is given by

$$v^2 = 2fs$$

i.e.
$$4\cdot5^2 = 2f \times 4\cdot5$$
$$\therefore f = 2\cdot25 \text{ m/s}^2$$

$$\therefore P_1 = m_1 g - m_1 f + F$$
$$= 1\,200(9\cdot81 - 2\cdot25) + 500 = \underline{9\,572 \text{ N}}$$

and
$$P_2 = m_2 g + m_2 f - F$$
$$= 500(9\cdot81 + 2\cdot25) - 500 = \underline{5\,530 \text{ N}}$$

8. *In a double reduction lifting gear, the moments of inertia of the motor armature and pinion, intermediate and drum shafts are 3·5, 45 and 1 000 kg m² respectively. The motor runs at 6 times the speed of the intermediate shaft and 6G times the speed of the drum shaft. The drum radius is 0·9 m and the load lifted is 1 t. For a constant motor torque of 550 N m, find the value of G for maximum acceleration and determine the value of this acceleration.*

The arrangement of the drive is shown in Fig. 11.16.

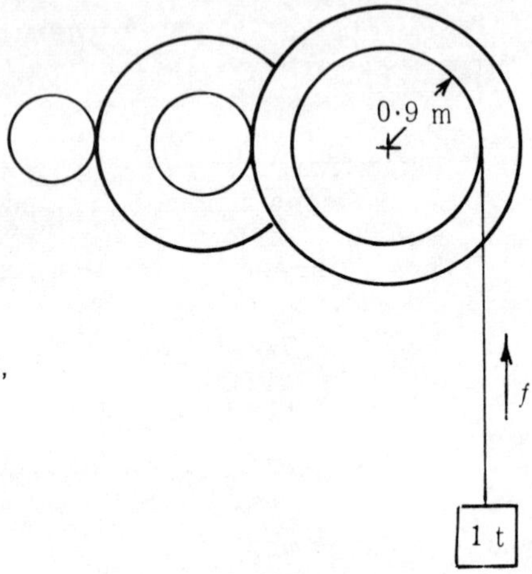

Effective moment of inertia of motor

$= 3·5 + \dfrac{45}{6^2} + \dfrac{1\,000}{(6G)^2}$ from Art. 11.16

$= 4·75 + \dfrac{27·78}{G^2}$

If the acceleration of the load is f,

rope tension $= 1\,000(9·81 + f)$

$\therefore$ drum torque $= 1\,000(9·81 + f) \times 0·9$

$\therefore$ motor torque $= \dfrac{900(9·81 + f)}{6G}$

Fig. 11.16

Angular acceleration of drum $= \dfrac{f}{0·9}$

$\therefore$ angular acceleration of motor $= \dfrac{f}{0·9} \times 6G$

Motor torque = torque to accelerate rotating parts + torque to accelerate load

i.e. $\quad 550 = \left(4·75 + \dfrac{27·78}{G^2}\right) \times \dfrac{6fG}{0·9} + \dfrac{900(9·81 + f)}{6G}$

from which $f = \dfrac{550G - 1\,472}{31·7G^2 + 335}$ \hfill (1)

For maximum acceleration, $\dfrac{df}{dG} = 0$

i.e. $\quad \dfrac{(31·7G^2 + 335) \times 550 - (550G - 1\,472) \times 63·4G}{(31·7G^2 + 335)^2} = 0$

from which $\quad G^2 - 5·35G - 10·56 = 0$

Hence $\quad G = 6·88$

$f_{max} = \dfrac{550 \times 6·88 - 1\,472}{31·7 \times 6·88^2 + 335}$ . . substituting in equation (1)

$= 1·26 \text{ m/s}^2$

DYNAMICS

9. *A car of mass 1 300 kg, with the engine at full throttle, can travel at 162 km/h on a level road with the engine developing 75 kW. The resistance to motion varies as the square of the speed.*

Determine the time taken for the speed of the car to rise from 72 km/h to 108 km/h at full throttle on an upgrade of 1 in 20, assuming that the engine torque remains constant.

$$1 \text{ km/h} = \frac{1\,000}{3\,600} = \frac{1}{3\cdot 6} \text{ m/s}$$

$$\therefore 162 \text{ km/h} = \frac{162}{3\cdot 6} = 45 \text{ m/s}$$

$$72 \text{ km/h} = \frac{72}{3\cdot 6} = 20 \text{ m/s}$$

and

$$108 \text{ km/h} = \frac{108}{3\cdot 6} = 30 \text{ m/s}$$

At 162 km/h, the whole of the engine power is absorbed in overcoming the resistance to motion,

i.e. $\quad$ tractive force $= \dfrac{75 \times 10^3}{45} = k \times 45^2$

from which $\quad k = 0\cdot 823$

At a speed v,

tractive force = mass × acceleration + resistance to motion +

+ component of weight down incline

i.e. $\quad \dfrac{75 \times 10^3}{45} = 1\,300 \dfrac{dv}{dt} + 0\cdot 823 v^2 + \dfrac{1\,300 \times 9\cdot 81}{20}$

from which $\quad dt = \dfrac{1\,300\, dv}{1\,030 - 0\cdot 823 v^2}$

$$\therefore t = \int_{20}^{30} \frac{1\,580\, dv}{35\cdot 4^2 - v^2}$$

$$= 1\,580 \times \frac{1}{2 \times 35\cdot 4} \left[\log_e \frac{35\cdot 4 + v}{35\cdot 4 - v} \right]_{20}^{30}$$

$$= 22\cdot 3 \log_e 3\cdot 37$$

$$= \underline{27 \text{ s}}$$

10. *The resistance to motion of a car of mass 800 kg is $(54 + 0 \cdot 8v^2)$ N, where v is the speed in m/s. The gear ratio between the engine and rear axle is $5 \cdot 3 : 1$, the moment of inertia of the rotating parts of the engine is $0 \cdot 3$ kg m² and that of the wheels is 1 kg m². The diameter of the wheels is $0 \cdot 52$ m.*

Find the distance travelled by the car up an incline of 1 in 40 whilst accelerating from 25 to 50 km/h assuming that the engine develops a constant torque of 40 N m during this period.

Effective moment of inertia of wheels

$$= 1 \cdot 0 + 5 \cdot 3^2 \times 0 \cdot 3 \quad \ldots \quad \text{from Art. 11.16}$$

$$= 9 \cdot 43 \text{ kg m}^2$$

Therefore effective mass of car for acceleration purposes

$$= 800 + \frac{9 \cdot 43}{0 \cdot 26^2} \quad \ldots \quad \text{from Art. 11.15}$$

$$= 939 \cdot 5 \text{ kg}$$

Tractive force $= \dfrac{40 \times 5 \cdot 3}{0 \cdot 26}$

$$= 815 \text{ N}$$

Component of weight down incline

$$= \frac{800 \times 9 \cdot 81}{40}$$

$$= 196 \cdot 2 \text{ N}$$

$$\therefore \; 815 = 939 \cdot 5 \, v \frac{dv}{dx} + (54 + 0 \cdot 8v^2) + 196 \cdot 2$$

from which $\quad dx = \dfrac{939 \cdot 5 \, v \, dv}{565 - 0 \cdot 8v^2}$

$$50 \text{ km/h} = \frac{50}{3 \cdot 6} = 13 \cdot 9 \text{ m/s}$$

and $\quad 25 \text{ km/h} = \dfrac{25}{3 \cdot 6} = 6 \cdot 95 \text{ m/s}$

$$\therefore \; x = \int_{6 \cdot 95}^{13 \cdot 9} \frac{1\,175 \, v \, dv}{706 \cdot 5 - v^2}$$

$$= -\frac{1\,175}{2} \Big[\log_e (706 \cdot 5 - v^2) \Big]_{6 \cdot 95}^{13 \cdot 9}$$

$$= -587 \cdot 5 \log_e 0 \cdot 779 = \underline{146 \text{ m}}$$

DYNAMICS

11. *A car is driven by the rear wheels and when it is stationary, 0·55 of its weight is supported on the rear wheels. The height of the centre of gravity above the road is one-fifth of the wheel base. In a test on a level road it was found that the greatest acceleration obtainable without skidding was 3 m/s².*

Calculate the coefficient of friction between the tyres and road. Using the same coefficient of friction, find the steepest gradient which the vehicle could climb.

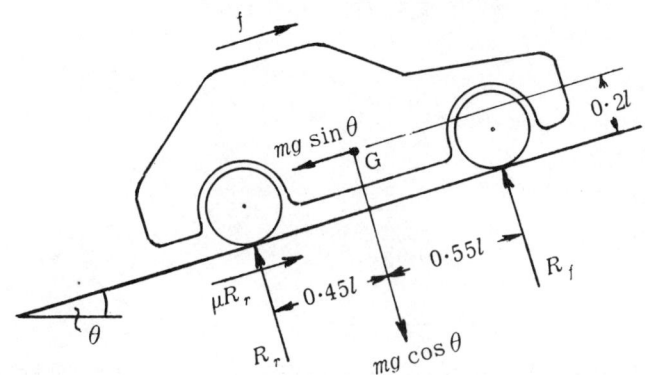

Fig. 11.17

Equating forces normal and parallel to the plane when $\theta = 0$, Fig. 11.17,

$$R_r + R_f = mg \qquad (1)$$

and
$$\mu R_r = mf \qquad (2)$$

Taking moments about G,

$$\mu R_r \times 0 \cdot 2l = R_r \times 0 \cdot 45l - R_f \times 0 \cdot 55l \qquad (3)$$

From equation (3) $R_f = \dfrac{0 \cdot 45 - 0 \cdot 2\mu}{0 \cdot 55} R_r$

Therefore in equation (1)

$$R_r \times \dfrac{1 - 0 \cdot 2\mu}{0 \cdot 55} = mg$$

Substituting for R_r in equation (2),

$$\mu \times \dfrac{0 \cdot 55 mg}{1 - 0 \cdot 2\mu} = mf$$

from which
$$\dfrac{f}{g} = \dfrac{0 \cdot 55\mu}{1 - 0 \cdot 2\mu} = \dfrac{3}{9 \cdot 81}$$

$$\therefore \mu = \underline{0 \cdot 5}$$

At a uniform speed on an inclined road,

$$R_f + R_r = mg \cos \theta \qquad (4)$$

$$\mu R_r = mg \sin \theta \qquad (5)$$

and $\quad \mu R_r \times 0.2l = R_r \times 0.45l - R_f \times 0.55l \qquad (6)$

From equation (6), $\quad R_f = \dfrac{0.35}{0.55} R_r$

From equations (4) and (5),

$$\tan \theta = \dfrac{0.5 R_r}{R_f + R_r}$$

$$= \dfrac{0.5}{\dfrac{0.35}{0.55} + 1} = 0.3055$$

$$\therefore \theta = \underline{17°}$$

12. A mass of 700 kg falling 0·2 m is used to drive a pile of mass 500 kg into the ground. Assuming there is no rebound, find the common velocity of the driver and pile at the end of the blow and the loss of kinetic energy. If the resistance of the ground is constant, find its value if the pile is driven 75 mm.
(*Ans.*: 1·155 m/s; 572 J; 22·45 kN)

13. A stationary truck of mass 9 t is set in motion by a shunting locomotive which provides an impulse of 30 kN s. The truck travels freely along a level track for a period of 15 s when it collides with a truck of mass 12 t which is moving at 0·6 m/s in the same direction. After collision, both trucks move on together. The track resistance is 65 N/t. Determine their common speed and the loss of energy at impact.
(*Ans.*: 1·353 m/s; 8 kJ)

14. A towing van of mass 2 300 kg is attached to a car of mass 1 350 kg by a rope 4·5 m long. The initial distance between the fixing points for the rope is 3 m so that the van can move 1·5 m before the rope tightened and the van is accelerated from rest under a tractive effort of 2·25 kN.

Determine the van speed (*a*) just before the rope starts to tighten, (*b*) at the instant when the rope ceases to stretch.

Determine also the impulsive force on the van during the stretching period if this takes place in 0·1 s.

Ignore frictional losses and the effect of the tractive effort during the tensioning period of 0·1 s. (*Ans.*: 1·713 m/s; 1·08 m/s; 14·55 kN)

15. A flywheel of mass 50 kg is mounted on a 75 mm diameter shaft in bearings on either side of the wheel. Due to friction, the speed of the flywheel falls from 200 rev/min to 150 rev/min in 14 s with uniform deceleration.

A plain cast iron ring, outside diameter 450 mm, inside diameter 350 mm, thickness 75 mm and density 7·2 Mg/m³ is now bolted concentrically on the side of the wheel and the effect of friction is to reduce the speed uniformly from 200 rev/min to 150 rev/min in 20 s.

Calculate the coefficient of friction and the radius of gyration of the flywheel.
(*Ans.*: 0·02; 140 mm)

16. A steel bar 2·4 m long and 75 mm diameter starts from rest and rolls without slipping down a plane inclined at 10° to the horizontal. The mass of the bar is 88 kg. Find the kinetic energy of the bar when it has rolled 6 m down the plane.
(*Ans.*: 900 J)

DYNAMICS

17. Two parallel shafts A and B are connected by gear wheels so that A rotates at four times the speed of B. The moments of inertia of the rotating parts on shafts A and B are respectively 160 and 500 kg m². Find the total kinetic energy of the system when B rotates at 200 rev/min and the driving torque required on shaft B to accelerate the system uniformly from rest so that the speed of B is 200 rev/min after 30 s.

If the shafts are 0·75 m apart, what is the tangential force between the teeth of the gears during this acceleration? (*Ans.*: 671 kJ; 2·14 kN m; 2·98 kN)

18. A flywheel A of mass 12·5 kg, outside diameter 450 mm and radius of gyration 175 mm, is initially rotating at 300 rev/min and second flywheel B of mass 7·5 kg, outside diameter 380 mm and radius of gyration 150 mm, is mounted on a parallel shaft and is initially stationary.

The two shafts are moved together so that the wheels make circumferential contact. Determine the speeds of the wheels when slipping ceases.

If the coefficient of friction between the wheels is 0·12 and the normal force between them is 60 N, calculate the time taken for slipping to cease.
 (*Ans.*: 185·4 rev/min; 219·5 rev/min; 2·84 s)

19. A rigid beam AB of uniform cross-section and of mass 40 kg is hinged at A to a fixed support and is maintained in a horizontal position by a vertical helical spring attached to B, AB being 1·8 m. A mass of 2·5 kg falls on to the beam with a striking velocity of 3 m/s, the point of impact being 1·2 m from A.

Assuming the mass and beam move together, determine the angular velocity of the beam immediately after impact. (*Ans.*: 0·1924 rad/s)

20. Fig. 11.18 shows a tilt hammer hinged at O with its head A resting on top of the pile B. The hammer and arm have a mass of 25 kg, the centre of gravity G is 400 mm from O and the radius of gyration about an axis through G parallel to the axis of the pin O is 75 mm. The pile has a mass of 135 kg. The hammer is raised through 45° from the horizontal and released. On striking the pile, there is no rebound.

Find the angular velocity of the hammer immediately before impact and the linear velocity of the pile immediately after impact.
 (*Ans.*: 5·79 rad/s; 0·343 m/s)

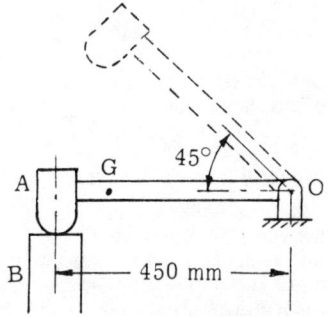

Fig. 11.18

21. A uniform rod AB, 0·75 m long and of mass 20 kg, is hinged at A and held in a horizontal position. It is allowed to fall, rotating about A in a vertical plane, until it strikes a horizontal spring C of stiffness 35 kN/m, whose centre line is 0·35 m below the level of A. When the spring force is a maximum, the rod is vertical.

Find (*a*) the maximum spring force, (*b*) the maximum force on the hinge A.
 (*Ans.*: 2270 N; 682 N)

22. The tailboard of a lorry is a uniform rectangle, 1·5 m long by 0·75 m high and has a mass of 27 kg. It is hinged along the bottom edge to the floor of the lorry. Chains are attached to the top corners of the board and to the sides of the lorry so that when the board is in the horizontal position, the chains are inclined at 45° to the horizontal. A tension spring is inserted in each chain to reduce the shock and these are adjusted to prevent the board from dropping below the horizontal. Each spring has a stiffness of 52 kN/m.

Find the greatest force in each spring and the resultant force at the hinges when the board falls freely from the vertical position. (*Ans.*: 2·27 kN; 3·59 kN)

23. A valve is opened by a cam and then released by a trip-gear, when it is closed by a helical spring concentric with the valve stem. The mass of the valve is 3 kg and the maximum opening is 16 mm. The stiffness of the spring is 25 kN/m and the compression when the valve is closed is 20 mm.

Determine (a) the time taken for the valve to close, (b) the velocity at the moment of impact. *(Ans.:* 0·010 8 s; 2·735 m/s)

24. The table of a machine tool slides on horizontal guides. The table has a mass of 100 kg and the frictional force opposing its motion is 180 N. When the table is moving at 0·9 m/s, the driving mechanism is disconnected and the table is brought to rest by a spring buffer which is initially unstressed.

Calculate the time required for the spring to attain its greatest compression of 40 mm. *(Ans.:* 0·072 3 s)

25. A winding drum raises a cage of mass 500 kg through a height of 120 m. The winding drum has a mass of 250 kg, a radius of 0·5 m and a radius of gyration of 0·36 m. The mass of the rope is 3 kg/m. The cage has at first an acceleration of 1·5 m/s^2 until a velocity of 9 m/s is reached, after which the velocity is constant until the cage nears the top, when the final retardation is 6 m/s^2.

Find (a) the time taken for the cage to reach the top, (b) the starting torque on the drum, (c) the power at the end of the acceleration period.

(Ans.: 17·08 s; 4 960 N m; 82·14 kW)

26. A lift of mass 900 kg is connected to a rope which passes over a drum 1 m diameter and then to a balance mass of 450 kg. The moment of inertia of the drum is 100 kg m^2 and it is driven through a reduction gear of 25 to 1, of 90 per cent efficiency. Neglecting the inertia of the gears, calculate the motor torque for a lift acceleration of 3 m/s^2. If the maximum output of the motor is 15 kW, what will be the maximum uniform speed of the lift? *(Ans.:* 215 N m; 3·06 m/s)

27. A truck of mass 5 t is hauled up an incline of 1 in 15 by a rope parallel to the track. The rope is wound on a drum driven by an electric motor. The drum is 1 m in diameter and has a mass of 1 t and a radius of gyration of 0·4 m. The efficiency of the drive from motor to drum is 88 per cent and the frictional resistance to the motion of the truck is 1·3 kN.

When the truck is moving up the incline with a speed of 3 m/s and an acceleration of 0·3 m/s^2, find the power output from the motor. *(Ans.:* 21·4 kW)

28. In a double reduction lifting gear driven by an electric motor, the gear ratio between the motor shaft and the intermediate shaft is 3·5 and between the intermediate shaft and drum shaft, it is 4·5. The moments of inertia of the three shafts are 5, 40 and 500 kg m^2 respectively. On the drum, 1·2 m diameter, is suspended a loaded cage of mass 6 t and a balance mass of 4·5 t which descends as the loaded cage is lifted. Determine the motor torque required to raise the cage with an acceleration of 0·4 m/s^2. *(Ans.:* 830 N m)

29. In a double reduction lifting gear, the speed of the motor shaft is G times the speed of the intermediate shaft and G^2 times the speed of the drum shaft. The moments of inertia of the three shafts are 3, 35 and 850 kg m^2 respectively. A load of 800 kg is suspended from the drum which is 1·2 m diameter and the motor exerts a constant torque of 570 N m.

Determine the value of G for maximum acceleration of the load when being raised and calculate this maximum acceleration. *(Ans.:* 5·59; 1·535 m/s^2)

30. A car has a mass of 1 t and the moment of inertia of the wheels are together 8·4 kg m^2. The diameter of the wheels is 0·62 m. Find the acceleration on the level when the engine output torque is 100 N m, the overall speed reduction is 14, the resistance to motion is 200 N and the transmission efficiency is 88 per cent.

(Ans.: 3·47 m/s^2)

31. A car has a maximum speed on a level road of 70 m/s, at which speed the engine develops 180 kW. The four road wheels have a rolling radius of 0·3 m, a radius of gyration of 0·2 m and the mass of each wheel is 23 kg. The rotating parts of the engine have a moment of inertia of 1 kg m^2 and the gear ratio, engine to wheels, is

DYNAMICS 147

4:1. The total mass of the car is 1 350 kg. The torque output of the engine may be regarded as constant over a wide speed range.

If the resistance to motion varies as the square of the road speed, determine the time taken for the speed to rise from 40 m/s to 60 m/s on a level road.

(*Ans.*: 27 s)

32. The resistance to motion of a train of mass 550 t is given by $R = 3\,800 + 900v$ where R is in newtons and v is in m/s. If the locomotive exerts a constant tractive force of 50 kN, find the distance travelled and the time taken to accelerate from 32 km/h to 48 km/h on an up gradient of 1 in 200. (*Ans.*: 271 s; 3·06 km)

33. A car has a mass of 2 t and the four wheels are each of mass 18 kg, diameter 0·75 m and radius of gyration 0·32 m. The engine develops a torque of 115 N m and the rotating parts have a moment of inertia of 0·47 kg m². Calculate the ratio of engine speed to back-axle speed for maximum acceleration up an incline of 1 in 120 against a wind resistance of 180 N. (*Ans.*: 25·9)

34. A vehicle is driven along a horizontal road by the rear wheels. The wheelbase is 3·3 m and the centre of gravity is 0·75 m above the ground and 1·35 m behind the front axle. The coefficient of friction between the tyres and ground is 0·3.

Determine (*a*) the maximum acceleration if the wheels are not to slip; (*b*) the maximum retardation when a braking torque is applied to the rear wheels.

(*Ans.*: 1·29 m/s²; 1·126 m/s²)

35. A car is driven by the rear wheels and when the car is stationary, 0·55 of the mass is supported by the rear wheels. The height of the centre of gravity above the ground is one-fifth of the wheel base. On a level road, the greatest acceleration possible without skidding the wheels is 3 m/s².

(*a*) Calculate the coefficient of friction between the tyres and the road.

(*b*) Using this coefficient of friction, find the steepest gradient which the car could climb. (*Ans.*: 0·5; $\tan^{-1} 0·305\,5$)

12 Velocity and acceleration diagrams

12.1 Introduction It is often necessary to determine the velocity and acceleration of points in a mechanism in order to obtain the forces involved. These can be found from velocity and acceleration diagrams, which give the velocity and acceleration of any point relative to any other point for one particular position of the mechanism.

12.2 Velocity of a rigid link The velocity of one point on a link relative to another must be perpendicular to the axis of the link, otherwise there would be a component along the axis which would involve a change in length. Thus, in the link shown in Fig. 12.1, the velocity of B relative to A is given by $v_{ba} = \omega \cdot AB$, perpendicular to AB; this is represented by the vector ab.

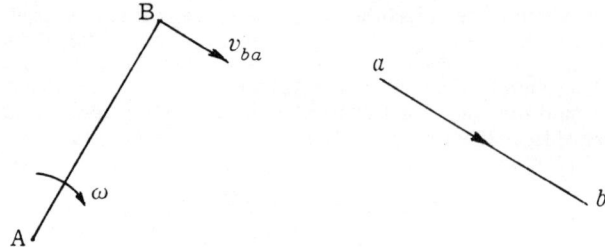

Fig. 12.1

If this link is part of a mechanism of several links such as the four-bar mechanism shown in Fig. 12.2 and the angular velocity of AB is given, the velocity diagram may be built up as follows:

The velocity of D relative to A is zero since AD is fixed and hence A and D are represented by a single point.

The velocity of B relative to A is $\omega \cdot AB$ and is perpendicular to AB.

The velocity of C relative to B is perpendicular to BC and passes through b.

The velocity of C relative to D is perpendicular to CD and passes through d.

Thus the intersection of the last two vectors gives the point c.

For a point on the mechanism such as P, the corresponding point on the velocity diagram is obtained by proportion,

i.e. $$\frac{bp}{bc} = \frac{BP}{BC}$$

and the velocity of P relative to the fixed points A and D is given by the vector ap.

The angular velocities of BC and CD are given by

$$\omega_{bc} = \frac{bc}{BC} \quad \text{and} \quad \omega_{cd} = \frac{cd}{CD}$$

VELOCITY AND ACCELERATION DIAGRAMS

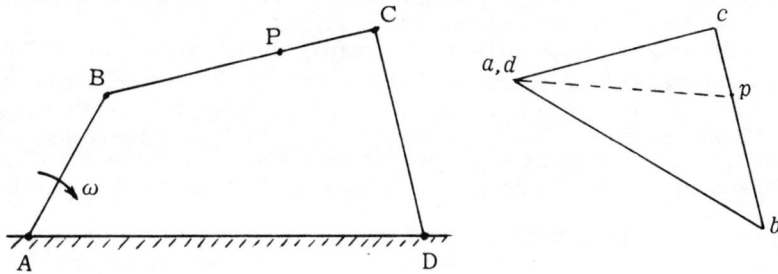

Fig. 12.2

12.3 Velocity of a block sliding on a rotating link
If a block A slides with velocity v along a link which rotates at a rate ω, Fig. 12.3, the velocity of the coincident link point A' is given by $v_{a'o} = \omega \cdot OA'$, perpendicular to OA'. This is represented by the vector oa'. The velocity of A relative to A' is v, parallel to OA' and this passes through a'. Hence the velocity of A relative to O is represented by oa.

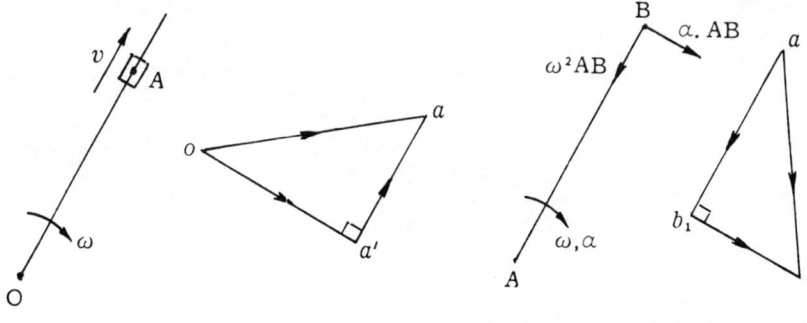

Fig. 12.3   Fig. 12.4

12.4 Acceleration of a rigid link
The acceleration of one point on a link relative to another has two components, (a) a centripetal component due to the angular velocity of the link, and (b) a tangential component due to the angular acceleration of the link.

If the link AB, Fig. 12.4, is rotating with angular velocity ω and angular acceleration α, the centripetal acceleration of B relative to A is $\omega^2 AB$, directed towards A, and the tangential acceleration is αAB, perpendicular to AB. These are represented in the acceleration diagram by vector ab_1 and $b_1 b$ respectively, the resultant acceleration of B relative to A being given by ab.

If this link is part of a mechanism such as the four-bar mechanism shown in Fig. 12.5, then A and D are represented by a single point, as in the velocity diagram since there is no relative motion between them.

The centripetal and tangential accelerations of B relative to A are given by ab_1 and $b_1 b$ respectively, as before. The centripetal acceleration of C relative to B is given by v_{cb}^2/BC and is directed towards B. The value of v_{cb} is obtained from the velocity diagram and this acceleration is represented by bc_1. The tangential acceleration of C relative to B is unknown in magnitude but its direction is perpendicular to BC, so that a vector is drawn through c_1 perpendicular to bc_1.

The centripetal acceleration of C relative to D is given by v_{cd}^2/CD and is directed towards D; this is represented by dc_2. The tangential acceleration of C relative to D is again unknown but its direction is perpendicular to CD and is represented by a line through c_2 perpendicular to dc_2. The intersection of the lines through c_1 and c_2 then give the point c.

The acceleration of a point such as P is again obtained by proportion, i.e. $\dfrac{bp}{bc} = \dfrac{\text{BP}}{\text{BC}}$, the absolute acceleration of P being represented by ap.

Also $a_{bc} = \dfrac{cc_1}{\text{BC}}$ and $a_{cd} = \dfrac{cc_2}{\text{CD}}$, only the tangential component of the acceleration of one end of the link relative to the other being relevant.

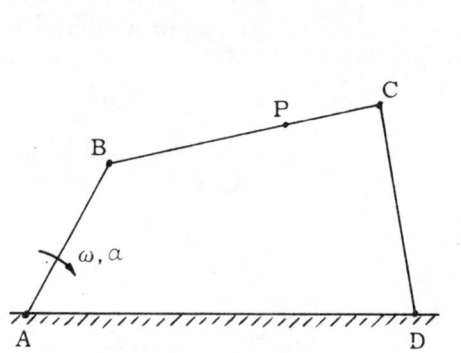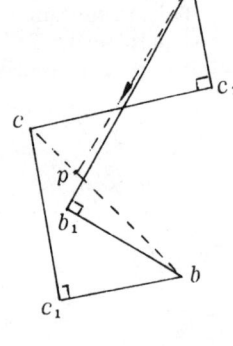

Fig. 12.5

12.5 Acceleration of a block sliding on a rotating link

Let a block A slide with velocity v and acceleration f along a link which has an angular velocity ω and an angular acceleration a, Fig. 12.6, and let the link turn through an angle $d\theta$ in time dt. The initial radial velocity of the block is v, represented in the vector diagram by oa_1 and the final velocity is $v + dv$, represented by oa_2. The change in velocity is therefore given by $a_1 a_2$ which has radial and tangential components dv and $v\, d\theta$ respectively.

In the tangential direction, the initial velocity of the block is ωr, represented by oa_3 in the vector diagram, and the final velocity is $(\omega + d\omega)(r + dr)$ represented by oa_4. The change in velocity is therefore given by $a_3 a_4$, which has radial and tangential components $\omega r\, d\theta$ and $\omega\, dr + r\, d\omega$ respectively, neglecting the product $d\omega\, dr$.

Thus the total change in the radial direction is $dv - \omega r\, d\theta$ so that

$$\text{radial acceleration} = \frac{dv}{dt} - \omega r \frac{d\theta}{dt}$$

$$= f - \omega^2 r \qquad (12.1)$$

This is evidently the outward acceleration of the block, f, less the centripetal acceleration of the coincident link point A' relative to O.

VELOCITY AND ACCELERATION DIAGRAMS

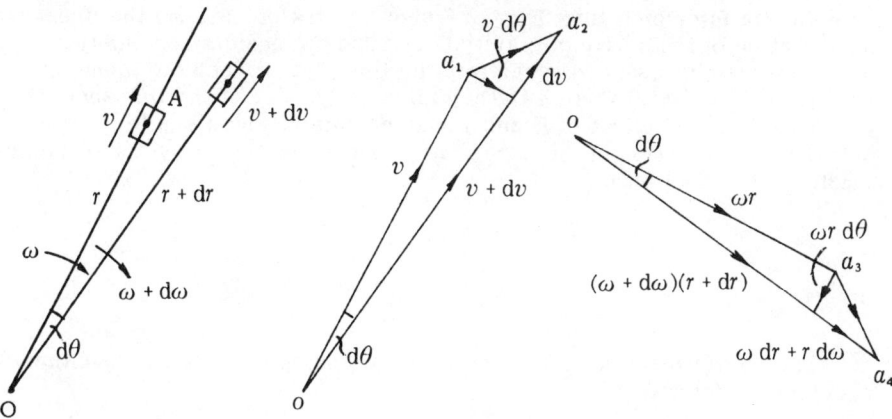

Fig. 12.6

The total change in the tangential direction is $v\,d\theta + \omega\,dr + r\,d\omega$ so that

$$\text{tangential acceleration} = v\frac{d\theta}{dt} + \omega\frac{dr}{dt} + r\frac{d\omega}{dt}$$

$$= v\omega + \omega v + r\alpha$$

$$= \alpha r + 2v\omega \qquad (12.2)$$

The term αr represents the tangential acceleration of the coincident link point A' relative to O and so the term $2v\omega$ represents the tangential acceleration of the block relative to the coincident link point. This is called the *Coriolis component* and arises whenever a block slides along a rotating link or when a link slides through a swivel block. The direction of the Coriolis component is obtained by rotating the sliding velocity vector through 90° in the direction of rotation of the link.

The acceleration diagram for the block is shown in Fig. 12.7, oa'_1 and $a'_1 a'$ are the radial and tangential accelerations of the link point A' relative to O and $a' a_1$ and $a_1 a$ are the sliding and tangential accelerations of A relative to A'. The absolute acceleration of A is then given by oa.

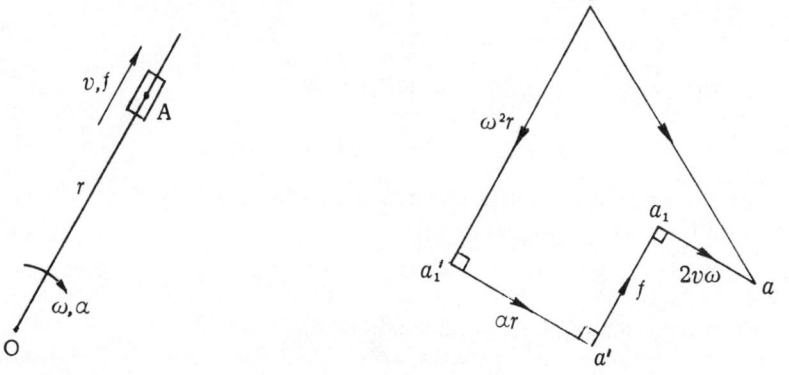

Fig. 12.7

12.6 Inertia force on a link Fig. 12.8 shows a link of mass m; the linear acceleration of the centre of gravity G is f and the angular acceleration is α, The force necessary to accelerate the link, $F = mf$ and the torque necessary, $T = I\alpha = mk^2\alpha$ where k is the radius of gyration of the link about G.

The combined effect of F and T may be obtained by a single force F acting in the direction of the acceleration but offset from it by a perpendicular distance l such that

$$Fl = mk^2\alpha$$

or
$$l = \frac{mk^2\alpha}{mf} = \frac{k^2\alpha}{f} \qquad (12.3)$$

The inertia reaction of the link is equal and opposite to the resultant accelerating force F.

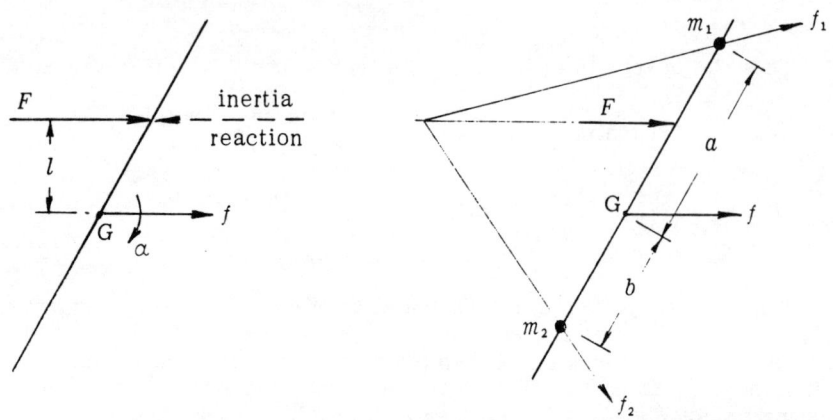

Fig. 12.8   Fig. 12.9

Alternatively the link may be replaced by two concentrated masses m_1 and m_2 provided that these masses are equivalent to that of the link, Fig. 12.9. The conditions for equivalence are:

(a) the total mass must be the same,

i.e. $\qquad\qquad\qquad m_1 + m_2 = m \qquad (12.4)$

(b) the centre of gravity must be at G,

i.e. $\qquad\qquad\qquad m_1 a = m_2 b \qquad (12.5)$

(c) the moment of inertia about G of the two-mass system must be the same as that of the link,

$$m_1 a^2 + m_2 b^2 = mk^2 \qquad (12.6)$$

From equations (12.4), (12.5) and (12.6),

$$m_1 = \frac{b}{a+b}m \,, \quad m_2 = \frac{a}{a+b}m \quad \text{and} \quad ab = k^2 \qquad (12.7)$$

VELOCITY AND ACCELERATION DIAGRAMS 153

Either a or b can be chosen arbitrarily and the other distance is then determined from the relation $ab = k^2$.

The directions of the accelerations of m_1 and m_2 are obtained from the acceleration diagram, shown by f_1 and f_2 and the line of action of F must then be parallel to the direction of f and pass through the intersection of the lines of f_1 and f_2, since the accelerating forces on the masses are in the directions of f_1 and f_2. It is then unnecessary to calculate the value of l or to decide on which side of G the force F must act. It is also unnecessary to calculate the magnitudes of m_1 and m_2; it is only their positions which are relevant.

1. *In the mechanism shown in Fig. 12.10, the crank OA rotates in a clockwise direction at 120 rev/min, the block B moves along the axis XX and the block D moves along the axis YY. AB and CD are connected by a pin-joint at C. AB = 210 mm, AC = 80 mm and CD = 240 mm. Draw the velocity diagram when the angle EOA is 75°, and find the velocity of D.*

If a force of 100 N acting along YY resists the motion of D, determine the forces acting on the pins at the ends of the bar AB and the torque required on the crank OA, neglecting friction.

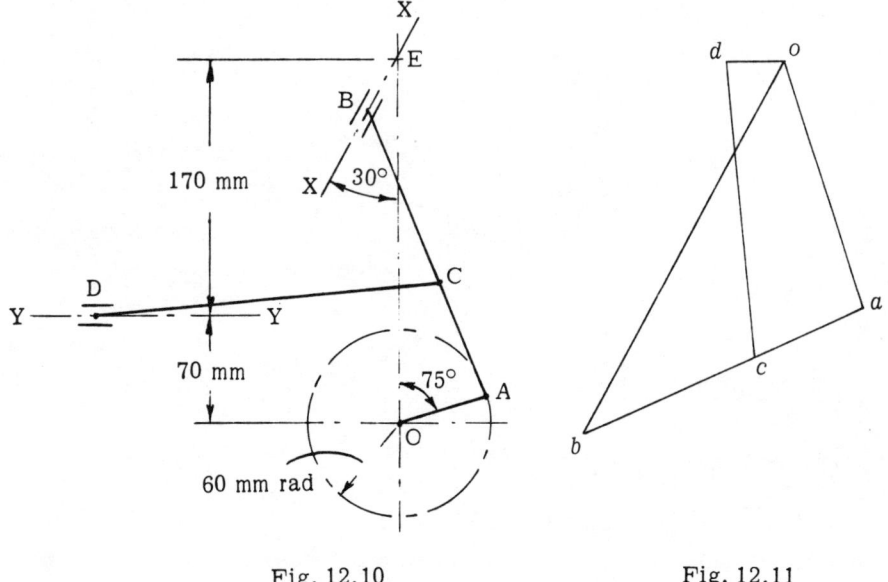

Fig. 12.10 Fig. 12.11

$$v_a = \omega r = 120 \times \frac{2\pi}{60} \times 0 \cdot 06 = 0 \cdot 754 \text{ m/s}$$

In the velocity diagram, Fig. 12.11, the absolute velocity of A is represented by oa and the velocity of B relative to A is perpendicular to AB, passing through the point a. The velocity of B relative to the fixed point O is inclined at 30° to the vertical and passes through the point o. The intersection of these lines through a and o then gives the point b.

The point c on ab is positioned such that $ac : ab :: AC : AB$. The velocity of D relative to C is perpendicular to DC and the velocity of D relative to the fixed point O is horizontal. Thus the intersection of a line through c perpendicular to DC and a horizontal line through o gives the point d.

From the diagram, $v_d = od = \underline{0 \cdot 187 \text{ m/s}}$.

The reaction at the slider D is perpendicular to YY, Fig. 12.12, and so, for a horizontal force of 100 N, the force in CD is 101 N, as determined by a triangle of forces. The forces acting on AB are then the force exerted at C by DC, the reaction at B which is perpendicular to XX and the force through A, which must be concurrent with the other two forces.

From the parallelogram of forces, force at A = pq = $\underline{62 \text{ N}}$

and force at B = pr = $\underline{59 \text{ N}}$

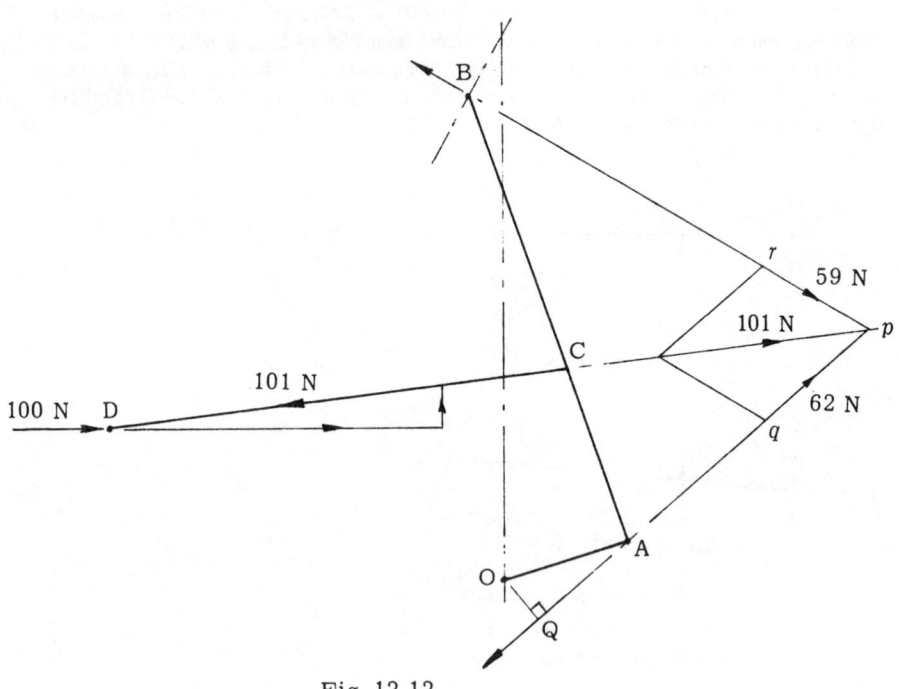

Fig. 12.12

The torque on OA is determined by equating the work done by this torque to the work done against the resistance at D,

i.e. $$T \times \left(120 \times \frac{2\pi}{60}\right) = 100 \times 0 \cdot 187$$

$$\therefore T = \underline{1 \cdot 49 \text{ Nm}}$$

Alternatively, the torque on OA is the product of the force through A and the perpendicular distance of its line of action from O, OQ.

Thus $$T = 62 \times 0 \cdot 024 = \underline{1 \cdot 49 \text{ Nm}}$$

VELOCITY AND ACCELERATION DIAGRAMS

2. *Fig. 12.13 shows a quick-return motion in which the driving crank OA rotates at 120 rev/min in a clockwise direction. For the position shown, determine the acceleration of the block D and the angular acceleration of the bar QB.*

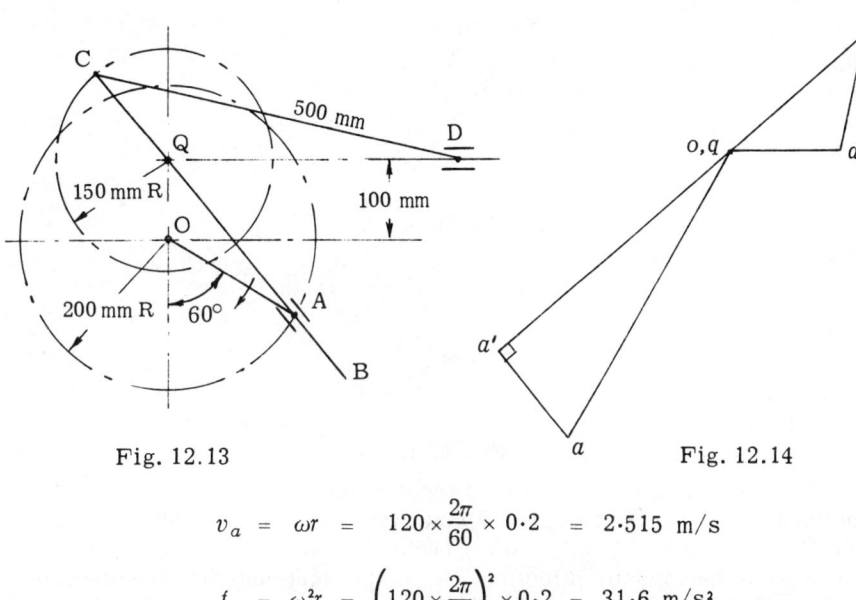

Fig. 12.13   Fig. 12.14

$$v_a = \omega r = 120 \times \frac{2\pi}{60} \times 0 \cdot 2 = 2 \cdot 515 \text{ m/s}$$

$$f_a = \omega^2 r = \left(120 \times \frac{2\pi}{60}\right)^2 \times 0 \cdot 2 = 31 \cdot 6 \text{ m/s}^2$$

Let A' be the point on QB which is coincident with the block A. Then, in the velocity diagram, Fig. 12.14, O and Q are represented by a single point o, q since there is no relative motion between them. oa represents the absolute velocity of a, aa' represents the sliding velocity of A' relative to A and qa' represents the velocity of A' relative to Q.

$a'q$ is extended to c such that $a'q : qc :: A'Q : QC$. cd then represents the velocity of C relative to D and the velocity of D relative to the fixed points is horizontal.

From the diagram, $\quad v_{aa'} = 0 \cdot 8 \text{ m/s}$

$$v_{a'q} = 2 \cdot 37 \text{ m/s}$$

and $\quad v_{dc} = 0 \cdot 9 \text{ m/s}$

In the acceleration diagram, Fig. 12.15, the fixed points O and Q are again represented by a single point o, q and oa is the absolute acceleration of A. The acceleration of A' relative to A has two components, a tangential (or Coriolis) component and a sliding component.

The tangential component is given by

$$2v_{a'a}\omega_{qb} = 2 \times v_{a'a} \times \frac{v_{a'q}}{A'Q}$$

$$= 2 \times 0 \cdot 8 \times \frac{2 \cdot 37}{0 \cdot 26} = 14 \cdot 6 \text{ m/s}^2$$

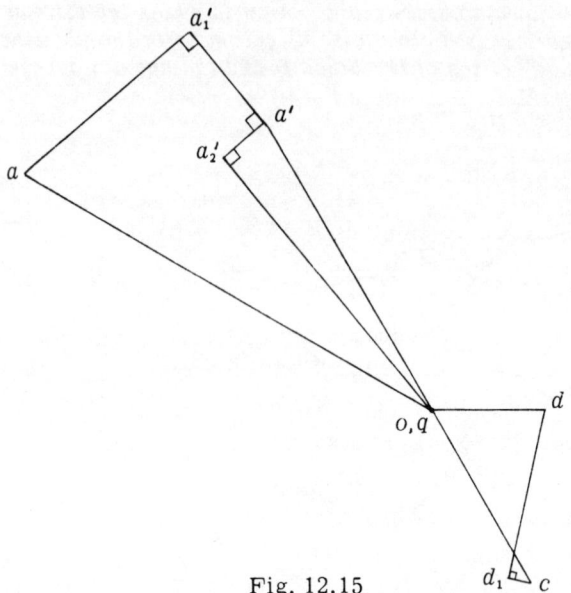

Fig. 12.15

From Art. 12.5, the direction of the acceleration of A relative to A' is obtained by turning the relative velocity vector aa' through 90° in the direction of ω_{qb}, i.e. to the left of QB. The acceleration of A' relative to A is therefore in the opposite direction, i.e. to the right and this is represented by aa'_1. The sliding acceleration of A' relative to A is parallel to QB and passes through a'_1 but its magnitude is unknown.

The acceleration of A' relative to Q also has two components, a centripetal component and a tangential component. The centripetal component is given by

$$\frac{v_{a'q}^2}{A'Q} = \frac{2 \cdot 37^2}{0 \cdot 26} = 21 \cdot 6 \text{ m/s}^2$$

Its direction is towards Q and this is represented by qa'_2. The tangential acceleration is perpendicular to QA' and passes through the point a'_2; the intersection of this line with the sliding acceleration through a'_1 then gives the point a'. $a'q$ is extended to c such that $a'q : qc :: A'Q : QC$. The centripetal acceleration of D relative to C is given by

$$\frac{v_{dc}^2}{DC} = \frac{0 \cdot 9^2}{0 \cdot 5} = 1 \cdot 62 \text{ m/s}^2$$

This is directed towards C and is represented by cd_1. The tangential acceleration of D relative to C is perpendicular to CD and passes through d_1. The acceleration of D relative to the fixed points is horizontal and thus the point d is obtained.

From the diagram, $f_d = od = \underline{7 \text{ m/s}^2}$

Angular acceleration of QB = $\dfrac{\text{tangential acceleration of A' relative to Q } (= a'_2 a')}{A'Q}$

$= \dfrac{4 \cdot 2}{0 \cdot 26} = \underline{16 \cdot 2 \text{ rad/s}^2}$

VELOCITY AND ACCELERATION DIAGRAMS

3. *Fig. 12.16 shows a mechanism in which the crank AB rotates anticlockwise about A at 70 rev/min. The link CD swings about D and is connected to the crankpin B by the link BC. The lengths are: AB, 0·25 m, BC, 1·0 m, CD, 0·75 m. The link BC has a mass of 14 kg, its centre of gravity G is at its mid-point and its radius of gyration about a transverse axis through G is 0·3 m.*

For the position in which AB is at 45° to the horizontal, find the torque required on AB to overcome the inertia of BC.

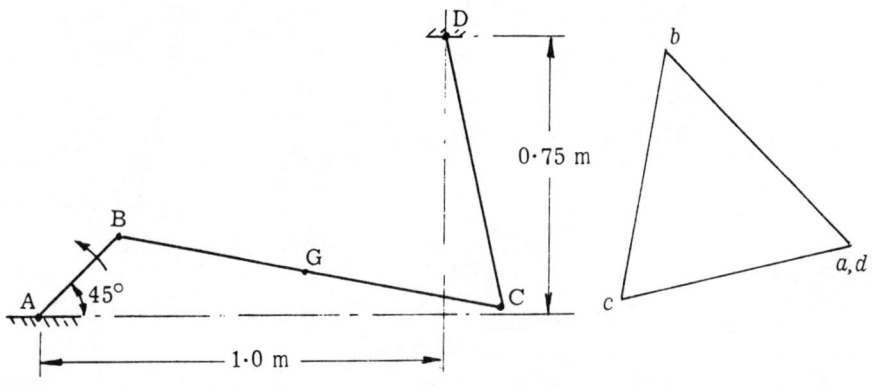

Fig. 12.16   Fig. 12.17

$$v_a = \omega r = 70 \times \frac{2\pi}{60} \times 0.25 = 1.83 \text{ m/s}$$

$$f_a = \omega^2 r = \left(70 \times \frac{2\pi}{60}\right)^2 \times 0.25 = 13.4 \text{ m/s}^2$$

In the velocity diagram, Fig. 12.17, ab represents the absolute velocity of B and cb and db represent the velocities of B relative to C and D respectively.

In the acceleration diagram, Fig. 12.18, ab represents the absolute acceleration of B, bc_1 represents the centripetal acceleration of c relative to B ($=v_{cb}^2/\text{CB}$) and dc_2 represents the centripetal acceleration of c relative to D ($=v_{cd}^2/\text{CD}$). The tangential accelerations of C relative to B and D are perpendicular to CB and CD respectively, passing through points c_1 and c_2; and the intersection of these lines gives the point c. The resultant acceleration of C relative to B is then given by bc.

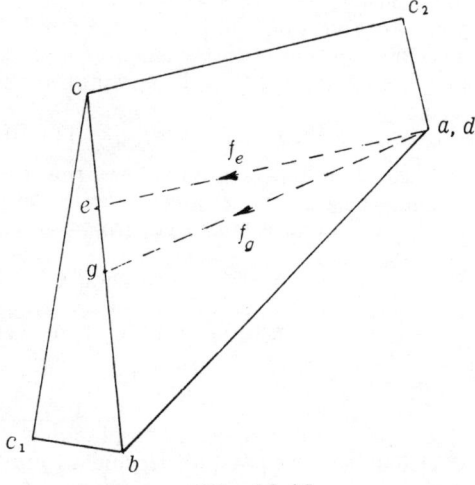

Fig. 12.18

To find the inertia force on BC, its mass is first replaced by an equivalent two-mass system (Art. 12.6), with one mass at B and the other at E, where $BG \times GE = k_G^2$, Fig. 12.19.

Thus $\qquad GE = \dfrac{0.3^2}{0.5} = 0.18$ m

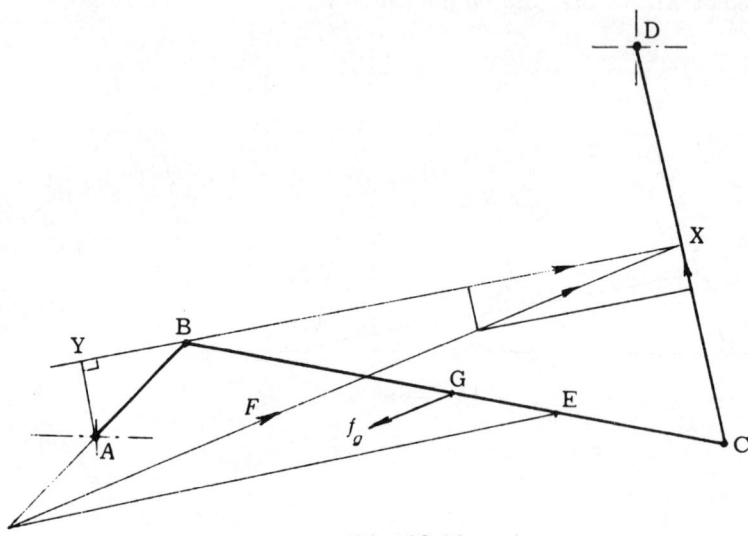

Fig. 12.19

The corresponding points in the acceleration diagram are g and e which divide bc in the same proportions as G and E divide BC; the absolute accelerations of G and E are then given by ag and ae respectively. The inertia force on the mass at B is directed along BA since the crank has only centripetal acceleration and the inertia force on the mass at E is parallel to ae. The intersection of these lines is a point on the line of action of the resultant inertia force on the link, which is parallel to ag and opposite in direction.

Inertia force, $\quad F = mf_g = 14 \times 10.9 = 152.6$ N

The forces acting on BC are the inertia force F and the reactions through the joints at B and C. In the absence of friction, the reaction at C is directed along CD and so, from the parallelogram of forces at X,

$\qquad$ force through B = 150 N

Therefore torque on A = $150 \times AY$

$\qquad\qquad\qquad\qquad\quad = 150 \times 0.14 = \underline{21 \text{ N m}}$

NOTE: It is not necessary to place one of the equivalent masses at B; any positions may be used for m_1 and m_2 provided that the condition $ab = k^2$ is satisfied.

VELOCITY AND ACCELERATION DIAGRAMS

4. The end A of a bar AB, Fig. 12.20, moves along the vertical path AD and the bar passes through a swivel bearing pivoted at C. When A has a velocity of 1·0 m/s towards D, find the velocity of sliding through the swivel and the angular velocity of the bar. (*Ans.*: 0·5 m/s; 7·5 rad/s)

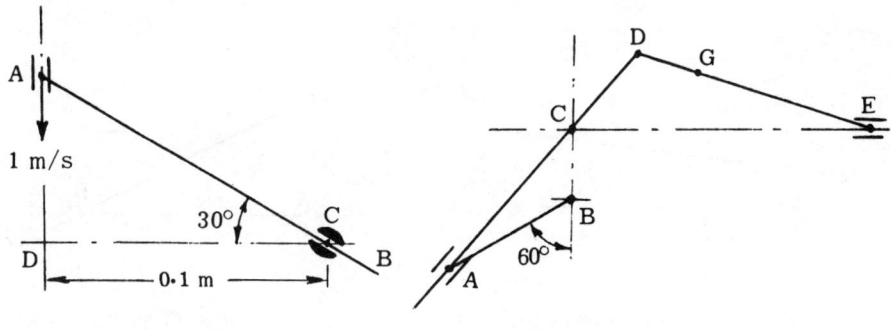

Fig. 12.20   Fig. 12.21

5. In the quick return mechanism shown in Fig. 12.21, the distance between the fixed centres BC = 90 mm, crank AB = 180 mm, CD = 180 mm and DE = 360 mm with its centre of gravity, G, 90 mm from D. If AB rotates clockwise at 120 rev/min, find the linear velocity of G and the angular velocity of DE.
(*Ans.*: 1·56 m/s; 3·16 rad/s)

6. Fig. 12.22 shows a crank OA, 72 mm long, which rotates anticlockwise about O at 150 rev/min. The bar DBC is pivoted at B which is 150 mm vertically below O. BC = 75 mm. The slider E moves in horizontal guides 30 mm below B and the rod CE is 240 mm long. A horizontal force of 150 N opposes the motion of E.

For the given position, where angle AOB = 120°, find
(*a*) the linear velocity of E and the angular velocity of DBC;
(*b*) the driving torque on the crank OA. (*Ans.*: 0·292 m/s; 4·34 rad/s; 2·79 N m)

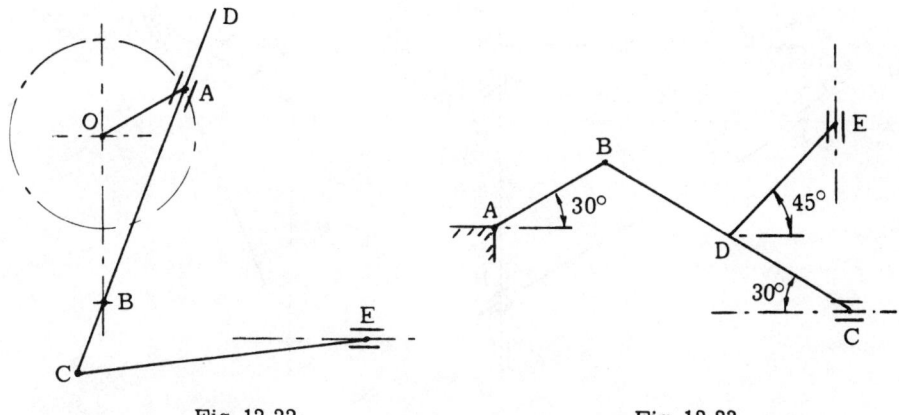

Fig. 12.22   Fig. 12.23

7. In the mechanism shown in Fig. 12.23, the crank AB is 75 mm long and rotates clockwise at 8 rad/s. BC = 300 mm and BD = DC = DE. Find the velocity and acceleration of pistons C and E.
(*Ans.*: 0·1875 m/s; 0·6 m/s; 4·16 m/s²; 9·06 m/s²)

8. In the mechanism shown in Fig. 12.24, the block P reciprocates along the line AB and the crank OC rotates at 240 rev/min. OC = 132 mm, CP = 732 mm, OD = 720 mm and angle COD = 60°. Find the velocity and acceleration of P.

(Ans.: 3·72 m/s; 63·9 m/s²)

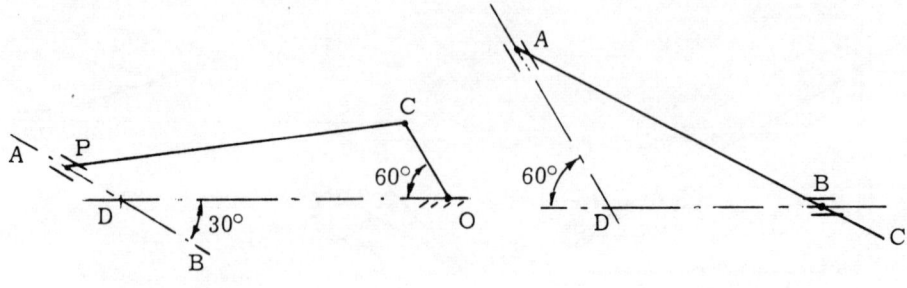

Fig. 12.24   Fig. 12.25

9. In the link ABC, Fig. 12.25, AB = 600 mm and BC = 225 mm. A and B are attached by pin-joints to the sliding blocks. When BD = 375 mm, A is sliding towards D with a velocity of 6 m/s and a retardation of 150 m/s². Find the acceleration of C and the angular acceleration of AC.   (Ans.: 259 m/s²; 294 rad/s²)

10. The rod OA shown in Fig. 12.26 rotates about O and lifts the vertical rod CD by means of the trunnion at B. At the instant when the angle EOA is 30°, OA has an anticlockwise angular velocity of 5 rad/s and zero angular acceleration.

Find the velocity and acceleration of CD at this instant.

(Ans.: ⅔ m/s; 2·5 m/s²)

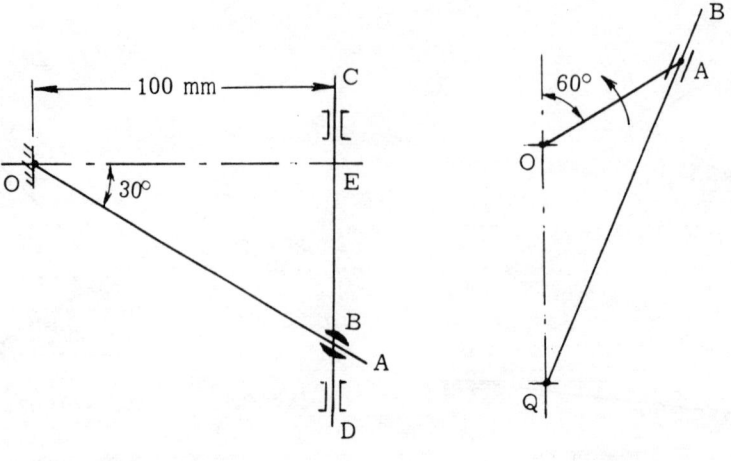

Fig. 12.26   Fig. 12.27

11. In part of a quick-return mechanism shown in Fig. 12.27, the crank OA rotates uniformly at 2·5 rad/s. OA = 225 mm and OQ = 300 mm. Determine the angular acceleration of the link QB.   (Ans.: 0·33 rad/s²)

VELOCITY AND ACCELERATION DIAGRAMS

12. In the mechanism shown in Fig. 12.28, crank OA rotates at 60 rev/min. The rod EF is pinned to AB at D and slides through a swivel block at Q. OA = 25 mm, AD = 75 mm, BC = 75 mm, DB = 75 mm and DE = 50 mm.

For the position shown, find the velocity and acceleration of E.

(*Ans.*: 0·208 m/s; 1·06 m/s²)

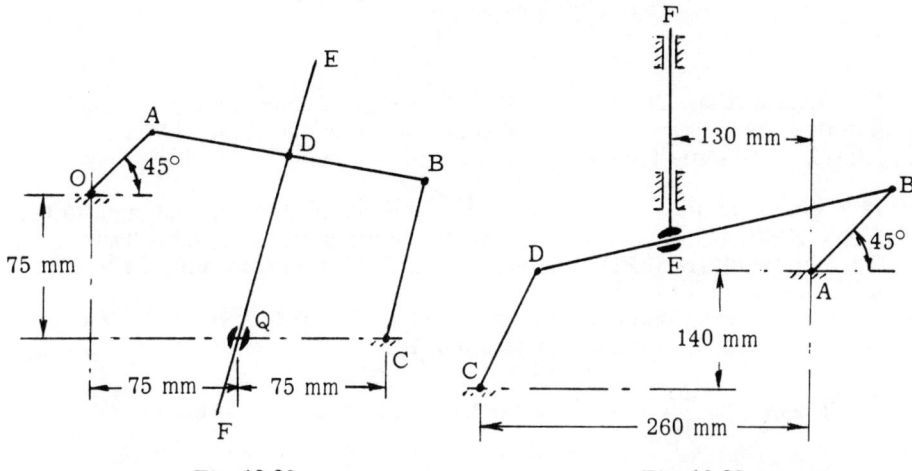

Fig. 12.28   Fig. 12.29

13. In the mechanism shown in Fig. 12.29, the crank AB rotates clockwise at 110 rev/min. AB = 70 mm, CD = 140 mm and BD = 260 mm. BD slides through a swivelling block at E at the lower end of EF and EF slides in vertical guides. For the position shown, find the linear velocity and linear acceleration of F and the angular velocity and angular acceleration of BD.

(*Ans.*: 0·452 m/s; 2·36 m/s²; 1·395 rad/s; 31·4 rad/s²)

14. In the mechanism shown in Fig. 12.30, the crank AB rotates at a uniform speed of 10 rad/s and CD oscillates about the fixed centre D. AB = 225 mm, BC = 600 mm and CD = 600 mm. CD is a uniform thin rod of mass 16 kg. Find the turning moment which must be applied to the crank to accelerate CD for the position where AB is vertical. (*Ans.*: 13·8 N m)

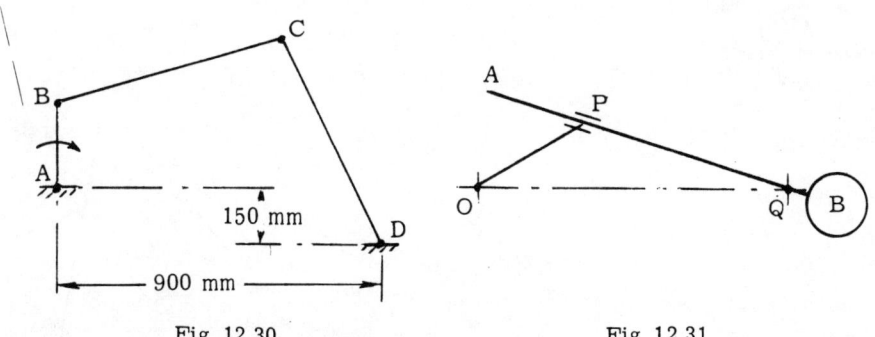

Fig. 12.30   Fig. 12.31

15. Fig. 12.31 shows a mechanism in which the crank OP revolves about O at 180 rev/min. The lever AB has a mass of 2·7 kg and its centre of gravity is at the pivot Q. OQ = 125 mm, OP = 50 mm and the radius of gyration of AB about Q is 100 mm.

When the angle POQ is 30°, find the torque on the crankshaft to overcome the inertia of AB. (*Ans.*: 3 N m)

13 Reciprocating mechanisms

13.1 Introduction The reciprocating mechanism is of such common application that special methods have been developed to determine velocity, acceleration and inertia forces.

13.2 Piston velocity The velocity of the crankpin C is perpendicular to OC, Fig. 13.1(a), and the piston P is constrained to move along OP. Thus, in the velocity diagram, Fig. 13.1(b), $oc = \omega \cdot OC$, cp is perpendicular to CP and op is parallel to OP.

The intersection of lines perpendicular to v_c and v_p gives the instantaneous centre, I, of the connecting rod, PC.

Then $$\frac{v_p}{IP} = \frac{v_c}{IC} = \Omega,$$ the angular velocity of PC

Thus $$v_p = \frac{IP}{IC} \cdot v_c = \frac{OM}{OC} \cdot v_c$$

$$= \omega \cdot OM \quad \text{since} \quad v_c = \omega \cdot OC \qquad (13.1)$$

Alternatively triangles OCM and ocp are similar, so that

$$\frac{op}{oc} = \frac{OM}{OC}$$

$$\therefore v_p = op = \omega \cdot OM \quad \text{since} \quad oc = \omega \cdot OC$$

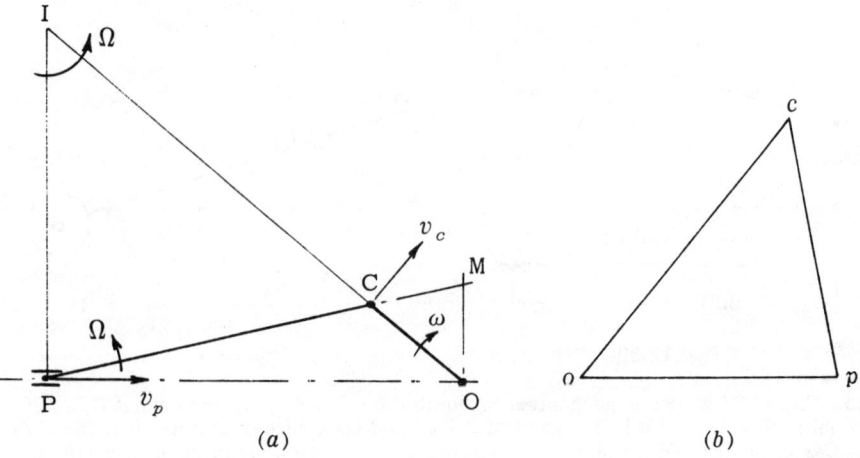

Fig. 13.1

13.3 Piston acceleration

Assuming the crank to be rotating at a uniform angular velocity, the acceleration diagram is shown in Fig. 13.2. oc is the centripetal acceleration of C ($= \omega^2 \cdot OC$) and cp_1 is the centripetal acceleration of P relative to C ($= v_{pc}^2/PC$). p_1p is the tangential acceleration of P relative to C and op is the acceleration of P relative to O.

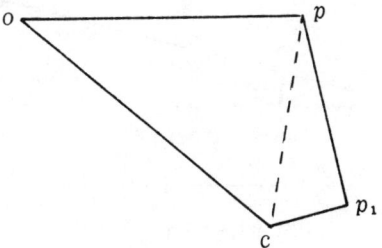

Fig. 13.2

Klein's construction Draw a circle with diameter PC, Fig. 13.3. Extend PC to cut the perpendicular through O at M and draw a circle of centre C and radius CM. Join H to K, H and K being the intersection of the two circles, and let HK intersect PC at L and PO at N.

The quadrilateral OCLN is then similar to the acceleration diagram ocp_1p, Fig. 13.2. Since $f_c = \omega^2 OC$,

then $$f_p = \omega^2 ON \tag{13.2}$$

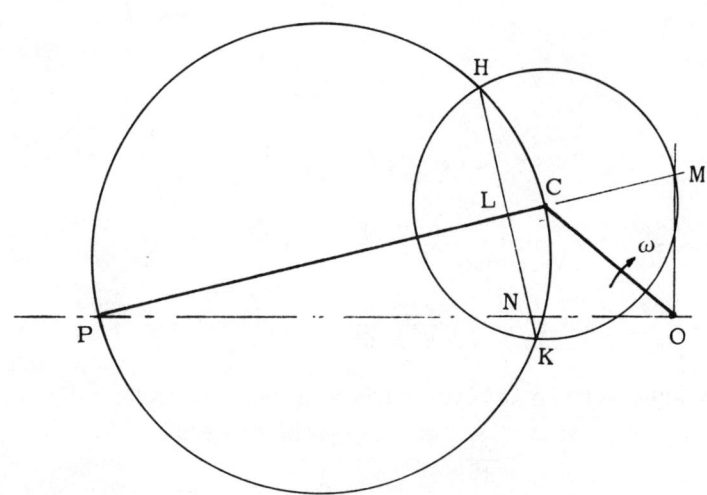

Fig. 13.3

13.4 Analytical method for piston velocity and acceleration The displacement of the piston P, Fig. 13.4, from the inner dead centre position is given by

$$x = (l + r) - (r \cos \theta + l \cos \phi)$$

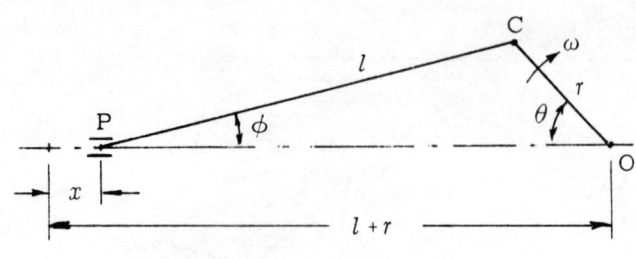

Fig. 13.4

But
$$l \sin \phi = r \sin \theta$$

$$\therefore \sin \phi = \frac{r}{l} \sin \theta$$

$$= \frac{\sin \theta}{n} \quad \text{where } n = \frac{l}{r}$$

$$\therefore \cos \phi = \sqrt{1 - \sin^2 \phi} = \sqrt{1 - \left(\frac{\sin \theta}{n}\right)^2}$$

$$\approx 1 - \frac{\sin^2 \theta}{2n^2} \quad \text{since } \frac{1}{n} \text{ is small and higher powers may be neglected}$$

Thus
$$x = (l + r) - \left(r \cos \theta + l \left[1 - \frac{\sin^2 \theta}{2n^2}\right]\right)$$

$$= r(1 - \cos \theta) + \frac{l \sin^2 \theta}{2n^2}$$

$$= r\left(1 - \cos \theta + \frac{\sin^2 \theta}{2n}\right)$$

$$\therefore v_p = \frac{dx}{dt} = r\left(\sin \theta + \frac{\sin 2\theta}{2n}\right)\frac{d\theta}{dt} = \omega r\left(\sin \theta + \frac{\sin 2\theta}{2n}\right) \quad (13.3)$$

and $\quad f_p = \frac{d^2 x}{dt^2} = \omega r\left(\cos \theta + \frac{\cos 2\theta}{n}\right)\frac{d\theta}{dt} = \omega^2 r\left(\cos \theta + \frac{\cos 2\theta}{n}\right) \quad (13.4)$

The angular velocity Ω of the connecting rod is given by

$$\Omega = \frac{d\phi}{dt}$$

Since
$$\sin \phi = \frac{\sin \theta}{n}$$

RECIPROCATING MECHANISMS

then
$$\cos\phi \frac{d\phi}{dt} = \frac{\cos\theta}{n} \frac{d\theta}{dt}$$

$$\therefore \Omega = \frac{\cos\theta}{n \cos\phi} \omega$$

$$\approx \frac{\omega \cos\theta}{n} \quad \text{since } \phi \text{ is small for usual values of } n \quad (13.5)$$

The angular acceleration α of the connecting rod is given by

$$\alpha = \frac{d\Omega}{dt}$$

$$= -\frac{\omega^2 \sin\theta}{n} \quad (13.6)$$

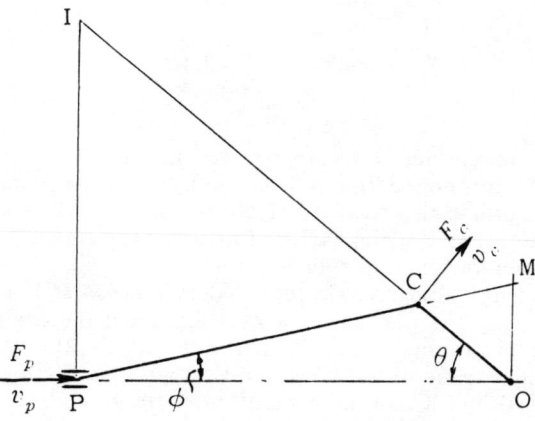

Fig. 13.5

13.5 Crankshaft torque due to piston force and mass Let the force acting on the piston in the direction of motion be F_p, Fig. 13.5, and the force on the crankpin perpendicular to the crank be F_c. Then, neglecting friction and the change in kinetic energy of the connecting rod,

i.e.   work input = work output

i.e.
$$F_p v_p = F_c v_c$$

$$\therefore F_c = F_p \times \frac{v_p}{v_c}$$

$$= F_p \times \frac{IP}{IC} = F_p \times \frac{OM}{OC}$$

$$\therefore \text{crankshaft torque} = F_c \times OC = F_p \times OM \quad (13.7)$$

If the piston is subjected to a gas pressure p acting on an area a, the force on the piston is pa.

If the mass of the piston is m, the force necessary to accelerate the piston is mf_p, so that the net piston force is given by

$$F_p = pa - mf_p$$
$$= pa - m\omega^2 r\left(\cos\theta + \frac{\cos 2\theta}{n}\right)$$

Hence crankshaft torque, $T = \left\{pa - m\omega^2 r\left(\cos\theta + \frac{\cos 2\theta}{n}\right)\right\} \times \text{OM}$

(13.8)

The side thrust S on the cylinder and the force in the connecting rod Q may be obtained from the equilibrium of the forces on the small end pin, Fig. 13.6.

Thus $S = F_p \tan\phi$ (13.9)

and $Q = F_p \sec\phi$ (13.10)

Fig. 13.6

13.6 Crankshaft torque due to connecting rod inertia Allowance may be made for the effect of connecting rod inertia by use of the equivalent two-mass system described in Art. 12.6. Klein's construction is used to construct the acceleration diagram OCLN, Fig. 13.7, the line CN representing the resultant acceleration of P relative to C.

The connecting rod mass m is replaced by masses at P and X such that PG.GX = k^2, where k is the radius of gyration about the centre of gravity, G. Lines are drawn through G and X, parallel with the line of stroke to intersect CN at g and x respectively. Og and Ox then give the directions of the accelerations of G and X and their magnitudes are given by $f_g = \omega^2 \text{O}g$ and $f_x = \omega^2 \text{O}x$ respectively.

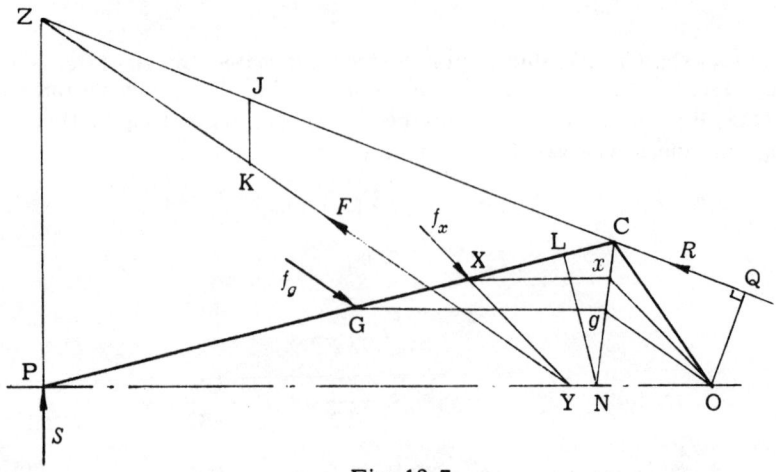

Fig. 13.7

RECIPROCATING MECHANISMS

The inertia force on the mass at P acts along OP and that due to the mass at X acts along YX. The resultant inertia force must pass through Y and be parallel to f_g, its direction being opposite to that of f_g.

The forces acting on the connecting rod are the inertia force ($m\omega^2 Og$), the side thrust S and the force at the crankpin, R. These are concurrent at Z and from the triangle of forces JKZ, JZ represents the force R and KJ represents the force S. The crankshaft torque is then given by

$$T = R \times OQ \qquad (13.11)$$

The *total* crankshaft torque is the algebraic sum of the torques given by equations (13.8) and (13.11).

1. *A horizontal steam engine has a stroke of 600 mm and a cylinder diameter of 225 mm. The piston rod diameter is 50 mm and the length of the connecting rod is 1 200 mm. The reciprocating parts have a mass of 100 kg. When the crank has rotated 60° from the inner dead centre, the steam pressure on the outside of the piston is 560 kN/m² and on the inside is 160 kN/m².*

For this position, calculate the thrust in the connecting rod, the thrust on the cross-head guide, the turning moment on the crankshaft and the radial force in the crank when the crankshaft rotates at 240 rev/min.

Crank radius, $r = 0.3$ m and $n = \dfrac{l}{r} = \dfrac{1.2}{0.3} = 4$

Net steam force on piston $= 560 \times 10^3 \times \dfrac{\pi}{4} \times 0.225^2 - 160 \times 10^3 \times \dfrac{\pi}{4}(0.225^2 - 0.05^2)$

$= 16\,200$ N

Inertia force on piston $= m\omega^2 r \left(\cos\theta + \dfrac{\cos 2\theta}{n}\right)$ from equation (13.4)

$= 100\left(\dfrac{2\pi}{60} \times 240\right)^2 \times 0.3 \left(\cos 60° + \dfrac{\cos 120°}{4}\right)$

$= 7\,100$ N

∴ net piston force, $F_p = 16\,200 - 7\,100 = 9\,100$ N

Referring to Fig. 13.5, $1.2\sin\phi = 0.3\sin 60°$

∴ $\phi = 12°\,30'$

$$\dfrac{OM}{\sin(\theta + \phi)} = \dfrac{OC}{\sin(90° - \phi)}$$

Hence $OM = 0.3\,\dfrac{\sin 72°\,30'}{\sin 77°\,30'} = 0.293$ m

Thus thrust in connecting rod, $Q = F_p \sec\phi$ from equation (13.10)

$= 9\,100 \sec 12°\,30' = \underline{9\,320\text{ N}}$

Side thrust, $S = F_p \tan\phi$ from equation (13.9)

$$= 9\,100\tan 12°30' = \underline{2\,015\text{ N}}$$

Turning moment, $T = F_p \times \text{OM}$. from equation (13.7)

$$= 9\,100 \times 0\cdot 293 = \underline{2\,670\text{ N m}}$$

Referring to Fig. 13.8,

$$\alpha = 60° + 12°30'$$
$$= 72°30'$$

∴ radial force in crank

$$= Q\cos\alpha$$
$$= 9\,320\cos 72°30'$$
$$= \underline{2\,800\text{ N}}$$

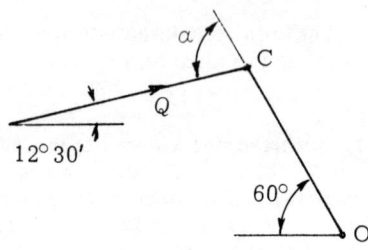

Fig. 13.8

2. *A marine engine has a stroke of 200 mm and a connecting rod of length 400 mm, its centre of gravity being at 175 mm from the crank-pin centre and radius of gyration about the centre of gravity being 125 mm. The connecting rod has a mass of 120 kg and the reciprocating mass is 90 kg. The crank rotates at 240 rev/min.*

Determine (a) the crankshaft torque due to the inertia of the reciprocating mass, and (b) the kinetic energy of the connecting rod for a crank angle of 45°.

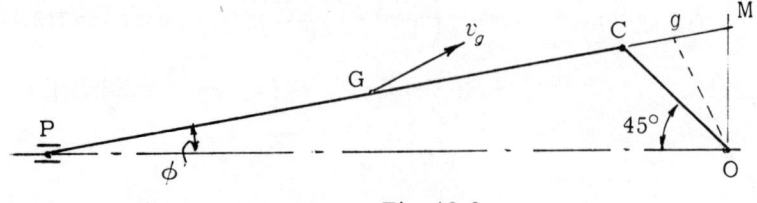

Fig. 13.9

By drawing or calculation, as in example 1,

$$\phi = 10°11' \quad \text{and} \quad \text{OM} = 83\cdot 3\text{ mm}, \quad \text{Fig. 13.9},$$

$$r = 0\cdot 1\text{ m} \quad \text{and} \quad n = \frac{400}{100} = 4$$

From equation (13.4), $F_p = m\omega^2 r\left(\cos\theta + \dfrac{\cos 2\theta}{n}\right)$

$$= 90 \times \left(\frac{2\pi}{60} \times 240\right)^2 \times 0\cdot 1\left(\cos 45° + \frac{\cos 90°}{4}\right)$$

$$= 4\,020\text{ N}$$

RECIPROCATING MECHANISMS 169

$\therefore$ crankshaft torque, $T = F_p \times OM$ . . . from equation (13.7)

$$= 4\,020 \times 0 \cdot 083\,3$$

$$= \underline{335\ \text{N m}}$$

Triangle OCM represents the velocity diagram for the mechanism, OC representing the velocity of C relative to O, OM the velocity of P relative to O and CM the velocity of P relative to C.

Thus the velocity of G is represented by Og, where

$$\frac{Cg}{CM} = \frac{CG}{CP}$$

Since $v_c = \omega \cdot OC$, then $v_g = \omega \cdot Og$

$$= \frac{2\pi}{60} \times 240 \times 0 \cdot 085 = 2 \cdot 14\ \text{m/s}$$

The direction of v_g is perpendicular to Og.

From equation (13.5), $\Omega = \dfrac{\omega \cos \theta}{n}$

$$= \frac{2\pi}{60} \times 240 \times \frac{\cos 45°}{4} = 4 \cdot 44\ \text{rad/s}$$

$\therefore$ total kinetic energy $= \tfrac{1}{2} m v_g^2 + \tfrac{1}{2} I \Omega^2$

$$= \tfrac{1}{2} \times 120 \times 2 \cdot 14^2 + \tfrac{1}{2} \times 120 \times 0 \cdot 125^2 \times 4 \cdot 44^2$$

$$= \underline{293 \cdot 5\ \text{J}}$$

3. *An engine of 120 mm stroke has a connecting rod 260 mm long between centres and of mass 1·25 kg. The centre of gravity is 80 mm from the big end centre and when suspended as a pendulum from the gudgeon pin axis, the rod makes 21 complete oscillations in 20 s.*

Determine (a) the radius of gyration of the rod about an axis through the centre of gravity, and (b) the inertia torque exerted on the crankshaft when the crank is 40° from the top dead centre and is rotating at 1500 rev/min.

The periodic time of a compound pendulum is given by

$$t = 2\pi \sqrt{\frac{k^2 + h^2}{gh}}$$

where k is the radius of gyration about an axis through the c.g. and h is the distance of the c.g. from the point of suspension.

$$h = 260 - 80 = 180\ \text{mm}$$

so that
$$\frac{20}{21} = 2\pi \sqrt{\frac{k^2 + 0 \cdot 18^2}{9 \cdot 81 \times 0 \cdot 18}}$$

from which $\quad\quad\quad k^2 = 0.0081 \text{ m}^2$

and $\quad\quad\quad k = 0.09 \text{ m} \quad \text{or} \quad \underline{90 \text{ mm}}$

Fig. 13.10 shows the required configuration of the engine. Replacing the connecting rod mass by an equivalent two-mass system with the masses placed at P and X, then

$$PG \times GX = k^2 \quad . \quad . \quad . \quad . \quad \text{from equation (12.7)}$$

i.e. $\quad\quad\quad GX = \dfrac{0.0081}{0.18} = 0.045 \text{ m} \quad \text{or} \quad \underline{45 \text{ mm}}$

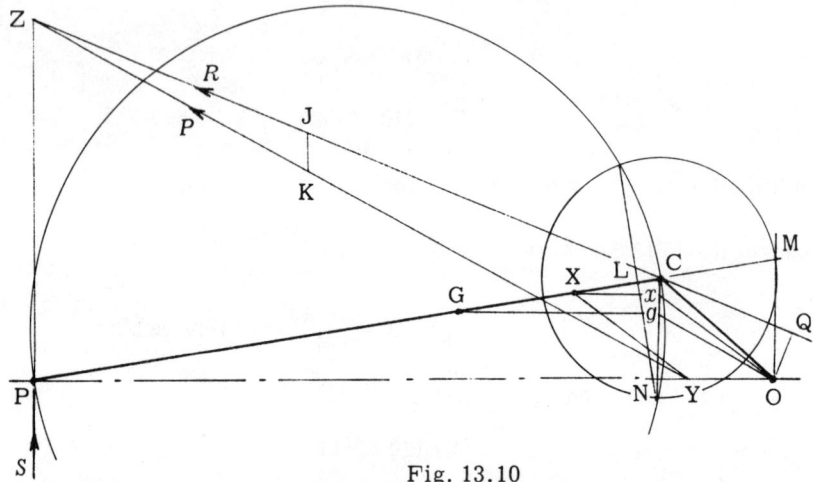

Fig. 13.10

Klein's construction is superimposed on the mechanism, giving the acceleration diagram OCLN. CN represents the acceleration of P relative to C and Og and Ox then give the directions of the accelerations of G and X respectively. The inertia forces on the mass at X is parallel to Ox and this intersects the inertia force on the mass at P at the point Y.

The resultant inertia force on the connecting rod passes through Y and is parallel to Og; its magnitude is given by

$$P = m\omega^2 \, Og$$

$$= 1.25 \times \left(1500 \times \dfrac{2\pi}{60}\right)^2 \times 0.057 = 1760 \text{ N}$$

The three forces acting on the rod are the inertia force, P, the side thrust at the piston, S, and the force at the crank pin, R. Forces P and S intersect at Z and so R must also pass through this point. Drawing the triangle of forces at Z, KZ represents P (= 1 760 N) and JZ represents R. This is found to be 1 620 N and the perpendicular distance of the line of action of R from O (OQ) is found to be 15 mm, so that the crankshaft torque,

$$T = R \times OQ$$

$$= 1620 \times 0.015 = \underline{24.3 \text{ N m}}$$

RECIPROCATING MECHANISMS

4. A vertical single-cylinder engine has a cylinder diameter of 250 mm and a stroke of 450 mm. The reciprocating parts have a mass of 180 kg, the ratio connecting rod/crank radius is 4 and the speed is 360 rev/min. When the crank has turned through an angle of 45° from t.d.c. the net pressure on the piston is 1·05 MN/m². Calculate the crankshaft torque for this position. (*Ans.*: 2 390 N m)

5. A vertical steam engine, 450 mm bore and 750 mm stroke, runs at 240 rev/min. The reciprocating parts of the engine have a mass of 70 kg and the connecting rod is 1·2 m long. When the piston is moving downwards and the crank is 90° beyond t.d.c., the steam pressure above the piston is 800 kN/m² and that below the piston is 120 kN/m².

Determine the instantaneous torque on the crankshaft, neglecting the piston rod area. (*Ans.*: 41·5 kN m)

6. An engine mechanism has a 150 mm crank radius and a 375 mm connecting rod with a piston mass of 10 kg. The crank rotates at 300 rev/min. Determine the acceleration of the piston and the crankshaft torque due to piston inertia when the crank is 45° from the o.d.c. position. (*Ans.*: 103 m/s²; 78·5 N m)

7. The crankshaft of a vertical single-cylinder engine, stroke 250 mm, rotates at 300 rev/min. The reciprocating parts have a mass of 100 kg. The connecting rod has a mass of 120 kg, it is 450 mm long, the c.g. is 300 mm from the gudgeon-pin axis and the radius of gyration about that axis is 363 mm.

When the crank is 30° from the t.d.c. position and moving downwards, find (*a*) the reaction at the cylinder walls due to the inertia of the reciprocating parts; (*b*) the total kinetic energy of the connecting rod. (*Ans.*: 1·74 kN; 739 J)

8. A petrol engine of cylinder diameter 100 mm and stroke 120 mm has a piston of mass 1·1 kg and a connecting rod of length 250 mm. When rotating at 2 000 rev/min, the gas pressure is 700 kN/m² when the crank is at 20° from the t.d.c. position.

Find (*a*) the resultant load on the gudgeon pin, (*b*) the thrust on the cylinder wall.

Determine also the speed at which the gudgeon pin load would be reversed in direction, the other conditions remaining constant.
(*Ans.*: 2 263 N; 186 N; 2 603 rev/min)

9. The connecting rod of an engine is 0·9 m long between centres, its mass is 25 kg and its centre of gravity is 0·3 m from the crank-pin. The radius of gyration about an axis through the c.g. is 0·375 m and the crank rotates at 300 rev/min. The crank radius is 0·3 m.

Determine the crankshaft torque due to the inertia of the connecting rod when the crank makes an angle of 45° with the t.d.c. position. (*Ans.*: 520 N m)

10. A connecting rod has a mass of 1·125 kg and the distance between the centres of the end bearings is 250 mm. The c.g. is 162·5 mm from the centre of the small end bearing and the moment of inertia about a transverse axis through the c.g. is 0·011 8 kg m². The crank radius is 62·5 mm and the speed is 200 rad/s. For a crank angle of 30° past the inner dead centre position, determine (*a*) the torque required at the crankshaft to accelerate the rod, assuming the rod to be equivalent to two mass particles, one at each end of the rod, (*b*) the percentage error in this assumption.
(*Ans.*: 37·1 N m; 11·8%)

14 Turning moment diagrams

14.1 Introduction The output torque from a reciprocating engine varies considerably over the working cycle and if the engine is driving a generator or machine which offers a constant resisting torque, the resulting speed will vary because the engine torque is at times greater than or less than the resisting torque. In order to reduce this fluctuation of speed, a flywheel is fitted to the engine to absorb energy at some points in the cycle and release it at others. The inertia of the flywheel required depends on the fluctuation of the energy available from the engine and the fluctuation of speed which is acceptable.

14.2 Crank effort diagrams If the output torque from the engine is plotted against crank angle, a turning moment or crank effort diagram is obtained. The net area under the graph represents the work done during the cycle and the average height represents the mean torque, which is equal to the resisting torque if the mean speed is to remain constant.

A typical crank effort diagram for a four stroke single cylinder engine is shown in Fig. 14.1. At points where the curve cuts the mean torque line, the engine speed is constant. Between points a and b, the engine torque is greater than the resisting torque and so the engine speeds up; similarly, between points b and c, the engine torque is less than the resisting torque and so the engine slows down.

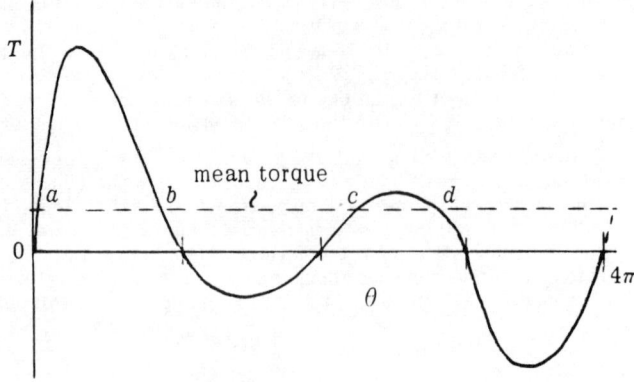

Fig. 14.1

A similar situation arises in the case of an electric motor driving a machine which offers a variable resistance. The turning moment then consists of a uniform input torque and a variable resisting torque but the analysis is identical with that for an engine application.

By inspection of the areas of the loops above and below the mean torque line, the points of maximum and minimum speed in the engine cycle may

TURNING MOMENT DIAGRAMS

be determined (see Ex. 1). The difference of energy supplied and energy required between these points is then responsible for the change of speed; this is termed the *fluctuation of energy*, U.

If the moment of inertia of the rotating parts of the system is I and ω_1 and ω_2 are respectively the maximum and minimum speeds during the cycle, the change of kinetic energy of the system is $\frac{1}{2}I\omega_1^2 - \frac{1}{2}I\omega_2^2$, which is brought about by the fluctuation of energy supplied by the engine,

i.e. $$U = \tfrac{1}{2}I\omega_1^2 - \tfrac{1}{2}I\omega_2^2 \qquad (14.1)$$

The fluctuation of speed, $\omega_1 - \omega_2$, is small in comparison with the mean speed ω and assuming that the variations above and below the mean speed are equal,

$$\omega_1 + \omega_2 = 2\omega$$

Thus equation (14.1) may be written

$$U = \tfrac{1}{2}I(\omega_1 + \omega_2)(\omega_1 - \omega_2)$$
$$= I\omega(\omega_1 - \omega_2)$$
$$\text{or } I\omega^2 \times \frac{\omega_1 - \omega_2}{\omega} \qquad (14.2)$$

The term $\dfrac{\omega_1 - \omega_2}{\omega}$ is called the *coefficient of fluctuation of speed*.

1. *The turning moment diagram for an engine is drawn on a base of crank angle and the mean resisting torque line added. The areas above and below the mean line are +4 400, -1 150, +1 300, -4 550 mm², the scales being 1 mm = 100 N m torque and 1 mm = 1° of crank angle.*

Find the mass of the flywheel required to keep the speed between 297 and 303 rev/min if its radius of gyration is 0·525 m.

The turning moment diagram is shown in Fig. 14.2.

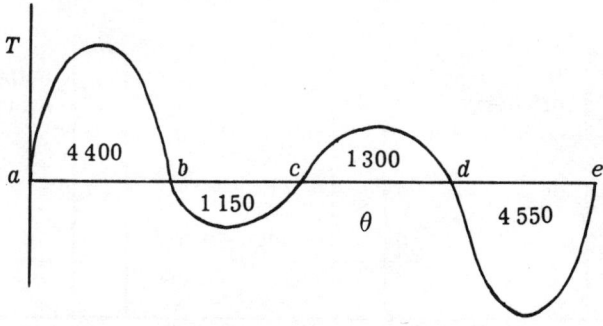

Fig. 14.2

1 mm on the torque scale represents 100 N m

1 mm on the crank angle scale represents $1 \times \dfrac{\pi}{180}$ rad

∴ 1 mm² of area represents $100 \times \frac{\pi}{180} = 1.745$ J

Commencing at point a, the engine speeds up to point b, slows down to point c, speeds up to point d and slows down to point e, when the cycle is repeated. Between b and c, less energy is abstracted than is added between a and b and so the engine is running faster at c than at a. It is running faster still at d but then a substantial amount of energy is abstracted between d and e, which restores the speed to that at a. Hence the minimum speed occurs at a and the maximum at d.

Thus the fluctuation of energy between points of minimum and maximum speeds is represented by *either* $4400 - 1150 + 1300$ mm² *or* 4550 mm², these being equal.

Hence fluctuation of energy $= 4550 \times 1.745 = 7940$ J

Therefore, from equation (14.1),
$$7940 = \tfrac{1}{2}I(\omega_1^2 - \omega_2^2)$$
$$= \tfrac{1}{2}I(303^2 - 297^2) \times \left(\frac{2\pi}{60}\right)^2$$
$$\therefore I = 402 \text{ kg m}^2$$
$$\therefore 402 = m \times 0.525^2$$
$$\therefore m = \underline{1459 \text{ kg}}$$

2. *A motor driving a punching machine exerts a constant torque of 675 Nm on the flywheel, which rotates at an average speed of 120 rev/min. The punch operates 60 times per minute, the duration of the punching operation being ⅕ s. It may be assumed that during the punching operation, the resisting torque on the flywheel is constant.*

Deduce the value of the resisting torque and find the moment of inertia of the flywheel if the speed variation between maximum and minimum is not to exceed 10 rev/min.

The turning moment diagram is shown in Fig. 14.3.

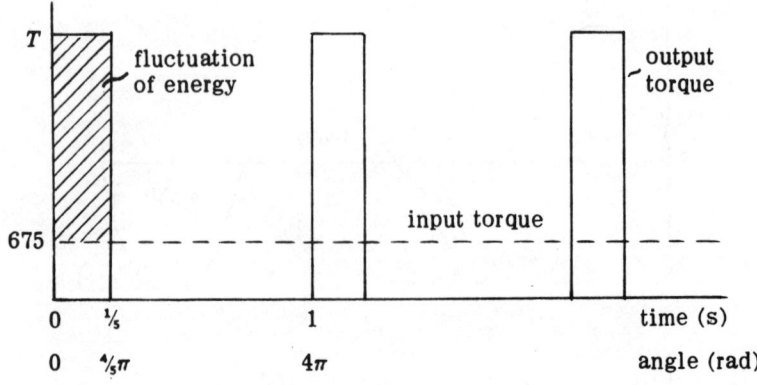

Fig. 14.3

TURNING MOMENT DIAGRAMS

Equating the areas under the input and output torques over a period of 1 s,

$$T \times \tfrac{1}{5} = 675 \times 1$$

$$\therefore T = 3\,375 \text{ N m}$$

Fluctuation of energy $= (3\,375 - 675) \times \dfrac{4\pi}{5}$

$$= 6\,786 \text{ J}$$

$\therefore 6\,786 = \tfrac{1}{2}I(\omega_1^2 - \omega_2^2)$ . . . from equation (14.1)

$$= \tfrac{1}{2}I(N_1^2 - N_2^2)\left(\dfrac{2\pi}{60}\right)^2 \quad \text{where } N_1 \text{ and } N_2 \text{ are the speeds in rev/min}$$

$$= \tfrac{1}{2}I(N_1 + N_2)(N_1 - N_2)\left(\dfrac{2\pi}{60}\right)^2$$

$$= \tfrac{1}{2}I \times 240 \times 10 \times \dfrac{\pi^2}{900}$$

$$\therefore I = \underline{516 \text{ kg m}^2}$$

3. *A single cylinder gas engine, working on the four-stroke cycle, develops 11 kW at 300 rev/min. The work done on the gas during the compression stroke is 0·7 times the work done by the gases during the power stroke. The turning moment diagram for the compression stroke may be taken as an isosceles triangle and that for the power stroke as another isosceles triangle. The turning moment during the suction and exhaust strokes is negligible.*

If the mass of the flywheel is 1500 kg and its radius of gyration is 400 mm, find the coefficient of fluctuation of speed.

The turning moment diagram is shown in Fig. 14.4.

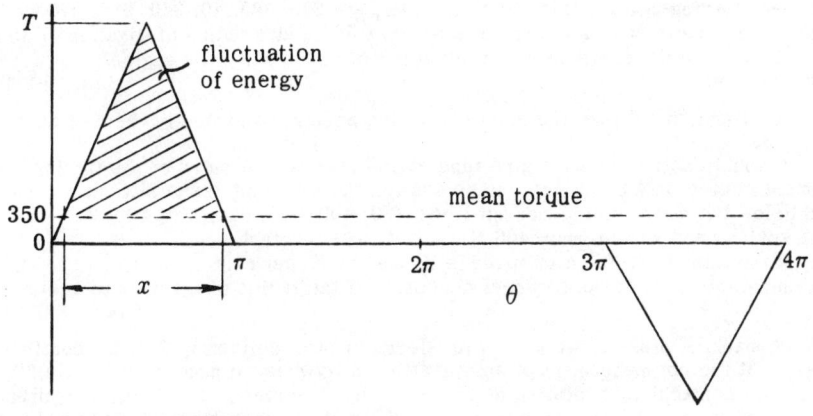

Fig. 14.4

$$\text{Power} = \text{mean torque} \times \text{speed}$$

$$\therefore \quad \text{mean torque} = \frac{11\,000}{300 \times \frac{2\pi}{60}}$$

$$= 350 \text{ Nm}$$

Equating areas under actual and mean torque lines,

$$\tfrac{1}{2}T\pi - 0.7 \times \tfrac{1}{2}T\pi = 350 \times 4\pi$$

$$\therefore \quad T = 9333 \text{ Nm}$$

From similar triangles, $\quad \dfrac{x}{\pi} = \dfrac{9333 - 350}{9333} = \dfrac{8983}{9333}$

$$\therefore \quad x = 3.025 \text{ rad}$$

Therefore, fluctuation of energy $= \tfrac{1}{2} \times 8983 \times 3.025$

$$= 13\,582 \text{ J}$$

$$\therefore \quad 13\,582 = I\omega^2 \, \frac{\omega_1 - \omega_2}{\omega} \quad \text{from equation (14.2)}$$

$$= 1500 \times 0.4^2 \times \left(300 \times \frac{2\pi}{60}\right)^2 \times \frac{\omega_1 - \omega_2}{\omega}$$

$$\therefore \quad \frac{\omega_1 - \omega_2}{\omega} = 0.0573 \quad \text{or} \quad \underline{5.73\%}$$

4. The turning-moment diagram for a petrol engine is drawn to the following scales: turning moment 1 mm = 5 Nm; crank angle 1 mm = 1°. The turning-moment diagram repeats itself at every half-revolution of the engine and the areas above and below the mean turning-moment line, taken in order, are 295, 685, 40, 340, 960, 270 mm². The rotating parts are equivalent to a mass of 36 kg at a radius of gyration of 150 mm. Determine the coefficient of fluctuation of speed when the engine runs at 1800 rev/min. *(Ans.: 0.299%)*

5. Distinguish between the functions of the governor and the flywheel of an engine.

A double-acting steam engine runs at 100 rev/min. A curve of the turning-moment plotted on a crank angle base showed the following areas alternately above and below the mean turning-moment line: 780, 400, 520, 620, 260, 460, 340, 420 mm². The scales used were 1 mm = 400 Nm and 1 mm = 1° crank angle.

If the total fluctuation in speed is limited to 1½ per cent of the mean speed, determine the mass of the flywheel necessary if the radius of gyration is 1.05 m.
(Ans.: 3464 kg)

6. A machine press is worked by an electric motor, delivering 2.25 kW continuously. At the commencement of an operation, a flywheel of moment of inertia 50 kg m² on the machine is rotating at 250 rev/min. The pressing operation requires 4.75 kJ of energy and occupies 0.75 s. Find the maximum number of pressings that can be made in 1 h and the reduction in speed of the flywheel after each pressing. Neglect friction losses. *(Ans.: 1705; 23.5 rev/min)*

TURNING MOMENT DIAGRAMS

7. A single-cylinder four-stroke internal combustion engine develops 30 kW at 300 rev/min. The turning-moment diagram for the expansion and compression strokes may be taken as two isosceles triangles, on bases 0 to π and 3π to 4π radians respectively, and the net work done during the exhaust and inlet strokes is zero. The work done during compression is negative and is one quarter of that during expansion.

Sketch the turning moment diagram for one cycle and find the maximum value of the turning moment during expansion.

If the load remains constant, mark on the diagram the points of maximum and minimum speeds. Also find the moment of inertia, in kg m², of a flywheel to keep the speed fluctuation within ±1·5 per cent of the mean speed.

(*Ans.*: 10·18 kN m; 444 kg m²)

8. A gas engine develops 22·5 kW at 270 rev/min. It has hit-and-miss governing and there are 125 explosions per minute. The flywheel has a mass of 900 kg and a radius of gyration of 0·675 m. If it is assumed that the work done is identical for each working cycle, that the work done by the gases on the explosion stroke is 2·4 times the work done on the gases during compression stroke, and that the work done on the other two strokes is negligible, find the maximum fluctuation of speed of the flywheel as a percentage of the mean speed. (*Ans.*: 4·82 per cent)

9. A shaft fitted with a flywheel rotates at 250 rev/min and drives a machine the resisting torque of which varies in a cyclic manner over a period of three revolutions. The torque rises from 675 N m to 2 700 N m in a uniform manner during ½ revolution and remains constant for the following 1 revolution. It then falls uniformly to 675 N m during the next ½ revolution and remains constant for 1 revolution, the cycle being then repeated.

If the driving torque applied to the shaft is constant and the flywheel has a mass of 450 kg and a radius of gyration of 0·6 m, find the power necessary to drive the machine and the percentage fluctuation of speed.

(*Ans.*: 44·2 kW; ±3·58 per cent).

15 Friction clutches, bearings and belt drives

15.1 Plate clutches A clutch is a device for connecting and disconnecting the drive between two co-axial shafts. In a friction clutch, the two parts are held in contact by springs and the torque is transmitted by friction between the contact surfaces. The drive may then be disconnected by separating the contact surfaces.

In a multi-plate clutch, a number of flat annular plates are held in contact by an axial force normal to the surfaces. Each pair of surfaces carries the full load and each side of each plate is effective, so that the total torque transmitted is then $2n$ times the torque transmitted by one pair of surfaces, where n is the number of plates.

Fig. 15.1 shows two surfaces pressed together with an axial force W; the outer and inner radii are r_1 and r_2 respectively and the pressure at a radius r is p.

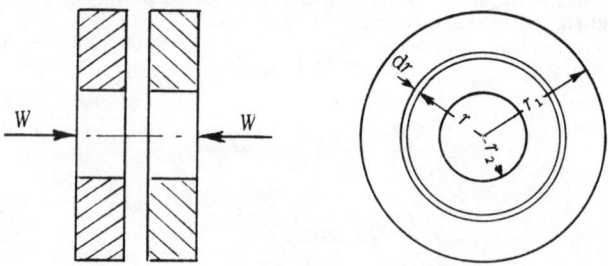

Fig. 15.1

Normal force on annular ring of radius r and width dr

$$= p \times 2\pi r \, dr$$

$$\therefore \text{total axial force, } W = 2\pi \int_{r_2}^{r_1} pr \, dr \tag{15.1}$$

Friction force on ring $= \mu \times 2\pi pr \, dr$, where μ is the coefficient of friction

$$\therefore \text{moment of friction force about axis} = \mu \times 2\pi pr \, dr \times r$$

$$\therefore \text{total torque transmitted, } T = 2\pi\mu \int_{r_2}^{r_1} pr^2 \, dr \tag{15.2}$$

Equations (15.1) and (15.2) may be integrated if some assumptions are made regarding the variation of pressure with radius.

FRICTION

For unworn clutches, it is assumed that p is constant, giving

$$W = 2\pi p \int_{r_2}^{r_1} r\, dr = \pi p\, (r_1^2 - r_2^2) \tag{15.3}$$

and
$$T = 2\pi\mu p \int_{r_2}^{r_1} r^2\, dr = \tfrac{2}{3} \pi\mu p\, (r_1^3 - r_2^3)$$

$$= \tfrac{2}{3}\mu W \frac{r_1^3 - r_2^3}{r_1^2 - r_2^2} \tag{15.4}$$

For worn clutches, it is assumed that wear is uniform over the contact area.

Since
$$\text{wear} \propto \text{pressure} \times \text{velocity}$$
$$\propto \text{pressure} \times \text{radius},$$

then
$$pr = c,$$

giving
$$W = 2\pi c \int_{r_2}^{r_1} dr = 2\pi c\, (r_1 - r_2) \tag{15.5}$$

and
$$T = 2\pi\mu c \int_{r_2}^{r_1} r\, dr = \pi\mu c\, (r_1^2 - r_2^2)$$

$$= \mu W \frac{r_1 + r_2}{2} \tag{15.6}$$

The torque transmissible by a worn clutch is less than by an unworn clutch and so this theory should always be used unless otherwise stated. If, however, the ratio $r_2/r_1 > \tfrac{1}{4}$, the difference between the two theories is very small.

15.2 Cone clutches A cone clutch consists of a single pair of friction faces arranged as the frustum of a cone, as shown in Fig. 15.2.

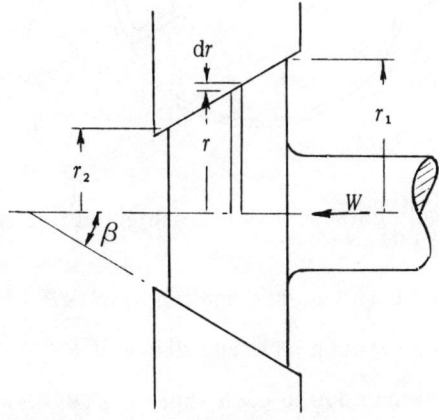

Fig. 15.2

Let p be the normal pressure between the surfaces at radius r.

Then normal force on elementary ring $= p \times 2\pi r \, dr \, \text{cosec}\, \beta$

∴ axial component of this force $= p \times 2\pi r \, dr \, \text{cosec}\, \beta \sin \beta$

$\qquad\qquad\qquad\qquad\qquad\qquad = p \times 2\pi r \, dr$

∴ total axial force, $W = 2\pi \int_{r_2}^{r_1} pr \, dr$ (15.7)

Friction force on ring $= \mu p \times 2\pi r \, dr \, \text{cosec}\, \beta$

∴ moment of friction force about axis $= \mu p \times 2\pi r \, dr \, \text{cosec}\, \beta \times r$

∴ total torque transmitted, $T = 2\pi\mu \, \text{cosec}\, \beta \int_{r_2}^{r_1} pr^2 \, dr$ (15.8)

If p is assumed constant, $\qquad T = \tfrac{2}{3}\mu W \dfrac{r_1^3 - r_2^3}{r_1^2 - r_2^2} \text{cosec}\, \beta$ (15.9)

If pr is assumed constant, $\qquad T = \mu W \dfrac{r_1 + r_2}{2} \text{cosec}\, \beta$ (15.10)

15.3 Centrifugal clutches A centrifugal clutch consists of a number of shoes which move outwards in guides due to centrifugal force. The shoes come into contact with a rim and cause the rim to rotate due to friction between shoes and rim. When the speed falls, the shoes are pulled in from the rim by springs.

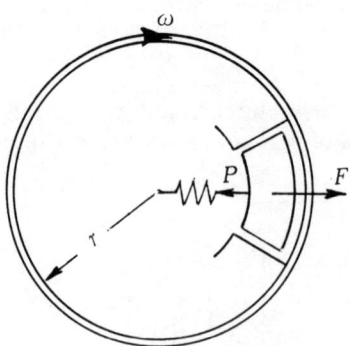

Fig. 15.3

Fig. 15.3 shows a clutch shoe which is subjected to a centrifugal force F and an inward spring force P.

Then radial force between shoe and rim $= F - P$

∴ friction force between shoe and rim $= \mu(F - P)$

∴ friction torque due to each shoe $= \mu(F - P)r$

If there are n shoes, total clutch torque, $T = n\mu r(F - P)$ (15.11)

FRICTION 181

15.4 Bearings Fig. 15.4(a), (b) and (c) shows three types of plain bearing, a collar bearing, a footstep bearing and a conical pivot respectively. In each case, the friction torque to be overcome is given by an identical equation to that for the corresponding flat plate or conical clutch.

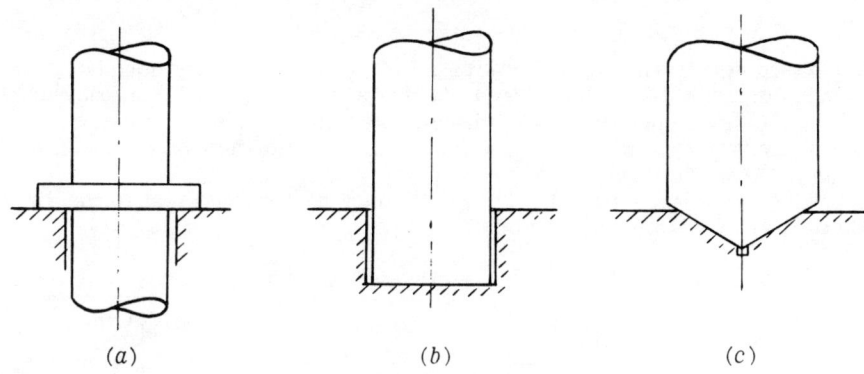

Fig. 15.4

In order to distribute the load over a larger area and hence reduce the bearing pressure, a multi-collar bearing, as shown in Fig. 15.5, may be used. The thrust pads between each pair of collars are horse-shoe shaped and are adjusted so that each takes an equal share of the axial thrust.

If the axial load is W and there are n collars, each collar carries a load $\frac{W}{n}$. The friction torque per collar can then be calculated for this load but since the total torque is n times that for a single collar, this is the same as that for one collar carrying the total load.

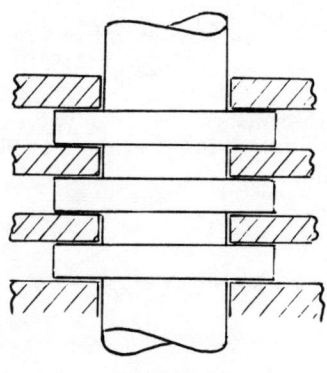

Fig. 15.5

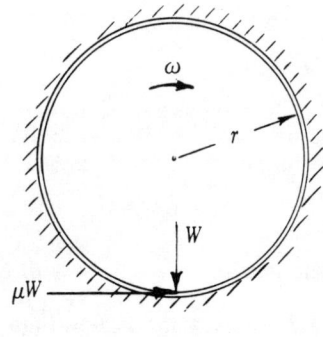

Fig. 15.6

15.5 Journal bearings Fig. 15.6 shows a shaft of radius r supported by a concentric bearing. If the radial load between the shaft and the bearing is W,

$$\text{tangential friction force} = \mu W$$
$$\therefore \text{ friction torque, } T = \mu W r$$
and $$\text{and power loss} = T\omega = \mu W r \omega \qquad (15.12)$$

15.6 Lubricated surfaces In the foregoing cases of friction torques, it has been assumed that the coefficient of friction, μ, has been constant, which is appropriate to unlubricated surfaces. If the bearing is lubricated, however, the effective coefficient of friction depends on the viscosity of the oil, the pressure between the surfaces, bearing dimensions and the shaft speed.

15.7 Viscosity When a fluid flows smoothly over a stationary boundary, the layer in contact with the boundary is at rest and subsequent layers move with increasing velocities as the distance from the boundary increases. Thus there is a velocity gradient across the section of flow and a shearing action between adjacent layers.

If the velocity changes by dv in a distance dy perpendicular to the direction of flow, Fig. 15.7, the viscous strain rate,

$$\phi = \frac{dv}{dy}$$

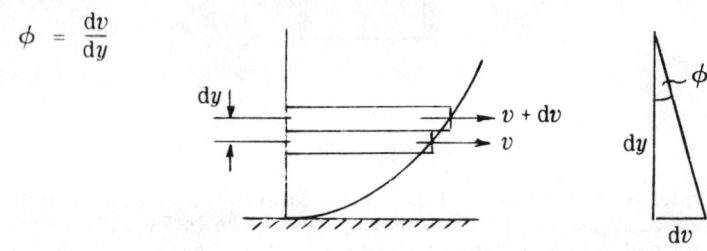

Fig. 15.7

The viscous stress τ is the viscous resistance per unit area and the coefficient of viscosity η is defined as the ratio viscous stress/viscous strain rate,

i.e.
$$\eta = \frac{\tau}{\phi} = \frac{\tau}{dv/dy}$$

or
$$\tau = \eta \frac{dv}{dy} \tag{15.13}$$

In the case of a thin film of lubricant, of thickness y, between two surfaces moving with a velocity v relative to one another, the gradient may be regarded as uniform, so that equation (15.13) becomes

$$\tau = \eta \frac{v}{y} \tag{15.14}$$

The units of η are kg/m s or N s/m² (1 poise = 0·1 kg/m s).

15.8 Application to bearings

(a) *Parallel bearings* Fig. 15.8 shows two flat surfaces of area a, between which is an oil film of thickness t. Then, if the upper surface is moving at a velocity v relative to the lower surface,

viscous force, P = stress × area

$$= \eta \frac{v}{t} \times a$$

Power loss = Pv = $\dfrac{\eta v^2 a}{t}$ (15.15)

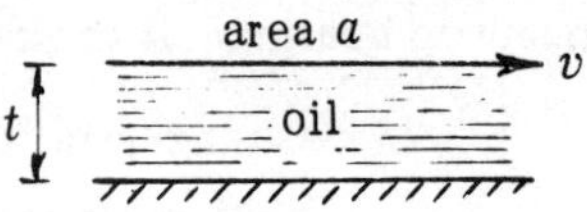

Fig. 15.8

FRICTION

(b) *Journal bearings* Fig. 15.9 shows a journal bearing of radius r and radial clearance c, rotating at ω rad/s.

Then $\quad \tau = \eta \dfrac{v}{c} = \eta \dfrac{\omega r}{c}$

∴ viscous force for a length l

$$= \text{stress} \times \text{area}$$

$$= \eta \frac{\omega r}{c} \times 2\pi r l$$

$$= \frac{2\pi \eta \omega r^2 l}{c}$$

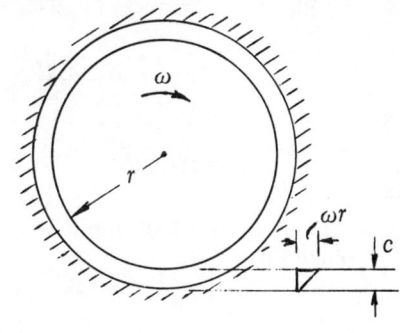

Fig. 15.9

∴ viscous torque $= \dfrac{2\pi \eta \omega r^2 l}{c} \times r = \dfrac{2\pi \eta \omega r^3 l}{c} \quad (15.16)$

and power loss $= \dfrac{2\pi \eta \omega r^3 l}{c} \times \omega = \dfrac{2\pi \eta \omega^2 r^3 l}{c} \quad (15.17)$

The coefficient of friction is defined by

$$\mu = \frac{\text{viscous friction force}}{\text{bearing load}}$$

$$= \frac{\dfrac{2\pi \eta \omega r^2 l}{c}}{p \times 2rl} = \frac{\pi \eta \omega r}{pc} \quad (15.18)$$

where p is the projected bearing pressure given by $\dfrac{\text{bearing load}}{\text{projected bearing area}}$

(c) *Collar and footstep bearings* Fig. 15.10 shows a collar bearing of external and internal radii r_1 and r_2 respectively. The shaft is supported on an oil film of thickness t and is rotating at ω rad/s.

Consider an annular element of radius x and thickness dx. Then

$$\tau = \eta \frac{v}{t} = \eta \frac{\omega x}{t}$$

Therefore shearing force on element

$$= 2\pi x \, dx \times \eta \frac{\omega x}{t}$$

Hence torque on element

$$= 2\pi x \, dx \times \eta \frac{\omega x}{t} \times x$$

$$= \frac{2\pi \eta \omega}{t} x^3 \, dx$$

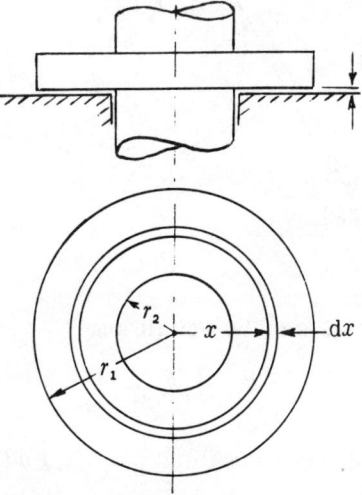

Fig. 15.10

so that total torque $= \dfrac{2\pi \eta \omega}{t} \displaystyle\int_{r_2}^{r_1} x^3 \, dx$

$$= \dfrac{\pi \eta \omega}{2t}(r_1^4 - r_2^4) \tag{15.19}$$

$$\text{Power loss} = T\omega = \dfrac{\pi \eta \omega^2}{2t}(r_1^4 - r_2^4) \tag{15.20}$$

In the case of a footstep bearing of radius r, Fig. 15.4(b), these equations reduce to

$$\text{torque} = \dfrac{\pi \eta \omega r^4}{2t} \tag{15.21}$$

and

$$\text{power} = \dfrac{\pi \eta \omega^2 r^4}{2t} \tag{15.22}$$

15.9 Belt drives When a belt transmits power to a pulley, there is a difference in tension between the tight and slack sides due to the friction force between the belt and pulley.

Consider the section of flat belt shown in Fig. 15.11, which has an angle of lap θ on the pulley and is about to slip in the direction of motion, the tight and slack side tensions then being T_1 and T_2 respectively. On a small arc subtending an angle $d\theta$, let T and $T + dT$ be the tensions at the element and R be the radial force between the pulley and belt.

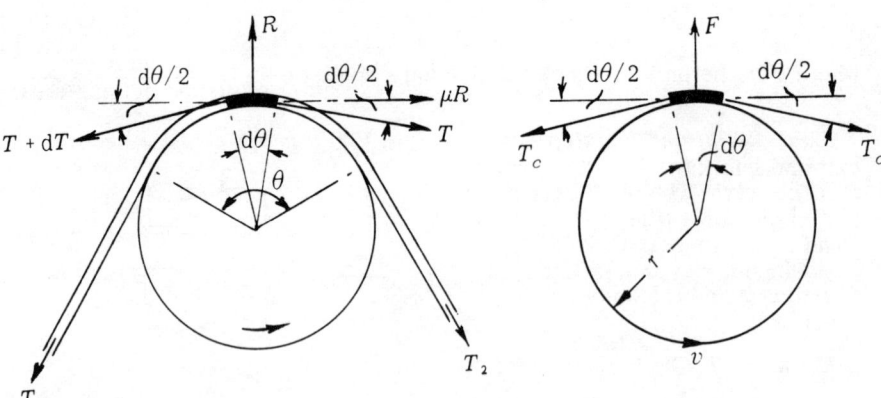

Fig. 15.11　　　　　　　Fig. 15.12

For radial equilibrium,

$$(T + dT)\dfrac{d\theta}{2} + T\dfrac{d\theta}{2} = R$$

which reduces to $\qquad T \, d\theta = R ,\qquad$ (15.23)

neglecting products of small quantities.

FRICTION

For tangential equilibrium,

$$(T + dT)\cos\frac{d\theta}{2} = T\cos\frac{d\theta}{2} + \mu R$$

which reduces to
$$dT = \mu R, \qquad (15.24)$$

assuming that $\cos\dfrac{d\theta}{2} = 1$.

From equations (15.23) and (15.24),

$$dT = \mu T\, d\theta$$

or
$$\frac{dT}{T} = \mu\, d\theta$$

$$\therefore \int_{T_2}^{T_1} \frac{dT}{T} = \int_0^\theta \mu\, d\theta$$

$$\therefore \log_e \frac{T_1}{T_2} = \mu\theta$$

or
$$\frac{T_1}{T_2} = e^{\mu\theta} \qquad (15.25)$$

If the belt is used to transmit power between two pulleys of unequal diameter, the belt will slip first on the pulley with the smaller angle of lap, i.e. on the smaller pulley.

If the belt moves with a velocity v, then

$$\text{power transmitted} = (T_1 - T_2)v = T_1\left(1 - \frac{1}{e^{\mu\theta}}\right)v \qquad (15.26)$$

15.10 Centrifugal tension As the belt passes over the pulley, each element is subjected to centrifugal force, which will increase the tension in the belt. Consider an element of mass m per unit length subtending an angle $d\theta$ and moving with velocity v, Fig. 15.12.

$$F = mr\, d\theta \times \frac{v^2}{r} = mv^2\, d\theta \qquad (15.27)$$

For radial equilibrium,

$$F = 2T_c \frac{d\theta}{2} = T_c\, d\theta \qquad (15.28)$$

where T_c is the centrifugal tension.

Therefore, from equations (15.27) and (15.28),

$$T_c = mv^2 \qquad (15.29)$$

This tension is additional to that due to the transmission of power. If allowance is to be made for this, equation (15.23) becomes

$$T\, d\theta = R + F = R + T_c\, d\theta$$

which leads to
$$\frac{dT}{T - T_c} = \mu \, d\theta$$

so that
$$\frac{T_1 - T_c}{T_2 - T_c} = e^{\mu\theta} \qquad (15.30)$$

T_1 and T_2 are the total tensions in the belt and $(T_1 - T_c)$ and $(T_2 - T_c)$ are the effective driving tensions.

The power transmitted is still $(T_1 - T_2)v$ but equation (15.26) becomes

$$\text{power} = (T_1 - T_c)\left(1 - \frac{1}{e^{\mu\theta}}\right)v \qquad (15.31)$$

For maximum power transmissible

$$\frac{d}{dv}[(T_1 - T_c)v] = 0$$

i.e.
$$\frac{d}{dv}[T_1 v - mv^3] = 0$$

from which
$$mv^2 = T_c = \frac{T_1}{3} \qquad (15.32)$$

This equation gives the velocity for maximum power, which may then be substituted in equation (15.31).

15.11 Initial tension A belt is assembled on pulleys with an initial tension T_0. To determine the tension required, it is assumed that the belt material is elastic and hence obeys Hooke's Law. Since the total length remains constant, the increase in length of the tight side is equal to the decrease in length of the slack side and it therefore follows that the increase in tension on the tight side is equal to the decrease in tension on the slack side.

i.e.
$$T_1 - T_0 = T_0 - T_2$$

or
$$T_1 + T_2 = 2T_0 \qquad (15.33)$$

15.12 V-belt drives For a V-grooved pulley, the normal reaction between the belt and pulley is given by

$R = 2N \sin\beta$, Fig. 15.13.

The friction force between the belt and pulley is then

$2\mu N = \mu R \operatorname{cosec} \beta$

The coefficient of friction is therefore effectively increased from μ to $\mu \operatorname{cosec} \beta$ and equation (15.25) then becomes

$$\frac{T_1}{T_2} = e^{\mu\theta \operatorname{cosec} \beta} \qquad (15.34)$$

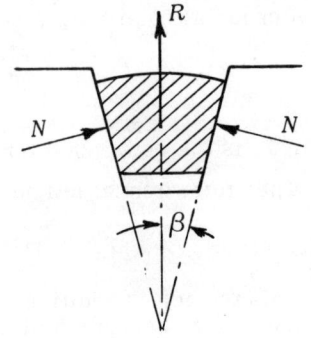

Fig. 15.13

FRICTION

1. *The rotating parts of a grinding mill have a mass of 2 t and rotate at 120 rev/min. They are supported by a conical bearing of 125 mm outer diameter and 50 mm inner diameter with an included angle of 120°. Assuming $\mu = 0.075$ and that the intensity of pressure varies inversely as the radius, determine the power wasted in friction.*

From equation (15.10), $T = \mu W \dfrac{r_1 + r_2}{2} \operatorname{cosec} \beta$

$$= 0.075 \times (2 \times 10^3 \times 9.81) \times \dfrac{62.5 + 25}{2 \times 10^3} \operatorname{cosec} 60°$$

$$= 74.3 \text{ N m}$$

∴ power wasted in friction $= T\omega$

$$= 74.3 \times 120 \times \dfrac{2\pi}{60}$$

$$= \underline{935 \text{ W}}$$

2. *An electric motor drives a co-axial rotor through a single-plate clutch which has two pairs of driving surfaces, each of 275 mm external and 200 mm internal diameter. The total spring load pressing the plates together is 500 N. The motor armature has a mass of 800 kg and a radius of gyration of 260 mm; the rotor has a mass of 1 350 kg and a radius of gyration of 225 mm.*

When the motor is running at 1 250 rev/min, the current is switched off and the clutch suddenly engaged. Determine the final speed of the motor and rotor, the time taken to reach that speed and the kinetic energy lost during clutch slip. $\mu = 0.35$.

Torque transmitted by clutch, $T = \mu W \dfrac{r_1 + r_2}{2} \times 2$ from equation (15.6)

$$= 0.35 \times 500 \times \dfrac{137.5 + 100}{10^3}$$

$$= 41.6 \text{ N m}$$

Let suffices m and r refer to the motor and rotor respectively.

Then $\quad I_m = 800 \times 0.26^2 = 54.1 \text{ kg m}^2$

and $\quad I_r = 1\,350 \times 0.225^2 = 68.3 \text{ kg m}^2$

Let N be the final speed in rev/min and t be the time of slipping.

Then, for the motor, $\quad T = I_m \alpha_m$

i.e. $\quad 41.6 = 54.1 \times \dfrac{1\,250 - N}{t} \times \dfrac{2\pi}{60}$

which reduces to $\quad 7.34 = \dfrac{1\,250 - N}{t} \quad\quad (1)$

and for the rotor, $T = I_r \alpha_r$

i.e. $41 \cdot 6 = 68 \cdot 3 \times \dfrac{N}{t} \times \dfrac{2\pi}{60}$

which reduces to $5 \cdot 82 = \dfrac{N}{t}$ (2)

Therefore, from equations (1) and (2),

$$N = \underline{552 \text{ rev/min}}$$

and $t = \underline{95 \text{ s}}$

Loss of K.E. $= \frac{1}{2} \times 54 \cdot 1 \times \left(1\,250 \times \dfrac{2\pi}{60}\right)^2 - \frac{1}{2} \times (54 \cdot 1 + 68 \cdot 3) \times \left(552 \times \dfrac{2\pi}{60}\right)^2$

$= \underline{259\,000 \text{ J}}$ or $\underline{259 \text{ kJ}}$

3. *A multi-plate friction clutch has to be designed to transmit 75 kW from an engine rotating at 2 000 rev/min. The inner and outer diameters are respectively 100 mm and 150 mm, the pressure is to be assumed uniform at 150 kN/m² and μ = 0·25. Determine the necessary end thrust and the number of plates required.*

If this clutch is then used to transmit power from a larger engine to a rotor which has a mass of 1 150 kg and a radius of gyration of 200 mm, determine the time required for this rotor to reach 1 500 rev/min from standstill, assuming that the clutch is transmitting the maximum possible torque.

Power $= T\omega$

i.e. $75 \times 10^3 = T \times 2\,000 \times \dfrac{2\pi}{60}$

$\therefore\ T = 358 \text{ N m}$

$W = \pi p\,(r_1^2 - r_2^2)$ . . . from equation (15.3)

$= \pi \times 150 \times 10^3 \times \dfrac{75^2 - 50^2}{10^6}$

$= 1\,474 \text{ N}$

From equation (15.4), $T = \tfrac{2}{3}\mu W \dfrac{r_1^3 - r_2^3}{r_1^2 - r_2^2} \times n$ where n is the number of contact surfaces

i.e. $358 = \tfrac{2}{3} \times 0 \cdot 25 \times 1\,474 \times \dfrac{75^3 - 50^3}{(75^2 - 50^2) \times 10^3} \times n$

from which $n = 15 \cdot 35$

Therefore 8 plates are required, since each side of each plate is effective.

FRICTION

When used with a larger engine, maximum torque transmissible

$$= \frac{16}{15 \cdot 35} \times 358 = 373 \text{ Nm}$$

Therefore the equation of motion of the rotor is

$$373 = 1\,150 \times 0 \cdot 2^2 \, \alpha$$

$$\therefore \alpha = 8 \cdot 1 \text{ rad/s}^2$$

$$\therefore t = \frac{1\,500 \times \frac{2\pi}{60}}{8 \cdot 1}$$

$$= \underline{19 \cdot 4 \text{ s}}$$

4. *A centrifugal clutch consists of a spider carrying four shoes which are kept from contact with the clutch case until increase in centrifugal force overcomes the spring force and power is transmitted by friction between the shoes and case.*

Determine the mass of each shoe if it is required to transmit 22·5 kW at 750 rev/min with engagement beginning at 75 per cent of the running speed. The inside diameter of the drum is 300 mm and the radial distance of the centre of gravity of each shoe from the shaft axis is 125 mm. $\mu = 0.25$.

$$\text{Power} = T\omega$$

i.e. $$22 \cdot 5 \times 10^3 = T \times 750 \times \frac{2\pi}{60}$$

$$\therefore T = 286 \cdot 5 \text{ Nm}$$

From equation (15.11),

$$T = n\mu r (F - P)$$

At 75% of 750 rev/min, (i.e. 562·5 rev/min)

$$P = F = m \left(562 \cdot 5 \times \frac{2\pi}{60} \right)^2 \times 0 \cdot 125$$

Therefore, at 750 rev/min,

$$T = 4 \times 0 \cdot 25 \times 0 \cdot 15 \times m \left[\left(750 \times \frac{2\pi}{60} \right)^2 - \left(562 \cdot 5 \times \frac{2\pi}{60} \right)^2 \right] \times 0 \cdot 125$$

$$= 50 \cdot 6 m$$

Hence $$m = \frac{286 \cdot 5}{50 \cdot 6} = \underline{5 \cdot 66 \text{ kg}}$$

5. *A piston, 50 mm diameter and 75 mm long, moves vertically in an open-ended lubricated cylinder with a radial clearance of 0·1 mm. When falling due to its own weight, the piston moves through 30 mm in 4·2 s at a uniform velocity. When a mass of 0·05 kg is added to the piston, it moves with uniform velocity through the same distance in 2·4 s. Calculate the viscosity of the oil.*

If the lubricated area is unwrapped, it becomes equivalent to the parallel bearing considered in Art. 15.8(a).

Thus viscous force $= \eta \dfrac{v}{t} a$ . . . . . from equation (15.15)

Let the weight of the piston be W.

Then $\qquad W = \eta \times \dfrac{0\cdot 03/4\cdot 2}{0\cdot 000\ 1} \times \pi \times 0\cdot 05 \times 0\cdot 075 = 0\cdot 841 \eta$

and $\qquad W + 0\cdot 05 \times 9\cdot 81 = \eta \times \dfrac{0\cdot 03/2\cdot 4}{0\cdot 000\ 1} \times \pi \times 0\cdot 05 \times 0\cdot 075 = 1\cdot 472 \eta$

Hence, by subtraction $\qquad 0\cdot 490\ 5 = 0\cdot 631 \eta$

or $\qquad \eta = \underline{0\cdot 777\ \text{N s/m}^2}$

6. *A journal bearing 50 mm diameter and 90 mm long carries a radial load of 4 500 N and the shaft runs at 1 400 rev/min. Calculate the power absorbed in friction.*
    (a) if the bearing is unlubricated and the coefficient of friction is 0·08;
    (b) if the bearing is lubricated with oil of viscosity 0·065 N s/m² and there is a radial clearance of 0·025 mm.

(a) $\qquad$ Power $= \mu W r \omega$ . . . . . . from equation (15.12)

$\qquad\qquad = 0\cdot 08 \times 4\ 500 \times 0\cdot 025 \times \left(1\ 400 \times \dfrac{2\pi}{60}\right)$

$\qquad\qquad = \underline{1\ 317\ \text{W}}$

(b) $\qquad$ Power $= \dfrac{2\pi \eta \omega^2 r^3 l}{c}$ . . . . . from equation (15.17)

$\qquad\qquad = \dfrac{2\pi \times 0\cdot 065 \times \left(1\ 400 \times \dfrac{2\pi}{60}\right)^2 \times 0\cdot 025^3 \times 0\cdot 09}{0\cdot 025 \times 10^{-3}}$

$\qquad\qquad = \underline{494\ \text{W}}$

Alternatively, $\mu = \dfrac{2\pi \eta \omega r^2 l}{cW}$ . . . . . from equation (15.18)

$\qquad\qquad = \dfrac{2\pi \times 0\cdot 065 \times \left(1\ 400 \times \dfrac{2\pi}{60}\right) \times 0\cdot 025^2 \times 0\cdot 09}{0\cdot 025 \times 10^{-3} \times 4\ 500} = 0\cdot 03$

∴ power $= \dfrac{0\cdot 03}{0\cdot 08} \times 1\ 317 = \underline{494\ \text{W}}$

FRICTION

7. *A belt drive transmits power from an electric motor to a machine. The diameter of the pulley on the motor shaft is 150 mm, that on the machine is 200 mm and the centre distance is 600 mm. If the motor speed is 1 440 rev/min and the maximum permissible belt tension is 900 N, then the maximum power transmissible is 6 kW.*

It is required to increase the power transmitted to 6·75 kW using the same pulleys, centre distance and motor speed. The belt material is to be treated with a preparation which increases the coefficient of friction by 10% and in addition, a jockey pulley is to be fitted.

Determine (a) the original coefficient of friction, (b) the new angle of lap.

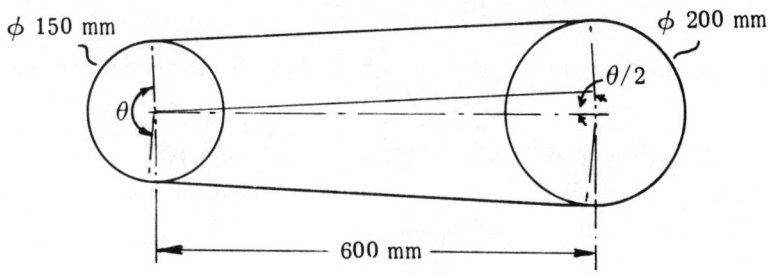

Fig. 15.14

The arrangement of the drive is shown in Fig. 15.14.

$$\cos\frac{\theta}{2} = \frac{\tfrac{1}{2}(200 - 150)}{600} = 0.041\,67$$

i.e. angle of lap on smaller pulley,

$$\theta = 175°\,14' = 3.055 \text{ rad}$$

$$v = 1\,440 \times \frac{2\pi}{60} \times 0.075 = 11.32 \text{ m/s}$$

$$\text{Power} = T_1\left(1 - \frac{1}{e^{\mu\theta}}\right)v \quad \ldots \quad \text{from equation (15.26)}$$

i.e. $6 \times 10^3 = 900\left(1 - \dfrac{1}{e^{3.055\mu}}\right) \times 11.32$

from which $e^{3.055\mu} = 2.43$

$$\therefore \mu = \underline{0.29}$$

New value of $\mu = 1.1 \times 0.29 = 0.319$

$\therefore 6.75 \times 10^3 = 900\left(1 - \dfrac{1}{e^{0.319\theta}}\right) \times 11.32$

from which $e^{0.319\theta} = 0.662$

$$\therefore \theta = 3.39 \text{ rad} = \underline{194°\,30'}$$

8. *An electric motor running at 1 400 rev/min transmits power by three V-belts, each of 320 mm² cross-sectional area, the total groove angle being 45°. The density of the belt material is 1·6 Mg/m³ and the maximum allowable stress in the belts is 2 MN/m². The angle of lap on the motor pulley is 145° and $\mu = 0·2$. Calculate the maximum power which can be transmitted and the diameter of the motor pulley.*

$$\text{Mass of 1 m of belt} = \frac{320}{10^6} \times 1·6 \times 10^3 = 0·512 \text{ kg}$$

$$T_1 = \frac{320}{10^6} \times 2 \times 10^6 = 640 \text{ N}$$

For maximum power, $T_c = mv^2 = \dfrac{T_1}{3} = 213$ N   from equation (15.32)

$$\therefore \text{ velocity for maximum power} = \frac{T_c}{\sqrt{m}} = \frac{213}{\sqrt{0·512}}$$

$$= 20·4 \text{ m/s}$$

$$\text{Effective coefficient of friction} = \mu \operatorname{cosec} \beta \quad . \text{ from Art.}(15.12)$$

$$= 0·2 \operatorname{cosec} 22\tfrac{1}{2}°$$

$$= 0·522\,6$$

$$\text{Angle of lap} = 145° \times \frac{\pi}{180} = 2·53$$

$$\therefore \ e^{\mu\theta \operatorname{cosec} \beta} = e^{0·522\,6 \times 2·53}$$

$$= 3·76$$

$$\text{Power transmitted by 3 belts} = (T_1 - T_c)\left(1 - \frac{1}{e^{\mu\theta \operatorname{cosec} \beta}}\right) v \times 3$$

from equation (15.31)

$$= (640 - 213)\left(1 - \frac{1}{3·76}\right) \times 20·4 \times 3$$

$$= 19\,200 \text{ W or } \underline{19·2 \text{ kW}}$$

$$\text{Angular speed of motor pulley, } \omega = 1\,400 \times \frac{2\pi}{60}$$

$$= 146·6 \text{ rad/s}$$

$$\therefore \ d = \frac{2v}{\omega} = 2 \times \frac{20·4}{146·6}$$

$$= \underline{0·278 \text{ m}}$$

FRICTION

9. *A belt drive connects two pulleys A and B, the centres of which are 4 m apart. The belt has a mass of 1·15 kg/m. Pulley A is 1 m diameter, has a mass of 25 kg and a radius of gyration of 420 mm. Pulley B is 0·5 m diameter, has a mass of 18 kg and a radius of gyration of 225 mm. When at rest, the tension in the belt is 700 N. Assuming that the belt obeys Hooke's Law, determine the tensions in the two portions of the belt between the pulleys when 1·5 kW is being transmitted, the speed of A being 180 rev/min.*

Find also the kinetic energy of the belt and pulleys under these conditions.

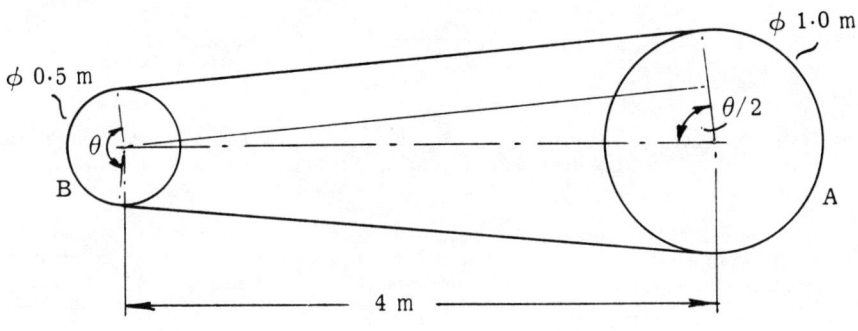

Fig. 15.15

In Fig. 15.15, $\quad \cos\dfrac{\theta}{2} = \dfrac{½(1 - 0·5)}{4} = 0·0625$

i.e. angle of lap on smaller pulley, $\theta = 172°50' = 3·02$ rad

$$v = \omega r = 180 \times \dfrac{2\pi}{60} \times 0·5 = 9·425 \text{ m/s}$$

Also $\quad T_1 + T_2 = 2T_0 = 1400$ N $\quad$ . . from equation (15.33)

$$\therefore T_2 = 1400 - T_1$$

Power $= (T_1 - T_2)v$ $\quad$ . . from equation (15.26)

i.e. $\quad 1500 = (T_1 - [1400 - T_1]) \times 9·425$

from which $\quad T_1 = \underline{779·6 \text{ N}}$

$$T_2 = 1400 - 779·6 = \underline{620·4 \text{ N}}$$

Angle of lap on larger pulley $= 2\pi - 3·02 = 3·26$ rad

$\therefore$ length of belt $= 0·25 \times 3·02 + 0·5 \times 3·26 + 2\sqrt{4^2 - 0·25^2}$

$\quad = 10·365$ m

$\therefore$ mass of belt $= 1·15 \times 10·365 = 11·93$ kg

$\therefore$ K.E. of belt $= ½ \times 11·93 \times 9·425^2 = \underline{529 \text{ J}}$

$$\text{Speed of pulley A} = \frac{9 \cdot 425}{0 \cdot 5} = 18 \cdot 85 \text{ rad/s}$$

$$\text{and speed of pulley B} = \frac{9 \cdot 425}{0 \cdot 25} = 37 \cdot 7 \text{ rad/s}$$

$$\therefore \text{ K.E. of pulleys} = \tfrac{1}{2} I_a \omega_a^2 + \tfrac{1}{2} I_b \omega_b^2$$

$$= \tfrac{1}{2} \{ 25 \times 0 \cdot 42^2 \times 18 \cdot 85^2 + 18 \times 0 \cdot 225^2 \times 37 \cdot 7^2 \}$$

$$= \underline{1434 \text{ J}}$$

10. A plate clutch has four pairs of contact surfaces, each of 240 mm external diameter and 120 mm internal diameter. Assuming uniform pressure, find the total spring load pressing the plates together to transmit 25 kW at 1575 rev/min. Take $\mu = 0 \cdot 3$.

If there are six springs each of stiffness 13 kN/m and each of the contact surfaces has worn away by 1·25 mm, what is the maximum power that can be transmitted at the same speed, assuming uniform wear and the same coefficient of friction?
(*Ans.*: 1·355 kN; 10·25 kW)

11. Determine the time required to accelerate a wheel of mass 500 kg and radius of gyration 0·2 m from rest to 250 rev/min through a single plate clutch of internal and external radii 125 mm and 200 mm, taking μ as 0·3 and the spring load as 600 N. Assume uniform pressure. (*Ans.*: 8·81 s)

12. A multi-plate clutch is to transmit 12 kW at 1500 rev/min. The inner and outer radii of the plates are 50 mm and 100 mm respectively and the maximum spring force is limited to 1 kN. If $\mu = 0 \cdot 35$, determine the necessary number of pairs of surfaces, assuming uniform wear.
What will be the necessary axial force? (*Ans.*: 3; 970 N)

13. A shaft running at a steady speed of 175 rev/min drives a countershaft through a single-plate clutch of external and internal diameters 375 mm and 225 mm respectively. The rotating parts on the countershaft have a mass of 350 kg and a radius of gyration of 250 mm. The axial spring load is 450 N and $\mu = 0 \cdot 3$. Determine the time required to reach full speed from rest and the energy dissipated due to clutch slip during this time. (*Ans.*: 9·9 s; 3·67 kJ)

14. A thrust of 30 kN along the axis of a shaft is taken by a pivot bearing consisting of the frustum of a cone. The outer and inner diameters are 200 mm and 100 mm and the semi-angle of the cone is 60°. The shaft speed is 200 rev/min and $\mu = 0 \cdot 02$. Assuming uniform pressure, determine (*a*) the magnitude of this pressure, (*b*) the power absorbed in friction. (*Ans.*: 1·272 MN/m²; 1·13 kW)

15. A cone clutch is required to transmit 30 kW at 1200 rev/min. The mean diameter of the bearing surface is 250 mm and the cone angle is 25°. Assuming that $\mu = 0 \cdot 3$ and the normal pressure is 140 kN/m², determine the axial width of the conical bearing surface and the axial load required. (*Ans.*: 54·1 mm; 1·32 kN)

16. Two co-axial rotors A and B are connected by a single-plate clutch with two pairs of friction surfaces, each of 300 mm external and 220 mm internal diameter. The total spring load pressing the plates together is 700 N. The masses and radii of gyration of A and B are 1100 kg, 200 mm and 800 kg, 350 mm respectively. $\mu = 0 \cdot 3$.

The rotor A is given a speed of 1200 rev/min while B is stationary and the clutch is then engaged. Determine the time for A and B to reach the same speed, the magnitude of that speed and the kinetic energy lost due to clutch slip. Assume constant wear. (*Ans.*: 70 s; 372 rev/min; 240 kJ)

FRICTION

17. A centrifugal clutch is fitted to a motor shaft to enable it to start without load. As the speed rises, the shoes move radially outward under centrifugal force against the inward pull of springs and press against the inner surface of a pulley. They then take up the drive by friction.

There are four shoes, each of mass 1·35 kg and the centre of gravity of each shoe is at 100 mm radius when it is just touching the pulley, which has a radius of 125 mm. Each spring then exerts a pull of 1·2 kN. $\mu = 0·25$.

Calculate the speed at which the shoes first touch the drum and the power which can be transmitted at 1 200 rev/min. (*Ans*.: 900 rev/min; 14·6 kW)

18. A centrifugal clutch has four blocks which slide radially in a spider keyed to the driving shaft and make contact with the internal surface of a drum keyed to the driven shaft. When the clutch is stationary, each block is pulled against a stop by a spring, leaving a radial clearance of 6 mm between the block and drum. The spring pull is then 450 N and the centre of gravity of each block is 200 mm from the axis of the clutch.

If the internal diameter of the drum is 500 mm, the mass of each block is 7 kg, the stiffness of each spring is 35 kN/m and $\mu = 0·3$, find the maximum power the clutch can transmit at 500 rev/min. (*Ans*.: 51·7 kW)

19. A bearing consists of a cylinder of outside diameter 150·00 mm which rotates at 100 rev/min in a fixed co-axial hollow cylinder of inside diameter 150·50 mm. The axial length of the surfaces is 0·3 m and the radial clearance is filled with oil of viscosity 0·12 N s/m². Calculate the power dissipated. (*Ans*.: 41·5 W)

20. A uniform film of oil 0·1 mm thick separates two 0·1 m diameter discs, one of which has a 10 mm diameter hole at the centre. Calculate the power required to rotate one disc at 5 rev/s relative to the other if the viscosity of the oil is 0·14 kg/m s. (*Ans*.: 13·57 W)

21. The thrust at the lower end of a vertical shaft is supported by a bearing consisting of a flat disc 100 mm diameter rotating with the shaft and a stationary housing. The disc is separated from the housing by a film of oil 0·25 mm thick and of viscosity 0·13 N s/m². Calculate the power absorbed in friction when the shaft rotates at 800 rev/min, ignoring the side effects. (*Ans*.: 35·7 W)

22. A ship is pulled through a lock by means of a capstan and rope. The capstan has a diameter of 0·5 m and turns at 30 rev/min. The rope makes three complete turns round the capstan and a pull of 100 N is applied at the free end. $\mu = 0·25$. Find (*a*) the pull on the ship, (*b*) the power required to drive the capstan.
(*Ans*.: 11·1 kN; 8·65 kW)

23. A pulley is driven by a flat belt, the angle of lap being 120°. The belt is 100 mm wide by 6 mm thick and has a mass of 1 Mg/m³. If $\mu = 0·3$ and the maximum stress in the belt is limited to 1·5 MN/m², find the greatest power the belt can transmit and the corresponding speed of the belt. (*Ans*.: 6·265 kW; 22·36 m/s)

24. Power is transmitted by a rope drive between two shafts 4·5 m apart. The pulleys are 3 m and 2 m diameter and the groove angle is 40°. If the rope has a mass of 4 kg/m and the maximum tension is 20 kN, determine the maximum power which the rope can transmit and the corresponding speed of the smaller pulley. $\mu = 0·2$. (*Ans*.: 446 kW; 390 rev/min)

25. Power is transmitted from a shaft rotating at 250 rev/min by five ropes running in grooves in the periphery of a wheel of effective diameter 1·65 m. The groove angle is 50° and the arc of contact is 180°. The maximum permissible load in each rope is 900 N and its mass is 0·55 kg/m. $\mu = 0·3$. What power can be transmitted under these conditions? (*Ans*.: 62 kW)

26. A 4 to 1 speed reduction between two shafts at 2 m centres is provided by four V-belts running on pulleys mounted on the shafts. The effective diameter of the driving pulley is 350 mm and it runs at 740 rev/min. The groove angle is 40° and each belt has a mass of 0·45 kg/m. $\mu = 0·28$. Determine the power that can be transmitted if the tension in each belt is not to exceed 800 N. (*Ans*.: 34·3 kW)

27. The drive from an electric motor to a shaft is by three parallel V-belts. The mean diameter of the motor pulley is 150 mm, the motor and shaft speeds are respectively 1 600 and 400 rev/min and the shafts are 1 m apart. Each belt has a mass of 1·4 Mg/m^3 and a cross-sectional area of 800 mm^2; the groove angle is 30° and $\mu = 0\cdot13$.

Find the power that can be transmitted by the drive if the tensile stress in the belt is not to exceed 8 MN/m^2. *(Ans.:* 173·6 kW)

28. A rope drive is required to transmit power from a pulley of 1·15 m effective diameter. The ropes each have a mass of 1·2 kg/m, the groove angle is 50° and the angle of lap is 170°. $\mu = 0\cdot3$.

(*a*) If the initial tension in each rope is 900 N, what is the maximum power which can be transmitted per rope? What will then be the load in the rope and its linear speed?

(*b*) If the permissible load is 1·6 kN per rope and this is to be utilized for maximum power, what then should be the initial tension in the rope?

(Ans.: 14 kW; 1·273 kN; 18·8 m/s; 1·13 kN)

16 Gyroscopic motion

16.1 Introduction Whenever the axis of a rotating body is caused to change direction, a couple is required, called a *gyroscopic couple*. The reaction to this couple is experienced in such cases as aircraft, marine and car engines when changing direction and gyroscopic effects may be usefully employed in navigating systems.

16.2 Gyroscopic couple Fig. 16.1 shows a rotor of polar moment of inertia I rotating at a rate ω about the axis OX, called the *spin* axis. The angular momentum, $I\omega$, is represented by the vector op, its direction being given by the forward movement of a corkscrew when turned in the direction of rotation.

If the spin axis is now rotated, or *precessed*, about the perpendicular axis OY at a rate Ω then, after a time dt, the vector op has moved to oq. The change of momentum is represented by oq and if the angle poq is dθ, then

$$pq = I\omega\,d\theta$$

Thus, rate of change of momentum $= I\omega \dfrac{d\theta}{dt}$

i.e. gyroscopic couple, $C = I\omega\Omega$ \hfill (16.1)

By applying the corkscrew rule to the vector pq, it will be seen that the couple required to cause precession is clockwise looking along axis OZ, mutually perpendicular to OX and OY.

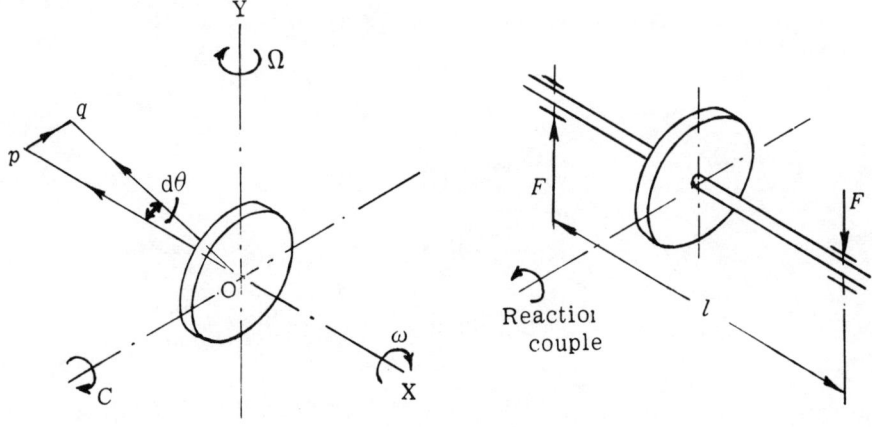

Fig. 16.1   Fig. 16.2

The reaction couple, which is supported by the shaft bearings, is equal and opposite to the gyroscopic couple. If the force at the bearings is F and the distance between the bearings is l, Fig. 16.2, then

$$F = \frac{I\omega\Omega}{l} \tag{16.2}$$

1. *A pair of locomotive driving wheels and axle have a moment of inertia of 400 kg m². The diameter of the wheels is 2 m and the distance between the wheel centres is 1·5 m. When the locomotive is travelling at 100 km/h, defective ballasting causes one wheel to fall 10 mm and rise again in a total time of 0·1 s. If the movement of the wheel takes place with simple harmonic motion, find the gyroscopic reaction on the locomotive.*

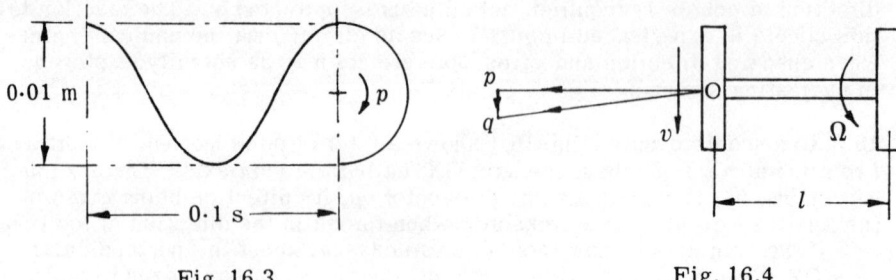

Fig. 16.3   Fig. 16.4

Referring to Fig. 16.3, the S.H.M. may be considered to be generated by a vector rotating at a uniform angular velocity p, (see Art. 18.2).

Periodic time of S.H.M. = 0·1 s

$$\therefore \; p = \frac{2\pi}{0\cdot 1} \text{ rad/s}$$

Amplitude of S.H.M., $a = 0\cdot 005$ m

Therefore maximum vertical velocity of wheel when descending,

$$v = pa = \frac{2\pi}{0\cdot 1} \times 0\cdot 005 = \frac{\pi}{10} \text{ rad/s}$$

Therefore maximum velocity of precession

$$\Omega = \frac{v}{l} = \frac{\pi}{10 \times 1\cdot 5} = \frac{\pi}{15} \text{ rad/s}$$

Spin velocity, $\omega = \dfrac{\text{velocity of loco}}{\text{wheel radius}}$

$$= \frac{100/3\cdot 6}{1} = \frac{100}{3\cdot 6} \text{ rad/s}$$

Therefore gyroscopic couple, $C = I\omega\Omega$ from equation (16.1)

$$= 400 \times \frac{100}{3\cdot 6} \times \frac{\pi}{15}$$

$$= \underline{2\,330 \text{ N m}}$$

GYROSCOPIC MOTION

Fig. 16.4 shows a view of the back of the wheels. If the left-hand wheel falls, the angular momentum vector op moves to oq, and the change of angular momentum, pq, shows that a clockwise couple looking downwards is required to cause precession. The reaction couple is opposite to this, i.e., anticlockwise downwards, and this tends to cause the locomotive to swing to the left.

2. *An electric motor on board ship is arranged with its rotor athwart the ship. Find the maximum load on its bearings due to gyroscopic action if the ship rolls with simple harmonic motion 30° on each side of the vertical and the time for one complete roll is 4 s. The mass of the rotor is 200 kg, its radius of gyration is 240 mm, the bearings are 1·2 m apart and the rotor speed is 3 000 rev/min, clockwise viewed from the starboard side.*

Fig. 16.5 shows a view from astern of the ship.

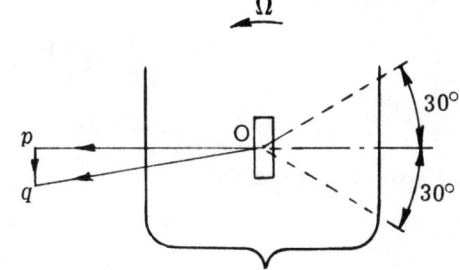

Fig. 16.5

Periodic time of S.H.M. = 4 s

Therefore angular velocity of vector generating S.H.M.,

$$p = \frac{2\pi}{4} \text{ rad/s}$$

Angular amplitude, $\phi = 30° = \frac{\pi}{6}$ rad

Therefore maximum angular velocity during rolling = maximum velocity of precession,

$$\Omega = p\phi = \frac{2\pi}{4} \times \frac{\pi}{6} = \frac{\pi^2}{12} \text{ rad/s}$$

Velocity of spin, $\omega = 3\,000 \times \frac{2\pi}{60} = 100\pi$ rad/s

Therefore gyroscopic couple, $C = I\omega\Omega$ . . . from equation (16.1)

$$= 200 \times 0\cdot 24^2 \times 100\pi \times \frac{\pi^2}{12}$$

$$= 2\,975 \text{ N m}$$

Therefore force on bearings, $F = \dfrac{I\omega\Omega}{l}$ . . . from equation (16.2)

$$= \frac{2\,975}{1\cdot 2} = \underline{2\,480 \text{ N}}$$

If the ship heels to port, the angular momentum vector moves from op to oq and hence a clockwise moment looking downwards must be applied by the shaft bearings to cause precession. The reaction couple is therefore anticlockwise looking downwards and tends to turn the ship to port.

3. *A truck with four wheels, each 750 mm diameter, travels on rails round a curve of 75 m radius at a speed of 50 km/h. The rails lie in a horizontal plane and are 1·4 m apart. The total mass of the truck is 5 t and its centre of gravity is midway between the axles, 1·2 m above the rails and midway between them. The moment of inertia of each pair of wheels is 15 kg m². Determine the load on each rail.*

Let v = velocity of truck,

r = radius of wheels

and R = radius of curve.

Then, referring to Fig. 16.6,

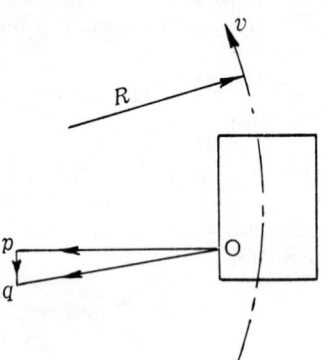

centrifugal force, $F = m\dfrac{v^2}{R}$

$= \dfrac{5 \times 10^3}{75}\left(\dfrac{50}{3\cdot 6}\right)^2$

$= 12\,850$ N

∴ overturning couple $= F \times h$

$= 12\,850 \times 1\cdot 2$

$= 15\,420$ N m

Fig. 16.6

Angular velocity of wheels,

$$\omega = \dfrac{v}{r} = \dfrac{50}{3\cdot 6 \times 0\cdot 375} = 37 \text{ rad/s}$$

Velocity of precession,

$$\Omega = \dfrac{v}{R} = \dfrac{50}{3\cdot 6 \times 75} = 0\cdot 185 \text{ rad/s}$$

Therefore gyroscopic couple due to wheel precession

$= I\omega\Omega$ . . . . . . from equation (16.1)

$= 2 \times 15 \times 37 \times 0\cdot 185 = \underline{205 \text{ N m}}$

As the truck turns, the angular momentum vector op turns to oq, so that a couple represented by pq is required to cause precession of the wheels. This is clockwise looking backwards, so that the reaction couple is clockwise looking forwards, tending to overturn the truck outwards in the same way as centrifugal force.

Therefore total overturning couple $= 15\,420 + 205 = 15\,625$ N m

Reaction at rails due to overturning couple $= \dfrac{15\,625}{1\cdot 4} = 11\,150$ N

This increases the load on the outer rail and reduces that on the inner rail.

GYROSCOPIC MOTION   201

Load on each rail due to dead weight $= \dfrac{mg}{2}$

$$= \dfrac{5 \times 10^3 \times 9\cdot 81}{2} = 24\,550 \text{ N}$$

Therefore resultant load on outer rail = 24 550 + 11 150 = <u>35 700 N</u>

and resultant load on inner rail = 24 550 − 11 150 = <u>13 400 N</u>

4. *A motor cyclist travels at 140 km/h round a curve of 120 m radius. The cycle and rider have a mass of 150 kg and their centre of gravity is 0·7 m above ground level when the machine is vertical. Each wheel is 0·6 m diameter and the moment of inertia about its axis of rotation is 1·5 kg m². The engine has rotating parts whose moment of inertia about their axis of rotation is 0·25 kg m² and it rotates at five times the wheel speed in the same direction.*

Find (a) the angle of banking so that there will be no tendency to side slip, (b) the angle of inclination of the cycle and rider to the vertical.

Let v = velocity of cycle,
 r = radius of wheels
and R = radius of curve.

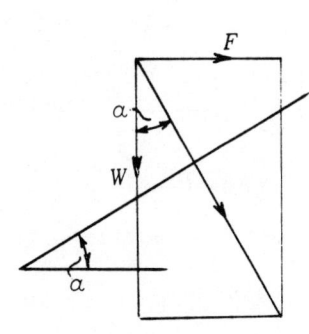

Fig. 16.7   Fig. 16.8

Front view

The forces acting on the cycle, Fig. 16.7, are

(i) its weight, $W = mg = 150 \times 9\cdot 81 = 1\,472$ N

(ii) the centrifugal force, $F = m\dfrac{v^2}{R}$

$$= \dfrac{150}{120}\left(\dfrac{140}{3\cdot 6}\right)^2 = 1\,890 \text{ N}$$

For no tendency to sideslip, the resultant force must be normal to the track,

i.e.
$$\tan\alpha = \frac{F}{W} = \frac{1890}{1472} = 1.284$$

$$\therefore \alpha = \underline{52°5'}$$

Angular velocity of wheels $= \dfrac{v}{r} = \dfrac{140}{3.6 \times 0.3} = 129.6$ rad/s

Angular velocity of engine $= 5 \times 129.6 = 648$ rad/s

If β is the inclination of the cycle to the vertical, Fig. 16.8, the velocity of precession, perpendicular to the axis of spin of the wheels and engine, is given by

$$\Omega = \frac{v}{r}\cos\beta = \frac{140}{3.6 \times 120}\cos\beta = 0.324\cos\beta$$

Therefore gyroscopic couple due to precession of wheels

$$= 2 \times 1.5 \times 129.6 \times 0.324\cos\beta = 125.9\cos\beta$$

and gyroscopic couple due to precession of engine

$$= 0.25 \times 648 \times 0.324\cos\beta = 52.5\cos\beta$$

Since the directions of rotation of the wheels and engine are the same,

total gyroscopic couple, $C = (125.9 + 52.5)\cos\beta = 178.4\cos\beta$ N m

As in example 3, the gyroscopic reaction tends to overturn the cycle outwards. Therefore, taking moments about the point of contact with the track, O,

$$C + F \times h\cos\beta = W \times h\sin\beta$$

i.e. $\quad 178.4\cos\beta + 1890 \times 0.7\cos\beta = 1472 \times 0.7\sin\beta$

from which $\quad\tan\beta = 1.458$

$$\therefore \beta = \underline{55°33'}$$

---

5. One of the driving axles of a locomotive, with its two wheels, has a moment of inertia of 350 kg m². The wheels are 1·85 m diameter and the distance between the planes of the wheels is 1·5 m. When travelling at 100 km/h, the locomotive passes over a defective rail which causes the right-hand wheel to fall 12 mm and rise again in a total time of 0·1 s, the vertical movement of the wheel being with S.H.M.

Find the maximum gyroscopic torque caused and state the effect this has on the locomotive. (*Ans.*: 2·64 kN m; loco tends to swing to right)

6. A generator is arranged on board ship with its axis parallel to the longitudinal centre-line of the ship. The revolving parts have a mass of 1 400 kg, a radius of gyration of 400 mm and revolve at 420 rev/min.

Find the magnitude and sense of the gyroscopic couple exerted on the ship when it steams at 36 km/h round a curve of 180 m radius. (*Ans.*: 547 N m; tends to lift bow when turning left if revolving clockwise looking forward)

GYROSCOPIC MOTION

7. The turbine rotor of a ship has a mass of 30 t, a radius of gyration of 0·6 m and rotates at 2 400 rev/min in a clockwise direction when viewed from aft. The ship pitches through a total angle of 15°, the motion being simple harmonic with a periodic time of 12 s.

Determine the maximum gyroscopic couple on the ship and its effect as the bow rises. *(Ans.:* 186·5 kN m; ship turns to starboard)

8. A car travels round a bend of 100 m radius. Each of the four wheels has a moment of inertia of 1·6 kg m² and a diameter of 0·6 m. The rotating parts of the engine have a moment of inertia of 0·85 kg m², the engine is parallel to the axles and the crankshaft rotates in the same sense as the wheels at three times the speed. The car has a mass of 1 400 kg and its centre of gravity is 0·45 m above road level. The width of the track of the car is 1·5 m.

Determine the limiting speed round the curve if the wheels are not to leave the road surface, which is horizontal. *(Ans.:* 142 km/h)

9. A diesel-electric locomotive has two axles 4 m apart. The wheels on each axle are 1·2 m diameter and 1·5 m apart and the moment of inertia of each axle is 70 kg m². The rotating parts of the engine and generator have a moment of inertia of 60 kg m² and they rotate about the longitudinal axis of the locomotive at 2 200 rev/min in an anticlockwise direction when looking forward. When the speed of the locomotive is 40 km/h, it enters a left-hand curve of 150 m radius. Find the change in the vertical reactions on each wheel due to gyroscopic action.
(Ans.: front outer, +192 N; front inner, +64 N; rear outer, −64 N; rear inner, −192 N)

10. A car has a mass of 1·5 t and the centre of gravity is 0·95 m above road level. The moment of inertia of the two front wheels is 10 kg m² and that of the two rear wheels is 15 kg m². The moment of inertia of the rotating parts of the engine is 2 kg m² and the gear ratio from engine to wheels is 10 to 1. The engine rotates in a clockwise direction when viewed from the front of the car. The wheel diameter is 0·64 m and when the car is travelling at 80 km/h, it enters a right-hand curve of 150 m radius.

Determine the magnitude and sense of the couples on the car due to (*a*) centrifugal effects, (*b*) gyroscopic effects of wheel rotation, (*c*) gyroscopic effects of engine rotation. *(Ans.:* 4·65 kN m, tending to overturn car outwards; 257 N m, tending to overturn car outwards; 205 N m, tending to lift nose)

11. A motor cycle and rider have a total mass of 320 kg, with the centre of gravity 0·525 m above ground level. The wheels each have a mass of 9 kg, a radius of gyration of 0·225 m and a rolling radius of 0·3 m. The rotating parts of the engine have a mass of 12 kg, a radius of gyration of 0·075 m and rotate in the same sense as the road wheels at 3·5 times the speed.

The machine travels round a banked curve of 60 m radius at 160 km/h. Determine the angle of banking necessary for the cycle to ride normal to the track.
(Ans.: 73°45′ to horizontal)

12. A solo motor cycle, complete with rider, has a mass of 225 kg, the centre of gravity being 0·6 m above ground level. The moment of inertia of each road wheel is 1 kg m² and the rolling diameter is 0·6 m. The engine crankshaft rotates, in the same sense as the wheels, at 5 times the speed of the wheels. The rotating parts of the engine are equivalent to a flywheel whose moment of inertia is 0·2 kg m².

Determine the heel-over angle required when the unit is travelling at 100 km/h in a curve of radius 60 m. *(Ans.:* 54° 36′)

13. A gyroscope turn indicator for an aeroplane consists of a uniform disc, 50 mm diameter, which rotates at 3 000 rev/min. Its bearings are carried in a frame which is free to turn in trunnions, the centre-line of the trunnions being at right angles to, and 5 mm above, the axis of rotation of the disc. On a straight course, the plane of the disc is vertical and at right angles to the fore-and-aft centre-line of the plane. Find the angle through which the frame will tilt when the speed is 240 km/h and the course is altered to a circular arc of 300 m radius. *(Ans.:* 24°)

17 Gear trains

17.1 Introduction Gears are used to transmit rotary motion from one shaft to another. The shafts may be parallel or inclined to one another and their speed ratio is determined by the numbers of teeth on the gears. Spur gears have teeth which are parallel with the shaft axis but helical gears have teeth which are cut on a helix; this gives a smoother and quieter drive as the engagement of mating teeth is gradual instead of instantaneous. The force on the helical tooth has a component in the axial direction which must be resisted by a thrust bearing; alternatively, double helical gears may be used in which the axial forces in the two parts balance each other.

17.2 Gear teeth definitions Fig. 17.1 shows two gears in mesh; the smaller one is termed the *pinion* and larger one the *wheel* or *spur*.

The *pitch circle diameters* (p.c.d.) are the diameters of discs which would transmit the same velocity ratio by friction as the gear wheels and the pitch point is the point of contact of the two pitch circles.

Let the numbers of teeth on the pinion and wheel be respectively t and T, the pitch circle diameters d and D and the speeds ω and Ω.

Then
$$\frac{\Omega}{\omega} = \frac{d}{D} = \frac{t}{T} \qquad (17.1)$$

The *circular pitch* (p) is the distance between a point on one tooth and the corresponding point on the next tooth, measured along the pitch circle,

i.e.
$$p = \frac{\pi d}{t} = \frac{\pi D}{T} \qquad (17.2)$$

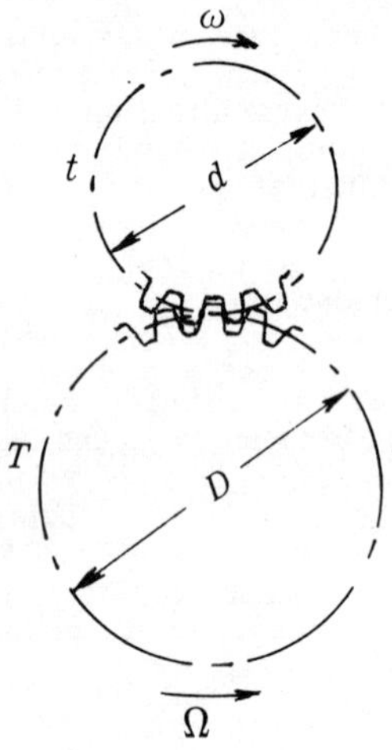

Fig. 17.1

The *diametral pitch* (P) is the number of teeth per mm of p.c.d.,

i.e.
$$P = \frac{t}{d} = \frac{T}{D} = \frac{\pi}{p} \qquad (17.3)$$

The *module* (m) is the number of millimetres of p.c.d. per tooth,

i.e.
$$m = \frac{d}{t} = \frac{D}{T} = \frac{1}{P} \qquad (17.4)$$

The *centre distance* $= \dfrac{d + D}{2} \qquad (17.5)$

GEAR TRAINS

Gear teeth are normally of involute profile, the involute curves being generated from a *base circle* and the angle which the generator makes with the common tangent to the pitch circles is termed the *pressure angle* (ψ) since the generator is the line of contact between the mating teeth, Fig. 17.2.

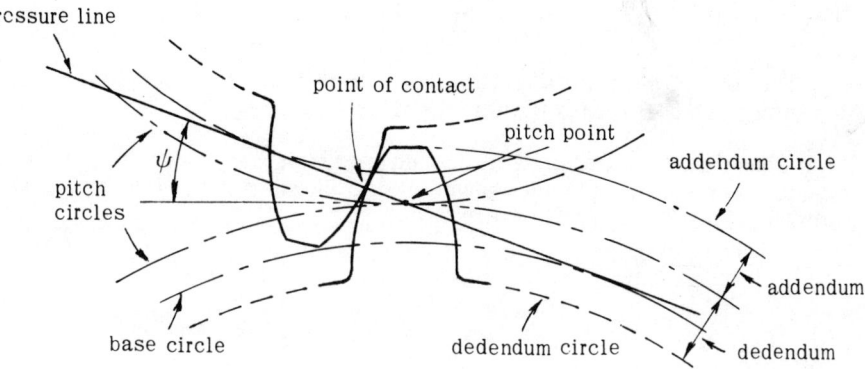

Fig. 17.2

The radial depth of the tooth above the pitch circle is termed the *addendum* and that below the pitch circle the *dedendum*. The diameter of the dedendum circle is smaller than that of the base circle to give clearance to the tips of the mating teeth and that part of the tooth profile below the base circle is non-involute.

The *working depth* is the sum of the addenda of two mating teeth.

Standard gear teeth proportions are:

$$\text{addendum} = 1/P = m$$
$$\text{dedendum} = 1\cdot 25/P = 1\cdot 25\,m$$
$$\text{pressure angle} = 20°$$

17.3 Simple gear trains A simple gear train is one in which all the gears are mounted on separate shafts. Fig. 17.3(a) shows a train of two wheels and Fig. 17.3(b) shows a train of three wheels. In the first case, the driving and driven shafts rotate in opposite directions and in the second case, they rotate in the same direction

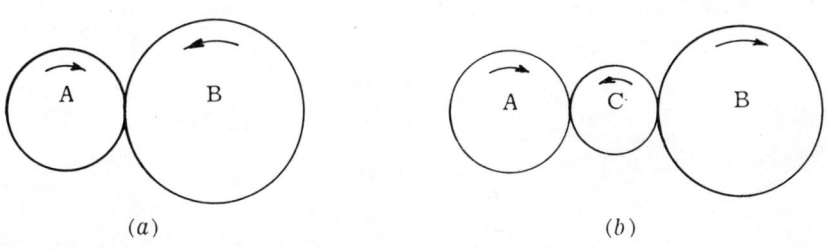

Fig. 17.3

Let N_a and N_b be the speeds of rotation of wheels A and B respectively. Then, allowing for changes in direction of rotation,

$$\frac{N_a}{N_b} = -\frac{T_b}{T_a} = -\frac{D_b}{D_a} \quad \text{in the first case}$$

and
$$\frac{N_a}{N_b} = \frac{T_b}{T_a} = \frac{D_b}{D_a} \quad \text{in the second case}$$

The idler C does not affect the velocity ratio between the driving and driven shafts but reverses the direction.

17.4 Compound gear trains A compound gear train has more than one wheel mounted on a shaft, a simple example being shown in Fig. 17.4(a). For this train, allowing for changes of direction,

$$\frac{N_a}{N_b} = -\frac{T_b}{T_a} \quad \text{and} \quad \frac{N_c}{N_d} = -\frac{T_d}{T_c}$$

Hence
$$\frac{N_a N_c}{N_b N_d} = \frac{N_a}{N_d} = \frac{T_b T_d}{T_a T_c}$$

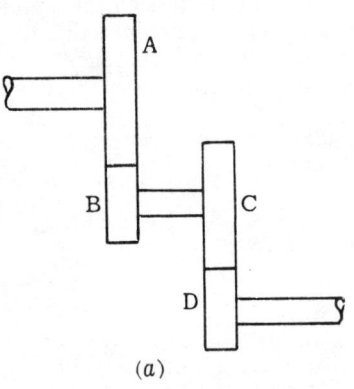
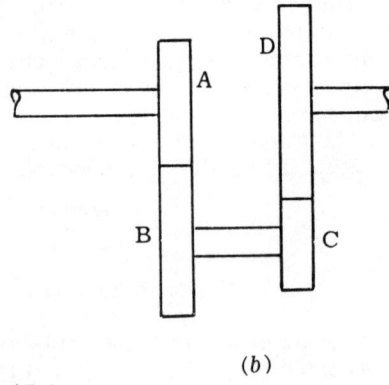

(a) (b)

Fig. 17.4

If the input and output shafts are mounted co-axially, as shown in Fig. 17.4(b), this is referred to as a co-axial or reverted train.

The gear ratio

$$\frac{N_a}{N_d} = \frac{T_b T_d}{T_a T_c}$$

as before but the equation

$$D_a + D_b = D_c + D_d$$

must also be satisfied to give equal centre distances.

GEAR TRAINS

If the pitch of each pair of wheels is the same, then

$$T_a + T_b = T_c + T_d$$

but the pitches of the two pairs may be made different to achieve a required velocity ratio and centre distance. Thus, if the modules of A and B are m_1 and those of C and D are m_2, then

$$(T_a + T_b)m_1 = (T_c + T_d)m_2 \qquad (17.6)$$

17.5 Epicyclic gear trains An alternative method of obtaining concentric motion of input and output shafts is to use an epicyclic gear train. This design can give a high velocity ratio within a small volume, leading to considerable saving in weight and it also has the capacity of giving a change of velocity ratio by locking or freeing one member of the gear train.

Fig. 17.5 shows a simple epicyclic train in which A is an annular wheel having internal teeth, P is a planet wheel, L is an arm, star or spider carrying pins on which the planets can rotate freely and S is a sun wheel which rotates about the same axis as A. The input and output shafts may be connected to any two of the sun, annulus or arm.

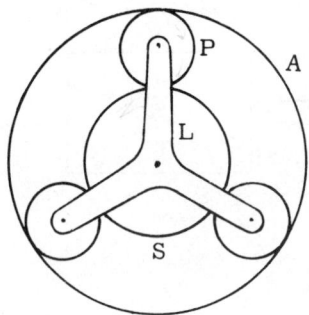

Fig. 17.5

Solution of an epicyclic gear problem is one of relative motion and a simple tabular method is used which involves three steps.

Number of revs. of wheels

		L	A	P	S
(i)	Rotate whole gear train $+m$ revs	$+m$	$+m$	$+m$	$+m$
(ii)	Fix arm and rotate annulus $+n$ revs	0	$+n$	$+n \cdot \dfrac{T_a}{T_p}$	$-n \cdot \dfrac{T_a}{T_s}$
(iii)	Add lines (ii) and (iii)	m	$m+n$	$m+n \cdot \dfrac{T_a}{T_p}$	$m-n \cdot \dfrac{T_a}{T_s}$

Line (iii) gives a general relation between the speeds of the various members and the constants m and n can then be solved from the given speeds of those members.

In many applications, one of the wheels is fixed so that the speed of that wheel is zero. This can be simply achieved by giving all members +1 rev in line (i) and then giving that fixed wheel −1 rev in line (ii) while the arm is held fixed. The resultant speed of the fixed wheel in line (iii) will then be zero.

If, as an example, the annulus is fixed, then, in tabular form:

		L	A	P	S
(i)	Rotate whole gear train +1 rev	+1	+1	+1	+1
(ii)	Fix arm and rotate annulus, −1 rev	0	−1	$-\dfrac{T_a}{T_p}$	$+\dfrac{T_a}{T_s}$
(iii)	Add lines (i) and (ii)	1	0	$1-\dfrac{T_a}{T_p}$	$1+\dfrac{T_a}{T_s}$

17.6 Torques in gear trains Fig. 17.6 shows a gear train in which A is the input shaft, B is the output shaft and C is the casing, usually connected to the annulus.

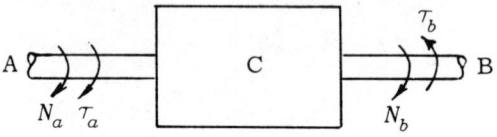

Fig. 17.6

If T_a, T_b and T_c are the external torques applied *to* the train, then T_a is in the *same* direction as the rotation N_a and T_b is in the *opposite* direction to the rotation N_b.

If there is no loss of power in the gears, the net power supplied to the train is zero.

i.e.
$$T_a N_a + T_b N_b = 0 \qquad (17.6)$$

If the direction of N_a is taken as positive, then either N_b or T_b will be negative.

The magnitude and direction of the torque on C will be determined from the equilibrium of the whole train, which is represented by the equation

$$T_a + T_b + T_c = 0 \qquad (17.7)$$

GEAR TRAINS 209

1. *Two shafts A and D in the same line are geared together through an intermediate parallel shaft carrying wheels B and C which mesh with wheels on A and D respectively. Wheels A and B have a module of 4 and wheels C and D have a module of 9. The number of teeth on any wheel is to be not less than 15, the speed of D is to be about but not greater than 1/12 the speed of A and the ratio of each reduction is the same. Find suitable wheels, the actual reduction and the distance of the intermediate shaft from shafts A and D.*

The arrangement of the gears is as shown in Fig. 17.4(b).

From equation (17.6), $(T_a + T_b)m_1 = (T_c + T_d)m_2$

i.e. $$T_a + T_b = 2 \cdot 25(T_c + T_d) \quad (1)$$

$$\frac{T_a}{T_b} = \frac{T_c}{T_d} \approx \frac{1}{\sqrt{12}}$$

$$\approx 0 \cdot 2885 \quad (2)$$

From equations (1) and (2) $T_b = 2 \cdot 25\, T_d \quad (3)$

and $T_a = 2 \cdot 25\, T_c \quad (4)$

$T = \dfrac{D}{m}$ and since m for C and D is 9 compared with 4 for wheels A and B the smallest number of teeth will be on wheel C.

Thus the lowest tooth numbers to satisfy equation (4) is

$$T_c = \underline{16} \quad \text{and} \quad T_a = \underline{36}$$

From equation (2), $T_b \geqslant \dfrac{T_a}{0 \cdot 2855} \geqslant 124 \cdot 8 \quad (5)$

The lowest tooth numbers to satisfy equations (3) and (5) are

$$T_b = \underline{126} \quad \text{and} \quad T_d = \underline{56}$$

$$\text{Actual gear ratio} = \frac{T_a}{T_b} \times \frac{T_c}{T_d}$$

$$= \frac{36}{126} \times \frac{16}{56} = \underline{\frac{1}{12 \cdot 25}}$$

$$\text{Centre distance} = \frac{D_a + D_b}{2}$$

$$= (T_a + T_b) \times \frac{m_1}{2}$$

$$= (36 + 126) \times \frac{4}{2} = \underline{324 \text{ mm}}$$

2. *In the epicyclic gear train shown in Fig. 17.7, A rotates at 1000 rev/min clockwise while E rotates at 500 rev/min anticlockwise. Determine the speed and direction of rotation of the annulus D and the shaft F. All gears are of the same pitch and the numbers of teeth are A, 30, B, 20, and E, 80.*

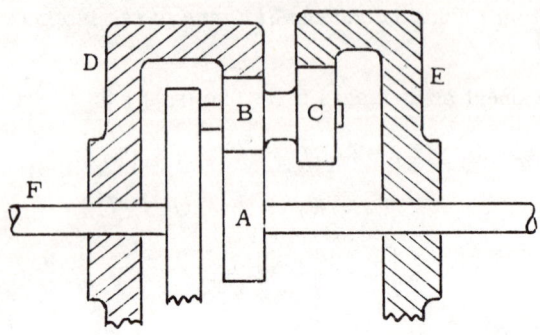

Fig. 17.7

Since diameters are proportional to numbers of teeth and all wheels are of the same pitch,

$$T_d = T_a + 2T_b$$
$$= 30 + 2 \times 20 = 70$$
$$T_e = T_a + T_b + T_c$$

i.e. $$T_c = 80 - 30 - 20 = 30$$

		F	A	D	E	B,C
(i)	Rotate whole gear train $+m$ revs	$+m$	$+m$	$+m$	$+m$	$+m$
(ii)	Fix arm and rotate wheel A $+n$ revs	0	$+n$	$-n \cdot \frac{30}{70}$	$-n \cdot \frac{30}{20} \cdot \frac{30}{80}$	$-n \cdot \frac{30}{20}$
(iii)	Add lines (i) and (ii)	m	$m+n$	$m - \tfrac{3}{7}n$	$m - \tfrac{9}{16}n$	$m - \tfrac{3}{2}n$

Taking clockwise rotation as positive,

$$\text{speed of A} = +1000 = m + n$$

and
$$\text{speed of E} = -500 = m - \tfrac{9}{16}n$$

$$\therefore m = +40 \quad \text{and} \quad n = +960$$

$$\therefore \text{speed of D} = +40 - \tfrac{3}{7} \times 960$$
$$= \underline{-371 \cdot 4 \text{ rev/min}}$$

and
$$\text{speed of F} = \underline{+40 \text{ rev/min}}$$

GEAR TRAINS 211

3. *In the epicyclic gear shown in Fig. 17.8, the driving wheel A has 14 teeth and the fixed annular wheel C, 100 teeth. The ratio of tooth numbers on wheels E and D is 98 : 41. If 2 kW at 1 200 rev/min is supplied to wheel A, find the speed and direction of rotation of E and the fixing torque required at C.*

$$T_b = T_c - T_a$$

$$= \frac{100 - 14}{2}$$

$$= 43$$

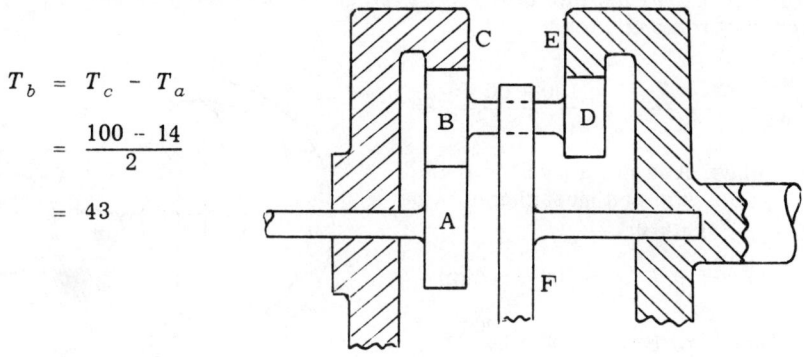

Fig. 17.8

		A	B,D	C	E	F
(i)	Rotate whole gear train +1 rev	+1	+1	+1	+1	+1
(ii)	Fix arm and rotate C −1 rev	$+\frac{100}{14}$	$-\frac{100}{43}$	−1	$-\frac{100}{43} \cdot \frac{41}{98}$	0
(iii)	Add lines (i) and (ii)	8·143	−1·326	0	0·027 1	1
	Multiply by $\frac{1\,200}{8 \cdot 143}$	1 200			3·99	

The speed of E is therefore 3·99 rev/min, in the same direction as A.

$$T_a = \frac{2 \times 10^3}{1\,200 \times \frac{2\pi}{60}} = 15 \cdot 92 \text{ N m}$$

From equation (17.6), $T_a N_a + T_e N_e = 0$

$$\therefore T_e = -\frac{15 \cdot 92 \times 1\,200}{3 \cdot 99}$$

$$= -4\,785 \text{ N m}, \text{ opposite in direction to } T_a.$$

The directions of the torques are similar to those shown in Fig. 17.6 and so, from equation (17.7),

$$T_a + T_c + T_e = 0$$

i.e. $15 \cdot 92 + T_c + 4\,785 = 0$

$$\therefore T_c = 4\,771 \text{ N m}, \text{ in the same direction as } T_a.$$

4. *In the epicyclic gear shown in Fig. 17.9, the gear B has 120 teeth externally and 100 teeth internally. The driver A has 20 teeth and the arm E is connected to the driven shaft. Gear D has 60 teeth. If A revolves at +100 rev/min and D revolves at +27 rev/min, find the speed of the arm, E.*

If D is now fixed and A transmits a torque of 10 N m at +100 rev/min, what will be the available torque on the arm E, assuming 96 per cent efficiency of transmission?

$$T_c = \frac{T_b - T_d}{2} = \frac{100 - 60}{2} = 20$$

Wheel A is not a member of the epicyclic train and must therefore be treated separately.

Speed of A = +100 rev/min

∴ speed of B = $-100 \times \frac{20}{120}$

$= -\frac{100}{6}$ rev/min

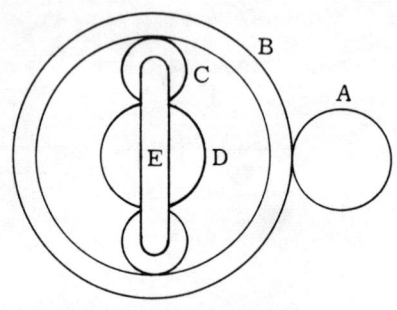

Fig. 17.9

		B	C	D	E
(i)	Rotate whole gear $+m$ rev	$+m$	$+m$	$+m$	$+m$
(ii)	Fix arm and rotate wheel B $+n$ rev	$+n$	$+n \cdot \frac{100}{20}$	$-n \cdot \frac{100}{60}$	0
(iii)	Add lines (i) and (ii)	$m+n$	$m+5n$	$m - {}^{10}\!/_{\!6}n$	m

Speed of B = $-\frac{100}{6} = m + n$

and speed of D = $+27 = m - {}^{10}\!/_{\!6}n$

∴ $m = -0.292$ and $n = -16.4$

∴ speed of E = $m = -0.29$ rev/min

When D is fixed, $m + n = -\frac{100}{6}$

and $m - {}^{10}\!/_{\!6}n = 0$

∴ $m = -\frac{1\,000}{96}$ and $n = -\frac{100}{16}$

∴ speed of E = $m = -\frac{1\,000}{96}$ rev/min

The ideal output torque is given by

$$\tau_a N_a + \tau_e N_e = 0$$

GEAR TRAINS

$$\therefore \quad T_e = \frac{-10 \times 100}{-\frac{1000}{96}} = 96 \text{ Nm}$$

Since, however, the gearing efficiency is only 96%,

$$\text{available torque} = 0.96 \times 96 = \underline{92 \cdot 16 \text{ Nm}}$$

5. Two parallel shafts X and Y are to be connected by toothed wheels; wheels A and B form a compound pair which can slide along, but rotate with, shaft X; wheels C and D are rigidly attached to shaft Y and the compound pair may be moved so that A engages with C or B with D.

Shaft X rotates at 640 rev/min and the speeds of shaft Y are to be 340 rev/min exactly, and 240 rev/min as nearly as possible. Using a module of 12 for all wheels find the minimum distance between the shaft axes, suitable tooth numbers for the wheels and the lower speed of Y.
(*Ans.*: 294 mm; A, 13; B, 17; C, 36; D, 32; 231·1 rev/min)

6. The first and third shafts of a double reduction gear are in line and a total reduction of approximately 10 to 1 is required. The module of the high speed pair is to be 5, that of the low speed pair is to be 8 and no wheel is to have fewer than 20 teeth. Obtain a suitable value for the centre distance between the first and second shafts and the numbers of teeth on the wheels to satisfy the above conditions. What is the actual gear ratio? (*Ans.*: 340 mm; A, 32; B, 104; C, 20; D, 65; 1:9·5)

7. In an epicyclic gear of the sun and planet type, the p.c.d. of the internally toothed ring is to be as nearly as possible 220 mm and the teeth are to have a module of 4. When the ring is stationary, the spider, which carries three planets, is to make one revolution for every five of the driving spindle carrying the sun wheel. Determine suitable numbers of teeth for all the wheels and the exact p.c.d. of the ring.

If a torque of 12 Nm is applied to the shaft carrying the sun wheel, what torque will be required to keep the ring stationary? (*Ans.*: 56, 21, 14; 224 mm; 48 Nm)

8. In the epicyclic gear shown in Fig. 17.10, the pinion B and the internal wheels E and F are mounted independently on the spindle O while C and D form a compound wheel which rotates on the pin P attached to the arm A. The wheels B, C and D have 15, 30 and 25 teeth respectively, all of the same pitch.
(*a*) If wheel E is fixed, what is the ratio of the speed of F to that of B?
(*b*) If wheel B is fixed, what are the ratios of the speeds of E and F to that of A?
(*Ans.*: 1:56; 6:5; 33:28)

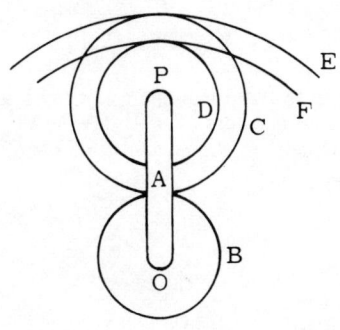

Fig. 17.10

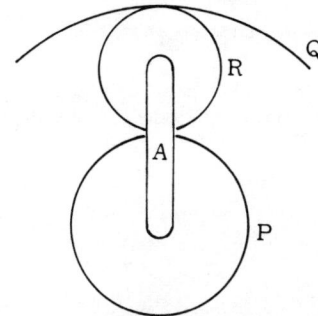

Fig. 17.11

9. Fig. 17.11 shows an epicyclic gear in which the wheel P, having 45 teeth of 15 mm pitch, is geared with Q through the intermediate wheel R at the end of the arm A. When P is rotating at 63 rev/min in a clockwise direction and A is rotating at 9 rev/min, also in a clockwise direction, Q is required to rotate at 21 rev/min anticlockwise. Find the necessary numbers of teeth on Q and R and the P.C.D. of Q. (*Ans.*: 81; 18; 386·9 mm)

10. An epicyclic gear consists of a sun wheel which has 24 teeth, planet wheels which have 28 teeth and an internally toothed annulus which is held stationary. Neglecting friction, find the torque required to hold the annulus fixed when 9 kW is being transmitted, the sun wheel rotating at 700 rev/min.

If the teeth have a module of 4, what is the diameter of the circle traced out by the centres of the planets? (*Ans.*: 409 N m; 208 mm)

11. An epicyclic gear consists of two sun wheels S_1 and S_2 with 24 and 28 teeth respectively, engaging with a compound planet with 26 and 22 teeth. S_1 is keyed to the driven shaft and S_2 is a fixed wheel co-axial with the driven shaft. The planet is carried on an arm fixed to the driving shaft. Find the velocity ratio of the gear.

If 750 W is transmitted when the output speed is 100 rev/min, what torque is required to hold S_2? (*Ans.*: −2·61 : 1; 98·7 N m)

12. An epicyclic train has a sun-wheel with 30 teeth and two planet wheels of 50 teeth, the latter meshing with the internal teeth of a fixed annulus. The input shaft, carrying the sun wheel, transmits 4 kW at 300 rev/min. The output shaft is connected to an arm which carries the planet wheels. Find the speed of the output shaft and the torque transmitted if the overall efficiency is 95%.

If the annulus is rotated independently, what should be its speed if the output shaft is to rotate at 10 rev/min? (*Ans.*: 56·25 rev/min; 645 N m; −56·9 rev/min)

13. In the epicyclic gear shown in Fig. 17.12, the input shaft A runs at 12 000 rev/min and the annular wheel B is fixed. Find the speed of the output shaft E and the speed of the planets relative to the spindle on which they are mounted. The tooth numbers are A, 15; B, 81; C, 41; D, 25. (*Ans.*: 1 216 rev/min; −3 946 rev/min)

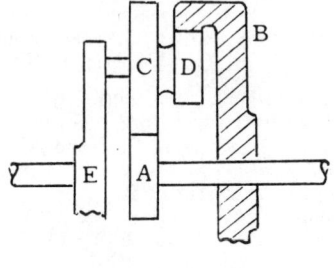

Fig. 17.12

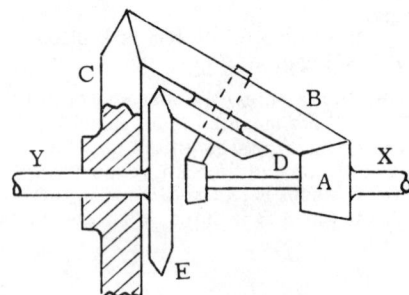

Fig. 17.13

14. In the gear shown in Fig. 17.13, the wheel C is fixed and shaft X rotates at 650 rev/min. Determine the speed of shaft Y. The numbers of teeth are A, 18; B, 55; C, 60; D, 24; E, 28. (*Ans.*: 9·75 rev/min)

18 Free vibrations

18.1 Introduction When an elastic system is displaced from its equilibrium position, the internal restoring force (and hence the acceleration) is proportional to the displacement and is directed towards the equilibrium position. Thus the body oscillates with simple harmonic motion, the rate being known as the natural frequency. In practice, internal damping will oppose the vibration, which will soon cease unless maintained by external excitement.

18.2 Simple harmonic motion If a point moves in a circular path with uniform velocity, the component of the motion along a diameter of the circle satisfies the conditions for simple harmonic motion. If the radius of the circle is a, Fig. 18.1, and the radius OP rotates at a rate ω rad/s, then measuring time from the point B, the angle BOP $= \omega t$.

The linear velocity and acceleration of P are respectively ωa and $\omega^2 a$ in the directions shown in Fig. 18.1 and the components of these quantities along AB give the corresponding velocity and acceleration of the point Q.

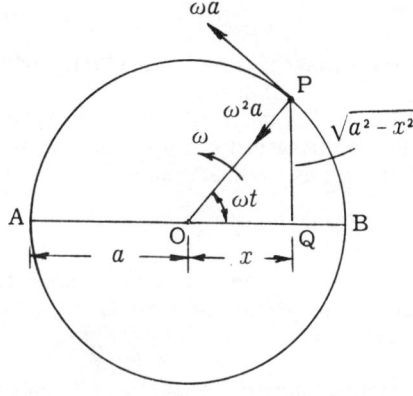

Thus $x = a \cos \omega t$ (18.1)

$v = \omega a \sin \omega t$

$ = \omega \sqrt{a^2 - x^2}$ (18.2)

and $f = \omega^2 a \cos \omega t$

$ = \omega^2 x$ (18.3)

Fig. 18.1

Thus the acceleration of Q is proportional to the displacement from the mid-position and is always directed towards that point, so that the motion of Q is simple harmonic.

Equations (18.1), (18.2) and (18.3) give merely numerical relations between displacement, velocity and acceleration without regard to direction.

The maximum displacement, a is termed the *amplitude* of the oscillation.
The maximum velocity is ωa and occurs at the mid-position when $x = 0$.
The maximum acceleration is $\omega^2 a$ and occurs at the end of the stroke.
The periodic time, T, is the time taken for one complete oscillation,

i.e. $$T = \frac{2\pi}{\omega}$$

But from equation (18.3), $\omega = \sqrt{\frac{f}{x}}$, so that

$$T = 2\pi \sqrt{\frac{x}{f}} \quad \text{or} \quad 2\pi \sqrt{\frac{\text{displacement}}{\text{acceleration}}} \quad (18.4)$$

The frequency, n, is the number of complete cycles per unit time,

i.e.
$$n = \frac{1}{T} = \frac{1}{2\pi}\sqrt{\frac{f}{x}} \qquad (18.5)$$

The unit of frequency is the Hertz (Hz), which is 1 cycle/s.

Simple harmonic motion also occurs in angular vibrations. If the angular amplitude of the motion is ϕ and the angular displacement, velocity and acceleration are θ, Ω and α respectively,

then
$$\theta = \phi \cos \omega t \qquad (18.6)$$

$$\Omega = \omega\phi \sin \omega t = \omega\sqrt{\phi^2 - \theta^2} \qquad (18.7)$$

$$\alpha = \omega^2 \phi \cos \omega t = \omega^2 \theta \qquad (18.8)$$

$$T = 2\pi \sqrt{\frac{\theta}{\alpha}} \qquad (18.9)$$

and
$$n = \frac{1}{2\pi}\sqrt{\frac{\alpha}{\theta}} \qquad (18.10)$$

18.3 Linear motion of an elastic system Consider the elastic system represented by the spring of stiffness S and the body of mass m shown in Fig. 18.2. If the mass if given a displacement x from the equilibrium position, the restoring force due to the spring stiffness is Sx. When released, this force gives the mass an acceleration f which is given by

$$Sx = mf$$

Thus the acceleration is proportional to the displacement and is always directed towards the equilibrium position, so that the mass moves with simple harmonic motion.

Thus the periodic time,
$$T = 2\pi\sqrt{\frac{x}{f}} = 2\pi\sqrt{\frac{m}{S}} \qquad (18.11)$$

If the static deflection of the spring under the action of gravity on the mass is δ, then

$$\delta = \frac{mg}{S}$$

so that
$$\frac{m}{S} = \frac{\delta}{g}$$

$$\therefore T = 2\pi\sqrt{\frac{\delta}{g}} \qquad (18.12)$$

The frequency $\quad n = \frac{1}{2\pi}\sqrt{\frac{g}{\delta}}\quad$ and if δ

is measured in metres, $\quad g = 9 \cdot 81$

so that $\quad n = \dfrac{1}{2 \cdot 006\, 5\sqrt{\delta}} \approx \dfrac{1}{2\sqrt{\delta}}$ Hz $\quad (18.13)$

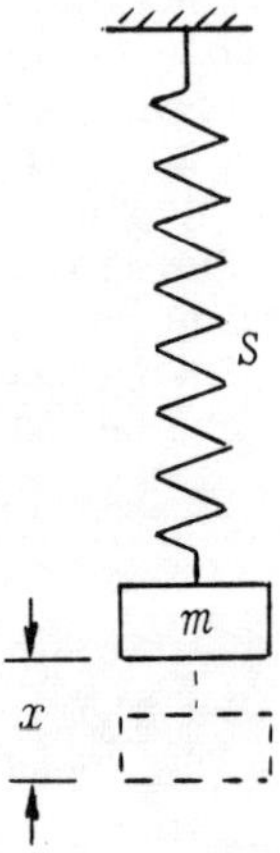

Fig. 18.2

FREE VIBRATIONS 217

18.4 Angular motion of an elastic system

Consider the elastic system represented by the rod of torsional stiffness q and the rotor of moment of inertia I shown in Fig. 18.3. If the rotor is given an angular displacement θ from the equilibrium position, the restoring torque due to the rod stiffness is $q\theta$. When released, this torque gives the rotor an acceleration α which is given by

$$q\theta = I\alpha$$

Thus the acceleration is proportional to the displacement and is always directed towards the equilibrium position, so that the rotor moves with simple harmonic motion.

Thus the periodic time,
$$T = 2\pi\sqrt{\frac{\theta}{\alpha}}$$
$$= 2\pi\sqrt{\frac{I}{q}} \qquad (18.14)$$

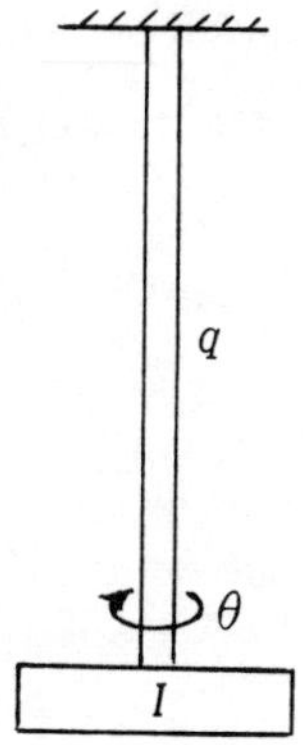

Fig. 18.3

If the length of the rod is l, the polar second moment of area of the cross-section is J and the modulus of rigidity of the material is G, then

$$q = \frac{GJ}{l} \quad . \quad . \quad . \quad . \text{ from equation (4.4)}$$

Thus
$$T = 2\pi\sqrt{\frac{Il}{GJ}} \qquad (18.15)$$

and
$$n = \frac{1}{2\pi}\sqrt{\frac{GJ}{Il}} \qquad (18.16)$$

18.5 Effect of mass of spring and inertia of shaft

Let the mass of the spring be m', Fig. 18.4. Then mass of element of length dx = $m'\dfrac{dx}{l}$. If the element is situated at a distance x from the support and the velocity of the free end is v, then

$$\text{velocity of element} = \frac{x}{l}v$$

$$\therefore \text{ K.E. of element} = \tfrac{1}{2}m'\frac{dx}{l}\left(\frac{x}{l}v\right)^2$$

$$= \tfrac{1}{2}m'\frac{v^2}{l^3}x^2\,dx$$

$$\therefore \text{ total K.E. of spring} = \tfrac{1}{2}m'\frac{v^2}{l^3}\int_0^l x^2\,dx$$

$$= \tfrac{1}{2}\frac{m'}{3}v^2 \qquad (18.17)$$

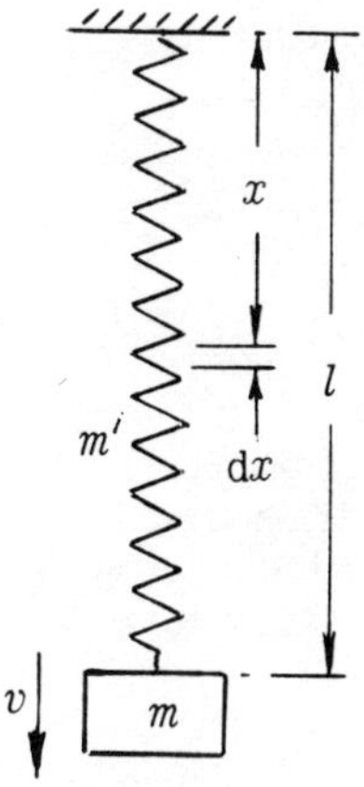

Fig. 18.4

The mass of the spring is therefore equivalent to a mass $\dfrac{m'}{3}$ added to the concentrated mass m at the free end.

It can similarly be shown that, in the case of angular motion, the moment of inertia of the rod, I', is equivalent to a body of moment of inertia $I'/3$ added to the rotor at the free end.

18.6 The motion of a pendulum Fig. 18.5 shows a simple pendulum consisting of a concentrated mass m at the end of a string of length l. If the string is given an angular displacement θ from the vertical, the restoring moment about O is $mgl\sin\theta$ and when released, the angular acceleration is given by

$$mgl\sin\theta = I_O\alpha = ml^2\alpha$$

Thus α is proportional to $\sin\theta$ and the motion is therefore not simple harmonic. If, however, θ is small, $\sin\theta \approx \theta$ and the equation of motion then becomes

$$mgl\theta = ml^2\alpha$$

from which
$$\frac{\theta}{\alpha} = \frac{l}{g}$$

whence
$$T = 2\pi\sqrt{\frac{l}{g}} \qquad (18.18)$$

The pendulum thus gives approximate S.H.M. provided that θ is kept small.

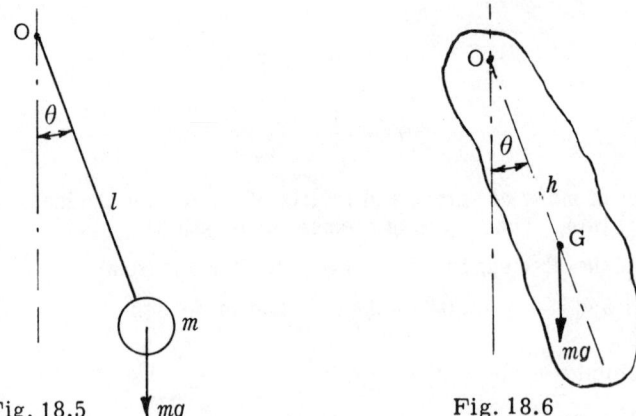

Fig. 18.5 Fig. 18.6

If the mass is not concentrated at a point, as shown in Fig. 18.6, the pendulum becomes *compound*. If the distance between the point of suspension O and the centre of gravity G is h, and the pendulum is displaced through a small angle θ, restoring moment about O $\approx mgh\theta$. When released, the angular acceleration is given by

$$mgh\theta = I_O\alpha$$

$$\therefore \frac{\theta}{\alpha} = \frac{I_O}{gh}$$

whence
$$T = 2\pi\sqrt{\frac{I_O}{mgh}} \qquad (18.19)$$

FREE VIBRATIONS

By the theorem of parallel axes, Art. 11.10,

$$I_O = I_G + mh^2$$
$$= m(k_G^2 + h^2)$$

so that the periodic time may be expressed in the form

$$T = 2\pi \sqrt{\frac{k_G^2 + h^2}{gh}} \qquad (18.20)$$

18.7 Differential equation of motion The motion of a vibrating body may be expressed mathematically by a differential equation. Referring to the system shown in Fig. 18.2, the equation of motion is given by

$$m\frac{d^2x}{dt^2} = -Sx$$

The negative sign indicates that the displacement and restoring force are in opposite directions.

This may be written as

$$\frac{d^2x}{dt^2} + \omega^2 x = 0 \qquad \text{where} \quad \omega^2 = \frac{S}{m}$$

The solution is *

$$x = A\cos\omega t + B\sin\omega t \qquad \text{where } A \text{ and } B \text{ are constants}$$

If the motion starts when $x = a$, corresponding with the point B of Fig. 18.1, then $x = a$ when $t = 0$ and $\frac{dx}{dt} = 0$ when $t = 0$.

Thus $\qquad\qquad A = a \quad \text{and} \quad B = 0$

so that $\qquad\qquad x = a\cos\omega t$

This is an oscillatory motion of period $\frac{2\pi}{\omega}$, so that

$$T = 2\pi\sqrt{\frac{m}{S}}$$

The corresponding differential equation for angular motion is

$$I\frac{d^2\theta}{dt^2} = -q\theta$$

or $\qquad\qquad \dfrac{d^2\theta}{dt^2} + \omega^2\theta = 0 \qquad \text{where} \quad \omega^2 = \dfrac{q}{I}$

This method is applicable to cases where the motion is not simple harmonic. Factors such as damping and forcing may be allowed for by adding the appropriate terms to the differential equation, as dealt with in Chapter 20.

* See Appendix

1. *In a mechanism, a crosshead moves in a straight guide with simple harmonic motion. At distances of 125 mm and 200 mm from its mean position, the crosshead has velocities of 6 and 3 m/s respectively. Determine (a) the amplitude of the motion, (b) the maximum velocity, and (c) the periodic time.*

If the crosshead has a mass of 0·2 kg, what is the maximum inertia force?

From equation (18.2), $v = \omega\sqrt{a^2 - x^2}$

$$\therefore 6 = \omega\sqrt{a^2 - 0\cdot125^2}$$

and $$3 = \omega\sqrt{a^2 - 0\cdot200^2}$$

$$\therefore \omega = 33\cdot3 \text{ rad/s}$$

and $$a = \underline{0\cdot2195 \text{ m}}$$

$$v_{max} = \omega a = 33\cdot3 \times 0\cdot2195 = \underline{7\cdot3 \text{ m/s}}$$

$$T = \frac{2\pi}{\omega} = \frac{2\pi}{33\cdot3} = \underline{0\cdot188 \text{ s}}$$

$$f_{max} = \omega^2 a = 33\cdot3^2 \times 0\cdot2195 = 243 \text{ m/s}$$

$$\therefore P_{max} = 0\cdot2 \times 243 = \underline{48\cdot7 \text{ N}}$$

2. *In the system shown in Fig. 18.7, the upper spring has a stiffness of 1 000 N/m and the lower spring 500 N/m. The suspended mass is 0·5 kg. Find the natural frequency of vertical vibration.*

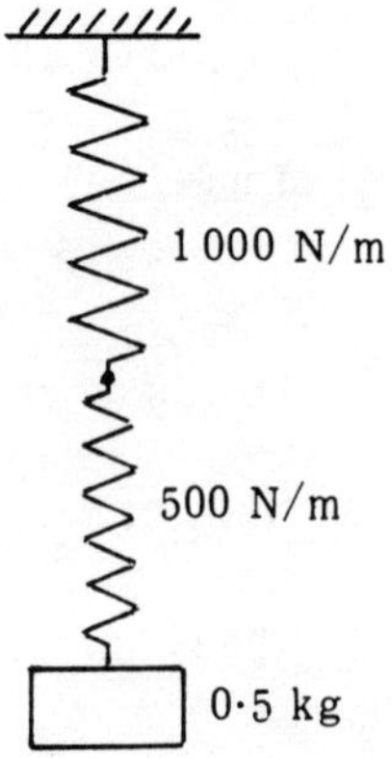

If a force P is applied to the compound spring, Fig. 18.7,

extension of upper spring $= \dfrac{P}{1\,000}$

and extension of lower spring $= \dfrac{P}{500}$

Therefore effective stiffness of compound spring

$$= \frac{P}{\dfrac{P}{1\,000} + \dfrac{P}{500}}$$

$$= \frac{500}{1\cdot5} \text{ N/m}$$

$$n = \frac{1}{T} = \frac{1}{2\pi}\sqrt{\frac{S}{m}} \quad . \quad . \quad \text{from equation (18.11)}$$

$$= \frac{1}{2\pi}\sqrt{\frac{500}{1\cdot5 \times 0\cdot5}}$$

$$= \underline{4\cdot11 \text{ Hz}}$$

Fig. 18.7

FREE VIBRATIONS

3. *A large gear wheel has a mass of 2 t and is suspended from a knife-edge so that it is free to swing in a vertical plane at right-angles to the gear axis. If the point of suspension is 0·7 m from the gear axis and the periodic time is 2·25 s, determine (a) the moment of inertia of the gear about its axis, and (b) the minimum possible periodic time if the point of suspension can be moved.*

From equation (18.20), $T = 2\pi\sqrt{\dfrac{k_G^2 + h^2}{gh}}$

i.e. $\quad 2 \cdot 25 = 2\pi\sqrt{\dfrac{k_G^2 + 0 \cdot 7^2}{9 \cdot 81 \times 0 \cdot 7}}$

from which $\quad k_G^2 = 0 \cdot 392$

$$\therefore I_G = m k_G^2$$

$$= 2 \times 10^3 \times 0 \cdot 392 = \underline{784 \text{ kg m}^2}$$

For minimum periodic time, $\dfrac{dT}{dh} = 0$

i.e. $\quad \dfrac{d}{dh}\left(\dfrac{k_G^2 + h^2}{h}\right) = 0$

from which $\quad h = k_G = 0 \cdot 626$ m

$$\therefore T_{min} = 2\pi\sqrt{\dfrac{2k_G^2}{g}}$$

$$= 2\pi\sqrt{\dfrac{2 \times 0 \cdot 626}{9 \cdot 81}} = \underline{2 \cdot 245 \text{ s}}$$

4. *A uniform bar AB, 2·5 m long and mass 100 kg, is supported on a hinge at one end A and on a spring support at the other end B so that it can vibrate in a vertical plane. The stiffness of the spring is 20 kN/m and when in static equilibrium, the bar is horizontal.*

The end B of the bar is depressed 10 mm and then released. Calculate (a) the frequency of the vibrations and (b) the maximum bending moment at the mid-point of the bar.

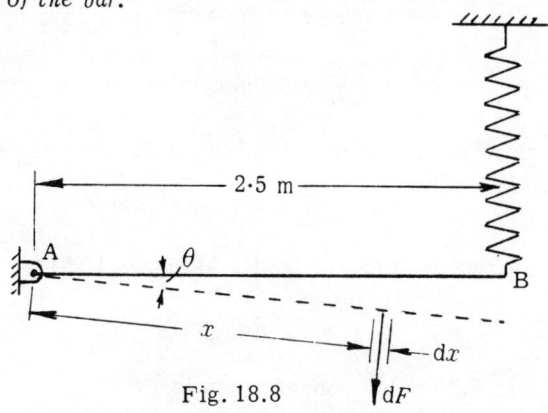

Fig. 18.8

If the beam is depressed through an angle θ, Fig. 18.8,

$$\text{then extension of spring} = 2 \cdot 5\theta \text{ m}$$

$$\therefore \text{ restoring force of spring} = 2 \cdot 5\theta \times 20$$

$$= 50\theta \text{ kN}$$

$$\therefore \text{ restoring moment about A} = 50\theta \times 2 \cdot 5$$

$$= 125\theta \text{ kN m}$$

$$\therefore 125 \times 10^3 \theta = I_A \alpha = \frac{ml^2}{3}\alpha$$

$$= \frac{100 \times 2 \cdot 5^2}{3}\alpha$$

$$\therefore \frac{\alpha}{\theta} = 600 \quad\quad\quad (1)$$

$$\therefore n = \frac{1}{2\pi}\sqrt{600} = \underline{3 \cdot 9 \text{ Hz}}$$

The maximum bending moment occurs when the inertia force is a maximum and this occurs at maximum displacement.

Thus, when extension of spring = 10 mm,

$$\theta = \frac{0 \cdot 010}{2 \cdot 5} = \frac{1}{250} \text{ rad}$$

$$\therefore \alpha = \frac{600}{250} \quad \cdot \quad \cdot \quad \cdot \quad \cdot \quad \cdot \quad \text{from equation (1)}$$

$$= 2 \cdot 4 \text{ rad/s}^2$$

$\therefore$ linear acceleration of element at distance x from A = $2 \cdot 4 x$ m/s²

$\therefore$ inertia force on element,

$$dF = dm \times f$$

$$= \frac{100}{2 \cdot 5} dx \times 2 \cdot 4 x = 96 x \, dx$$

$$\therefore \text{ moment of } dF \text{ about centre} = dF(x - 1 \cdot 25)$$

$$= 96x(x - 1 \cdot 25) dx$$

$$\therefore \text{ total inertia moment about centre} = \int_{1 \cdot 25}^{2 \cdot 5} 96x(x - 1 \cdot 25) dx$$

$$= 156 \cdot 3 \text{ N m}$$

This tends to bend the bar convex upwards.

FREE VIBRATIONS 223

Upward force at B due to spring tension

$$= 20 \times 10^3 \times 0.01 = 200 \text{ N}$$

$$\therefore \text{ moment about centre } = 200 \times 1.25 = 250 \text{ Nm}$$

This tends to bend the bar convex downwards

Static bending moment at centre due to dead weight of bar

$$= \frac{wl^2}{8} \quad . \quad . \quad . \quad \text{from Fig. 2.4}(d)$$

$$= \frac{100 \times 9.81 \times 2.5}{8} = 306.3 \text{ Nm}$$

This tends to bend the beam convex downwards.

Thus resultant bending moment at centre

$$= 250 + 306.3 - 156.3$$

$$= \underline{400 \text{ Nm}} \text{ , tending to bend bar convex downwards}$$

5. *A vertical steel wire, 2 mm diameter and 2 m long, is fixed at its upper end and at the lower end, a solid steel cylinder is attached centrally so that its axis is horizontal. The cylinder is 75 mm diameter and its density is 7.8 Mg/m³. Find the length of the cylinder to give 0.6 torsional oscillation per second.*

Calculate the amplitude of the vibrations when the maximum shearing stress is 120 MN/m². G = 80 GN/m².

For a solid cylinder of length L and radius R, the moment of inertia about the axis of rotation AB, Fig. 18.9, is given by

$$I = m\left(\frac{L^2}{12} + \frac{R^2}{4}\right)$$

$$= \frac{\rho \pi R^2 L}{12}(L^2 + 3R^2)$$

$$= \frac{7.8 \times 10^3 \pi \times 0.0375^2 L (L^2 + 3 \times 0.0375^2)}{12}$$

$$= 2.87L (L^2 + 0.00422)$$

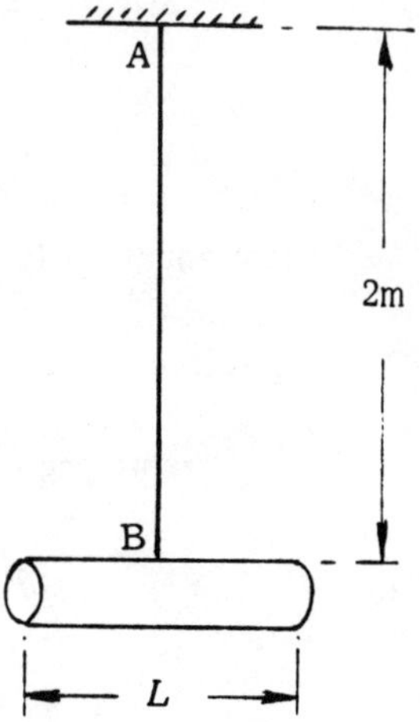

For the wire, $J = \dfrac{\pi d^4}{32}$

$$= \frac{\pi \times 2^4 \times 10^{-12}}{32} = 1.571 \times 10^{-12} \text{ m}^4$$

Fig. 18.9

$$n = \frac{1}{2\pi}\sqrt{\frac{GJ}{Il}} \quad . \quad . \quad . \quad . \quad \text{from equation (18.16)}$$

i.e.
$$0 \cdot 6 = \frac{1}{2\pi} \sqrt{\frac{80 \times 10^9 \times 1 \cdot 571 \times 10^{-12}}{2 \cdot 87 L (L^2 + 0 \cdot 004\,22) \times 2}}$$

i.e. $L(L^2 + 0 \cdot 004\,22) = 0 \cdot 001\,54$

By trial or plotting, $L = 0 \cdot 103\,5$ m

From equation (4.4), $\dfrac{\tau}{r} = \dfrac{G\theta}{l}$

$$\therefore \theta = \frac{120 \times 10^6 \times 2}{80 \times 10^9 \times 0 \cdot 001}$$

$$= \underline{3 \text{ rad} \quad \text{or} \quad 172°}$$

6. *A rotor has a mass of 225 kg and a radius of gyration of 0·4 m. It is bolted between the ends of two shafts, one of which is 75 mm diameter, 0·9 m long and the other is 65 mm diameter, 0·45 m long. The other ends of the shafts are rigidly fixed in position. Find the frequency of natural torsional vibration of the rotor. G = 80 GN/m².*

The arrangement is shown in Fig. 18.10.

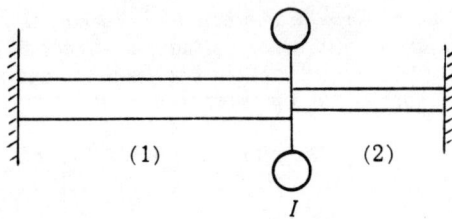

Fig. 18.10

From equation (4.5), $\dfrac{T}{J} = \dfrac{G\theta}{l}$

Therefore when the rotor is twisted through an angle θ,

restoring torque, $T = \dfrac{GJ_1\theta}{l_1} + \dfrac{GJ_2\theta}{l_2}$

Hence $G\theta\left(\dfrac{J_1}{l_1} + \dfrac{J_2}{l_2}\right) = I\alpha$

$$\therefore n = \frac{1}{2\pi}\sqrt{\frac{\alpha}{\theta}} = \frac{1}{2\pi}\sqrt{\frac{G}{I}\left(\frac{J_1}{l_1} + \frac{J_2}{l_2}\right)}$$

$$= \frac{1}{2\pi}\sqrt{\frac{80 \times 10^9}{225 \times 0 \cdot 4^2} \times \frac{\pi}{32}\left(\frac{75^4}{0 \cdot 9} + \frac{65^4}{0 \cdot 45}\right) \times 10^{-12}}$$

$$= \underline{20 \cdot 3 \text{ Hz}}$$

FREE VIBRATIONS

7. *The flywheel of an engine driving a dynamo has a mass of 135 kg and a radius of gyration of 250 mm. The armature has a mass of 100 kg and a radius of gyration of 200 mm. The driving shaft is 450 mm long and 50 mm diameter and a spring coupling is incorporated at one end, having a stiffness of 28 kN m/rad. Determine the natural frequency of torsional vibration of the system.* $G = 80 \text{ GN}/m^2$.

The arrangement is shown in Fig. 18.11.

The torsional stiffness of a shaft is given by

$$q = \frac{T}{\theta} = \frac{GJ}{l}$$

$$\therefore l = \frac{GJ}{q}$$

Thus the spring coupling is equivalent to a length of shaft of diameter 50 mm given by

$$l = \frac{80 \times 10^9 \times \frac{\pi}{32} \times 0.05^4}{28 \times 10^3} = 1.753 \text{ m}$$

Therefore the actual shaft and coupling is equivalent to a uniform shaft of length $0.45 + 1.753 = 2.203$ m, as shown in Fig. 18.12.

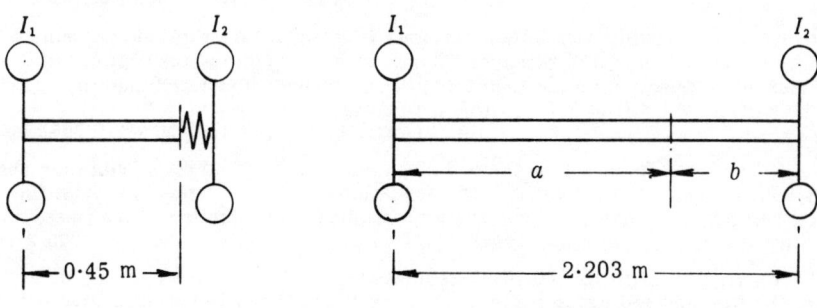

Fig. 18.11   Fig. 18.12

When oscillating, the two rotors will be moving in opposite directions at any instant and there will be a stationary point (called a node) somewhere between the two rotors. Let the node be at distances a and b from I_1 and I_2 respectively.

Then, for I_1,
$$n = \frac{1}{2\pi}\sqrt{\frac{GJ}{I_1 a}} \quad (1)$$

and for I_2,
$$n = \frac{1}{2\pi}\sqrt{\frac{GJ}{I_2 b}} \quad (2)$$

treating each as a rotor on the end of a shaft attached to a fixed point. The frequency of each rotor must be the same,

hence
$$I_1 a = I_2 b$$

or
$$\frac{a}{b} = \frac{I_2}{I_1} = \frac{100 \times 0.20^2}{135 \times 0.25^2} = 0.474$$

Also $\quad a + b = 2.203$

$\therefore a = 0.708$ m and $b = 1.495$ m

Substituting in equation (1) gives

$$n = \frac{1}{2\pi}\sqrt{\frac{80 \times 10^9 \times \frac{\pi}{32} \times 0.05^4}{135 \times 0.25^2 \times 0.708}}$$

$$= \underline{14.45 \text{ Hz}}$$

8. A mass of 25 kg is suspended from a spring of stiffness 14 kN/m and vibrates with an amplitude of 12 mm. Find the periodic time, the velocity and acceleration when displaced 8 mm from the equilibrium position and the time taken to move from this position to the position of maximum displacement.
(*Ans.*: 0.266 s; 0.212 m/s; 4.48 m/s^2; 0.035 6 s)

9. A body of mass 5 kg oscillates in a horizontal straight line with S.H.M. The amplitude of the oscillation is 0.9 m and the maximum horizontal force acting on the body is 300 N. Determine (*a*) the frequency of the oscillations, (*b*) the velocity of the body at a point 0.45 m from the mid-position, (*c*) the time taken for the body to travel 0.45 m from one extreme end of the oscillation.
(*Ans.*: 1.3 Hz; 6.363 m/s; 0.128 s)

10. A particle moving with S.H.M. performs 10 complete oscillations per minute and its speed, when at a distance of 200 mm from the centre of oscillation, is ⅗ of the maximum speed. Find the amplitude, the maximum acceleration and the speed of the particle when it is 150 mm from the centre of oscillation.
(*Ans.*: 0.25 m; 0.274 2 m/s^2; 0.209 4 m/s)

11. A 2 kg mass is hung from the end of a helical spring and is set vibrating vertically. The mass makes 100 complete oscillations in 55 s. Determine the stiffness of the spring. Also calculate the maximum amplitude of vibration if the mass is not to leave the hook during its motion. (*Ans.*: 261 N/m; 75.2 mm)

12. A spring of stiffness 200 N/m has a mass of 0.75 kg. A mass of 5 kg is attached to the free end and set in motion. Find the time of oscillation (*a*) neglecting the mass of the spring; (*b*) allowing for the mass of the spring.
(*Ans.*: 1.005 Hz; 0.982 Hz)

13. A horizontal shaft, supported in bearings at the ends, deflects at the centre by 0.005 mm per 100 N of load applied there. When a wheel of mass 300 kg is centrally fitted, the system responds to an external disturbance and free vertical vibrations of amplitude 0.25 mm are set up. Calculate the frequency, the maximum velocity and the maximum acceleration for this vibration.
(*Ans.*: 41.1 Hz; 0.064 5 m/s; 16.67 m/s^2)

14. A thin uniform rod AB, of mass 1 kg and length 0.6 m, carries a concentrated mass of 2.5 kg at B. The rod is hinged at A and is maintained in a horizontal position by a vertical spring of stiffness 1.8 kN/m attached at its mid-point. Find the frequency of oscillation in the vertical plane. (*Ans.*: 2.01 Hz)

15. A bar has a mass of 5 kg and is pivoted at one end. The radius of gyration of the bar about the pivot is 0.6 m. The bar is supported in the horizontal position by a vertical spring of stiffness 500 N/m and mass 3 kg which is attached to the bar at a point 0.4 m from the pivot. Determine the periodic time of oscillation in the vertical plane. (*Ans.*: 0.98 s)

FREE VIBRATIONS

16. A wheel is mounted on a knife-edge on the inside surface of its rim at a distance of 400 mm from its centre of gravity. It is found to make 100 complete vibrations in 2 min 40 s. Calculate its radius of gyration about an axis through the centre of gravity. *(Ans.:* 307 mm)

17. In order to determine the radius of gyration of a wheel, it is swung from a knife-edge as a compound pendulum. When the knife-edge is 900 mm from the centre of gravity, it is found that the wheel makes twice as many swings in a given time as it does when the knife-edge is 125 mm from the centre of gravity. What is the radius of gyration of the wheel? *(Ans.:* 988·5 mm)

18. A torsional pendulum consists of a wire 0·5 m long, 10 mm diameter, fixed at its upper end and attached at its lower end to a disc of moment of inertia 0·06 kg m^2. The modulus of rigidity of the wire is 44 GN/m^2. Find the frequency of torsional oscillation of the disc.

If the maximum displacement to one side of the rest position is 5°, find the maximum angular velocity and acceleration of the disc.

(Ans.: 6·04 Hz; 3·31 rad/s; 125·5 rad/s^2)

19. A solid metal cylinder 450 mm diameter is suspended with its axis vertical by means of a wire coaxial with the cylinder and rigidly attached to it. The stiffness of the wire is 22 N m per radian of twist. Find the necessary mass of the cylinder so that when it is given a small angular displacement about its axis, it will make 40 vibrations per minute. *(Ans.:* 49·5 kg)

20. One end of a shaft, 0·9 m long and 25 mm diameter, is fixed and a coaxial cylinder, 150 mm long and 100 mm diameter, is attached at the other end. Both the shaft and cylinder are made of steel of density 7·8 Mg/m^2. Calculate the frequency of torsional oscillation of the system, allowing for the inertia of the shaft.
$G = 85 \text{ GN/m}^2$. *(Ans.:* 84·6 Hz)

21. A uniform shaft is rigidly fixed at both ends and a rotor is attached to the shaft at some intermediate point. Show that the frequency of torsional oscillation is a minimum when the rotor is attached at the mid-point.

22. A rotor of mass 34 kg and radius of gyration 1·15 m is fixed to one end of a shaft and another, of mass 16 kg and radius of gyration 1·4 m is fixed to the other end. The shaft is 0·45 m long and 45 mm diameter. Calculate the frequency of torsional vibrations if $G = 80 \text{ GN/m}^2$. *(Ans.:* 9·9 Hz)

23. The moving parts of a radial engine have a total moment of inertia of 0·8 kg m^2 and are concentrated in the plane of the single crank pin. The engine is connected to a propeller of moment of inertia 15 kg m^2 by a hollow shaft of length 250 mm and outer and inner diameters 75 and 32 mm respectively. The stiffness of the crank alone is 2·5 MN m/rad. Determine the frequency of torsional vibration of the system if $G = 80 \text{ GN/m}^2$. *(Ans.:* 152 Hz)

24. Two rotors A and B are fixed to the ends of a shaft of length 530 mm and the node is to be at a section C distant 330 mm from A. Rotor A has a mass of 40 kg and radius of gyration 140 mm while rotor B has a mass of 18 kg and radius of gyration 160 mm. If the diameter of part AC is 45 mm, find the diameter of part CB and the frequency of torsional oscillation. $G = 80 \text{ GN/m}^2$. *(Ans.:* 34·7 mm; 56·1 Hz)

19 Transverse vibrations and whirling speeds

19.1 Light beam with single load The beam shown in Fig. 19.1 may be considered in the same way as a mass attached to the end of a spring, Art. 18.3. If the mass m is displaced a distance y from the equilibrium position, the restoring force is given by Sy, where S is the stiffness of the beam, i.e., the force required at the load point per unit deflection at that point. This is of the form $S = kEI/l^3$ where k is a constant depending on the loading and method of support (see Chap. 5)

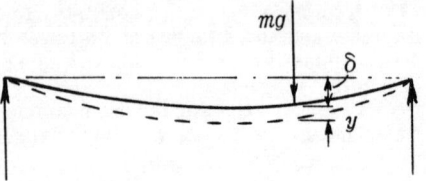

Fig. 19.1

When released, the acceleration of the mass is given by

$$Sy = mf$$

Thus the acceleration is proportional to the displacement of the load, so that the motion is simple harmonic. The frequency is then given by

$$n = \frac{1}{2\pi}\sqrt{\frac{f}{y}} = \frac{1}{2\pi}\sqrt{\frac{S}{m}} \quad *$$

If the static deflection at the load point is δ, then $\delta = \frac{mg}{S}$

or
$$\frac{S}{m} = \frac{g}{\delta}$$

Hence
$$n = \frac{1}{2\pi}\sqrt{\frac{g}{\delta}}$$

If δ is measured in metres, $g = 9 \cdot 81$ m/s²

so that
$$n \approx \frac{1}{2\sqrt{\delta}} \text{ Hz} \tag{19.1}$$

19.2 Uniformly distributed load Exact analysis of this case is beyond the scope of this book but a very close approximation may be obtained by assuming that the vibrating beam is of the same shape as the static deflection curve, i.e., that the amplitude a at any point is proportional to the static deflection y at that point

or $\quad a = cy \quad$ where c is a constant.

If the mass of the beam is m per unit length then, considering an element of length dx, Fig. 19.2, the additional load required to deflect the beam a further distance a

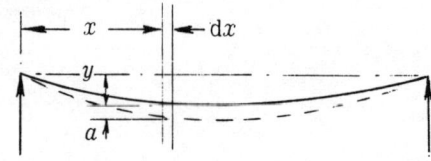

Fig. 19.2

* cf. Art. 18.3

TRANSVERSE VIBRATIONS

$$= mg\,dx \times \frac{a}{y}$$

Work done by this load in moving through the distance a

$$= \tfrac{1}{2}(mg\,dx \times \frac{a}{y}) \times a$$

$$= \tfrac{1}{2} mg\, c^2 y\, dx$$

Therefore, for the whole beam,

$$\text{work done} = \tfrac{1}{2} mg\, c^2 \int_0^l y\, dx \qquad (19.2)$$

This represents the strain energy stored in the beam at maximum displacement, which must equal the kinetic energy of the beam as it passes through the mid-position.

At the mid-point, velocity of element $= \omega a$. from equation (18.2)

Hence, K.E. of element in mid-position $= \tfrac{1}{2} m\,dx\,(\omega a)^2$

$$= \tfrac{1}{2} m\, \omega^2 c^2 y^2 \,dx$$

Therefore total K.E. of beam $= \tfrac{1}{2} m\, \omega^2 c^2 \int_0^l y^2\, dx \qquad (19.3)$

Equations (19.2) and (19.3) give $\quad \omega^2 = g\, \dfrac{\displaystyle\int_0^l y\,dx}{\displaystyle\int_0^l y^2\,dx}$

so that

$$n = \frac{\omega}{2\pi} = \frac{1}{2\pi} \sqrt{g\, \frac{\displaystyle\int_0^l y\,dx}{\displaystyle\int_0^l y^2\,dx}}$$

If y is measured in metres,

$$n \approx \tfrac{1}{2} \sqrt{\frac{\displaystyle\int_0^l y\,dx}{\displaystyle\int_0^l y^2\,dx}} \qquad (19.4)$$

For a simply supported beam, $\quad y = \dfrac{mg}{24EI}(l^3 x - 2l x^3 + x^4)$ *,

taking the origin at one end,

which gives $\quad n = 4 \cdot 935 \sqrt{\dfrac{EI}{mg l^4}}$

---

* It is only *magnitudes* of deflections which are relevant to the subsequent analysis.

$$= \frac{0 \cdot 564}{\sqrt{\Delta}} \qquad (19.5)$$

where Δ is the maximum static deflection of the beam.

For a cantilever, $y = \dfrac{mg}{24EI}(6l^2 x^2 - 4lx^3 + x^4)$ *,

taking the origin at the fixed end,

which gives
$$n = 1 \cdot 755 \sqrt{\frac{EI}{mgl^4}}$$

$$= \frac{0 \cdot 624}{\sqrt{\Delta}} \qquad (19.6)$$

This is known as the energy method. The error resulting from the initial assumption is extremely small and quite accurate results may also be obtained by assuming the shape to be a sine wave or parabola.

19.3 Several loads (a) *Energy method* Assuming that the vibrating beam is similar in shape to the static deflection curve, the frequency may be obtained in the same way as for a distributed load by equating the strain energy at maximum displacement to the kinetic energy in the mid-position. This leads to the equation

$$n = \tfrac{1}{2}\sqrt{\frac{\Sigma my}{\Sigma my^2}} \text{ Hz} \qquad (19.7)$$

The deflection y under each load must be calculated with all the loads *acting together*.

Any distributed mass may be allowed for by adding to each concentrated mass the mass of the beam between the centres of the sections into which the loads divide the beam.

A good approximation can again be obtained by assuming the shape of the vibrating beam to be a sine wave or parabola instead of the static deflection curve.

(b) *Dunkerley's method* This approximate method relates the frequency with all the loads acting together with those of each of the loads when acting alone. If n_0 is the frequency of the beam due to its own distributed mass and n_1, n_2, n_3, etc., are the frequencies of the concentrated masses m_1, m_2, m_3, etc., when each acts alone on the beam, then

$$\frac{1}{n^2} = \frac{1}{n_0^2} + \frac{1}{n_1^2} + \frac{1}{n_2^2} + \frac{1}{n_3^2} + \ldots \qquad (19.8)$$

This is an empirical relation and gives results which are less accurate than those given by the energy method.

---

* It is only *magnitudes* of deflections which are relevant to the subsequent analysis.

TRANSVERSE VIBRATIONS

19.4 Whirling speed of shafts Fig. 19.3 shows a shaft carrying a rotor of mass m, the c.g. of which has an eccentricity e relative to the shaft axis O. When the shaft rotates at a rate ω a centrifugal force $m\omega^2(y + e)$ will act upon the rotor, causing a deflection y relative to the static deflection δ. The resulting force due to the shaft stiffness S is Sy, so that, for equilibrium,

$$m\omega^2(y + e) = Sy$$

from which
$$y = \frac{e}{\dfrac{S}{m\omega^2} - 1} \qquad (19.9)$$

When $\dfrac{S}{m\omega^2} = 1$, y will become infinite and the shaft is said to whirl.

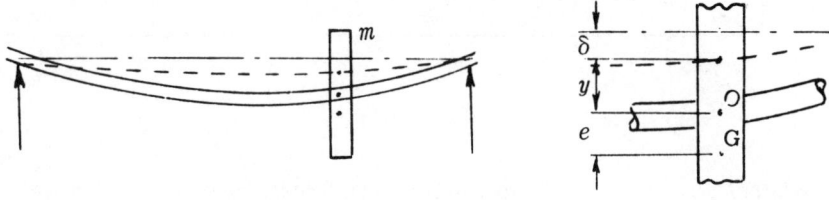

Fig. 19.3

The whirling, or critical, speed ω_c is therefore given by

$$\omega_c^2 = \frac{S}{m} = \frac{g}{\delta}$$

so that the whirling speed
$$n_c = \frac{\omega_c}{2\pi} = \frac{1}{2\pi}\sqrt{\frac{g}{\delta}} \text{ rev/s}$$

$$\approx \frac{1}{2\sqrt{\delta}} \text{ where } \delta \text{ is in metres} \quad (19.10)$$

Thus the whirling speed is identical with the frequency of transverse vibration of the same shaft, carrying the same load and this applies equally to other systems of distributed and multiple loads.

At speeds other than the critical speed, the deflection of the shaft relative to the static deflection position may be expressed in the form

$$y = \frac{e}{\dfrac{\omega_c^2}{\omega^2} - 1} = \frac{\omega^2 e}{\omega_c^2 - \omega^2} \qquad (19.11)$$

Equation (19.11) shows that when $\omega < \omega_c$, e and y are of the same sign, i.e. G is to the outside of O but when $\omega > \omega_c$, e and y are of opposite sign, so that G lies between O and the static deflection curve. As the speed increases, $y \to -e$ so that eventually G lies on the static deflection curve.

Fig. 19.4 shows the numerical relation between y and ω. In practice, internal damping in the shaft material will prevent the deflection form becoming infinite when $\omega = \omega_c$ but the speed should be kept well away from the critical speed to prevent damage to the shaft.

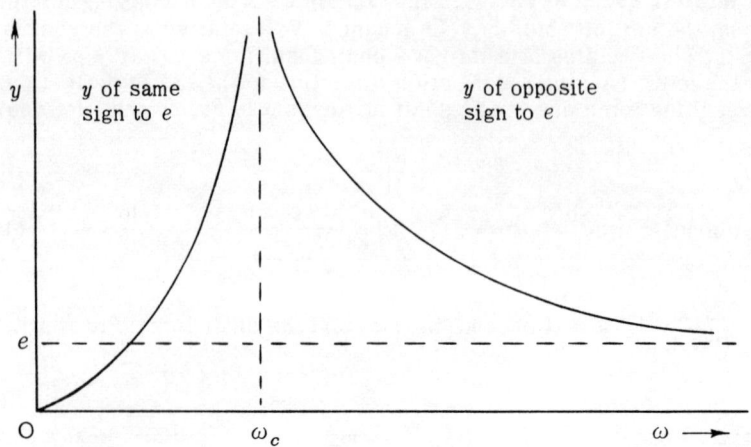

Fig. 19.4

1. *A steel strip 10 mm wide and 0·8 mm thick rests on supports 200 mm apart and a mass of 0·15 kg is fixed to the strip at mid-span. Find the frequency of transverse vibration.*

If the greatest bending stress in the strip is 100 MN/m², calculate the amplitude of movement of the mass and also the least force on the supports. Neglect the mass of the strip. E = 200 GN/m².

$$\text{Static deflection under load} = \frac{Wl^3}{48EI} \quad . \quad . \quad \text{from equation (5.15)}$$

$$= \frac{0 \cdot 15 \times 9 \cdot 81 \times 0 \cdot 2^3}{48 \times 200 \times 10^9 \times \frac{10 \times 0 \cdot 8^3}{12 \times 10^{12}}}$$

$$= 0 \cdot 002\,875 \text{ m}$$

$$\therefore n = \frac{1}{2\sqrt{0 \cdot 002\,875}} \quad . \quad \text{from equation (19.1)}$$

$$= \underline{9 \cdot 32 \text{ Hz}}$$

$$\text{Greatest force on strip} = \text{weight} + \text{maximum inertia force}$$

$$= mg + m\omega^2 a \quad \text{where } a \text{ is the amplitude}$$

$$= m[g + (2\pi n)^2 a]$$

$$= 0 \cdot 15[9 \cdot 81 + (2\pi \times 9 \cdot 32)^2 a]$$

$$= 0 \cdot 15(9 \cdot 81 + 3430\,a)$$

$$\sigma_{max} = \frac{M}{Z} \quad . \quad . \quad . \quad \text{from equation (3.2)}$$

TRANSVERSE VIBRATIONS

$$= \frac{Wl}{4} \bigg/ \frac{bd^2}{6} = \frac{3Wl}{2bd^2}$$

$$\therefore \; 100 \times 10^6 = \frac{3 \times 0.15\,(9.81 + 3\,430\,a) \times 0.2}{2 \times 10 \times 0.8^2 \times 10^{-9}}$$

from which $\quad a = 0.001\,29$ m $\quad$ or $\quad 1.29$ mm

Least force on each support $= \tfrac{1}{2}(mg - m\omega^2 a)$

$$= \frac{0.15}{2}(9.81 - 3\,430 \times 0.001\,29)$$

$$= \underline{0.404 \text{ N}}$$

2. *A simply supported beam of length 3a carries two loads W situated at distances a and 2a from one end. Calculate the frequency of transverse vibration.*

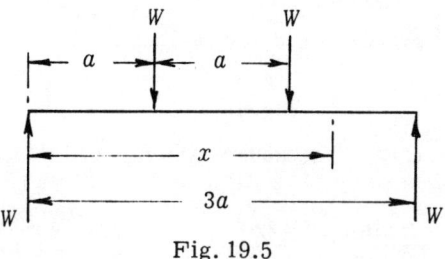

Fig. 19.5

(a) *Energy method* Referring to Fig. 19.5, the deflections under the loads are obtained by Macaulay's method (Art. 5.6). Taking the origin at the left-hand end,

$$EI\frac{d^2y}{dx^2} = Wx - W[x-a] - W[x-2a]$$

$$\therefore \; EI\frac{dy}{dx} = \frac{W}{2}x^2 - \frac{W}{2}[x-a]^2 - \frac{W}{2}[x-2a]^2 + A$$

$$\therefore \; EIy = \frac{W}{6}x^3 - \frac{W}{6}[x-a]^3 - \frac{W}{6}[x-2a]^3 + Ax + B$$

When $x = 0$, $y = 0$, $\quad \therefore B = 0$

When $x = 3a$, $y = 0$, $\quad \therefore A = -Wa^2$

Therefore, when $x = a$, $\quad EIy = \dfrac{Wa^3}{6} - Wa^3$

$$\therefore \; y = -\frac{5Wa^3}{6EI} \; {}^{*}$$

---

* It is only the *magnitude* of the deflection which is relevant to the subsequent analysis.

Since the beam is symmetrical, the deflection under the second load is identical.

Hence
$$\Sigma my = 2 \times \frac{W}{g}\left(\frac{5Wa^3}{6EI}\right)$$

and
$$\Sigma my^2 = 2 \times \frac{W}{g}\left(\frac{5Wa^3}{6EI}\right)^2$$

$$\therefore n = \tfrac{1}{2}\sqrt{\frac{\Sigma my}{\Sigma my^2}} = \tfrac{1}{2}\sqrt{\frac{6EI}{5Wa^3}} \qquad \text{from equation (19.7)}$$

$$= \underline{0\cdot 548 \sqrt{\frac{EI}{Wa^3}}}$$

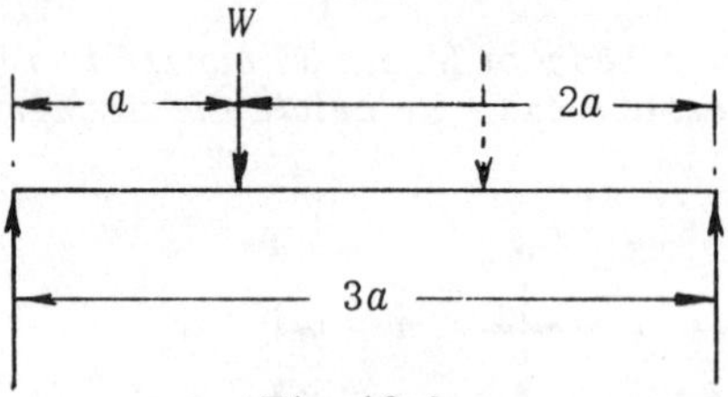

Fig. 19.6

(b) *Dunkerley's method* The numerical value of the deflection under one load assumed to be acting alone, Fig. 19.6, is given by

$$y = \frac{Wa^2(2a)^2}{3EI(3a)} = \frac{4Wa^3}{9EI} \qquad \text{. . . from equation (5.25)}$$

Thus
$$n_1 = n_2 = \frac{1}{2\sqrt{\frac{4Wa^3}{9EI}}} \qquad \text{. . from equation (19.1)}$$

$$= \tfrac{3}{4}\sqrt{\frac{EI}{Wa^3}}$$

$$\therefore \frac{1}{n^2} = \frac{1}{n_1^2} + \frac{1}{n_2^2} \qquad \text{. . from equation (19.8)}$$

$$= \frac{32}{9}\frac{Wa^3}{EI}$$

$$\therefore n = \underline{0\cdot 531 \sqrt{\frac{EI}{Wa^3}}}$$

3. *A shaft of diameter 40 mm is supported in flexible bearings 0·6 m apart. It carries a rotor of mass 100 kg at its centre, the c.g. of the rotor being 0·01 mm eccentric to the shaft axis.*

(a) Determine the critical speed of rotation, allowing for the mass of the shaft which has a density of 7·8 Mg/m³.

(b) Find the speed range over which the maximum deflection of the shaft relative to its static position will exceed 0·5 mm.

$E = 200 \ GN/m^2.$

TRANSVERSE VIBRATIONS

The arrangement is shown in Fig. 15.7.

Static deflection due to mass of rotor

$$= \frac{Wl^3}{48EI} \quad \text{from equation (5.15)}$$

$$= \frac{100 \times 9\cdot 81 \times 0\cdot 6^3}{48 \times 200 \times 10^9 \times \frac{\pi}{64} \times 0\cdot 04^4}$$

$$= 0\cdot 000\ 175\ 6 \text{ m}$$

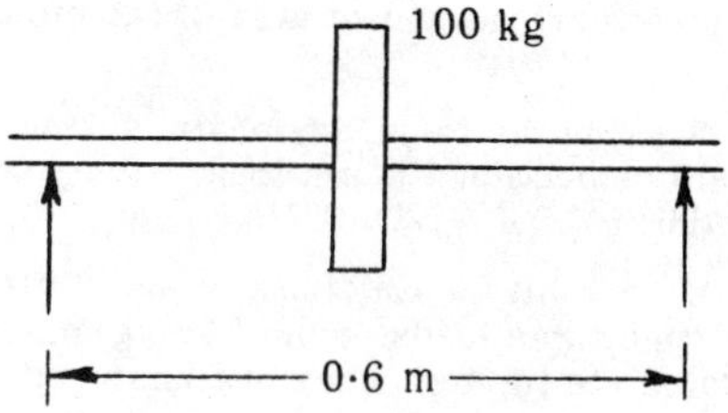

Fig. 19.7

$$\therefore n_1 = \frac{1}{2\sqrt{0\cdot 000\ 175\ 6}} = 37\cdot 7 \text{ rev/s} \quad . \quad . \quad \text{from equation (19.1)}$$

Static deflection due to mass of shaft

$$= \frac{5}{384} \frac{wl^4}{EI} \quad . \quad . \quad . \quad . \quad . \quad \text{from equation (5.17)}$$

$$= \frac{5}{384} \times \frac{\frac{\pi}{4} \times 0\cdot 04^2 \times 7\cdot 8 \times 10^3 \times 9\cdot 81 \times 0\cdot 6^4}{200 \times 10^9 \times \frac{\pi}{64} \times 0\cdot 04^4}$$

$$= 0\cdot 000\ 006\ 45 \text{ m}$$

$$\therefore n_0 = \frac{0\cdot 564}{\sqrt{0\cdot 000\ 006\ 45}} = 222 \text{ rev/s} \quad . \quad \text{from equation (19.5)}$$

$$\frac{1}{n^2} = \frac{1}{n_0^2} + \frac{1}{n_1^2}$$

$$= \frac{1}{222^2} + \frac{1}{37\cdot 7^2}$$

$$\therefore n = \underline{37\cdot 1 \text{ rev/s}}$$

From equation (19.11), $y = \dfrac{e}{\dfrac{\omega_c^2}{\omega^2} - 1} = \dfrac{e}{\dfrac{n_c^2}{n^2} - 1}$

Since y may be of the same or opposite sign to e, depending on whether ω is less than or greater than ω_c,

$$0\cdot 5 = \pm \frac{0\cdot 01}{\dfrac{37\cdot 1^2}{n^2} - 1}$$

from which $\qquad n = \underline{36\cdot 8 \text{ and } 37\cdot 5 \text{ rev/s}}$

4. Determine the frequency of transverse vibration of an I-section beam 6 m long simply supported at its ends which carries a uniformly distributed load of 750 kg/m. The second moment of area of the cross-section is 140×10^{-6} m^4 and $E = 200$ GN/m^2.
(*Ans.*: 8·46 Hz)

5. Find the frequency of transverse vibration of a cantilever turbine blade of uniform section 125 mm long, having a mass of 2 kg/m and second moment of area 2 700 mm^4. $E = 200$ GN/m^2. (*Ans.*: 590 Hz)

6. A shaft 50 mm diameter and 0·8 m long is simply supported at its ends and carries three loads each of 36 kg, one at the centre and one 0·2 m from each end. Calculate the frequency of transverse vibration, given that the deflection under the central load is $\dfrac{19}{384} \dfrac{mgl^3}{EI}$ and under each end load it is $\dfrac{9}{256} \dfrac{mgl^3}{EI}$. $E = 200$ GN/m^2.
(*Ans.*: 45·5 Hz)

7. A shaft is simply supported on bearings 3 m apart and carries five equal concentrated loads equally spaced with the end loads 0·3 m from each bearing. If the maximum static deflection is 2·5 mm, estimate the frequency of transverse vibration of the shaft when the static deflection is assumed to be (*a*) a sine wave, (*b*) a parabola. (*Ans.*: 11·37 Hz; 11·30 Hz)

8. A beam of mass 18 kg/m is simply supported on a span of 3·6 m and carries a body of mass 250 kg at the centre. The second moment of area of the cross-section is 8×10^{-6} m^4 and $E = 200$ GN/m^2. Find the frequency of transverse vibrations.
(*Ans.*: 12·2 Hz)

9. A beam, simply supported on a span of 6 m, has a mass of 52 kg/m. The second moment of area of the cross-section is 120×10^{-6} m^4 and $E = 200$ GN/m^2. A load of 3 t is carried at a point 2·4 m from one support.

Find the frequency of transverse vibration of the beam, neglecting the mass of the beam. Then find the approximate frequency if the mass of the beam is taken into account. (*Ans.*: 7·01 Hz; 6·82 Hz)

10. A light shaft carries a rotor in which the centre of gravity is 0·5 mm from the axis of rotation. If the whirling speed is 750 rev/min, find the speed range over which the deflection of the rotor, relative to the static position, will exceed 1·25 mm.
(*Ans.*: 633·8 to 968·2 rev/min)

11. A light shaft carrying a rotor is to be run at 1½ times its critical speed. If the centrifugal deflection is not to exceed 0·25 mm, find the greatest permissible displacement of the centre of gravity from the axis of rotation. (*Ans.*: 0·139 mm)

12. A light shaft carries a single disc whose centre of gravity is at a distance e from the shaft axis. The whirling speed is 3 600 rev/min and at a speed of 3 240 rev/min, the centre of gravity revolves in a circle of radius 3·8 mm. Calculate the distance e. (*Ans.*: 0·722 mm)

20 Damped and forced vibrations

20.1 Introduction All vibrations are damped, either externally by a dash-pot or eddy current damper, or by internal hysteresis forces within the system. The damping forces or torques are assumed to be proportional to the velocity of the vibration.

Damping will cause the vibration to die away unless sustained by an externally applied harmonic force or torque, which may be applied directly to the body or through its support. Examples of such a disturbance are an out-of-balance rotor or the reciprocating masses in an engine.

Immediately on starting, the motion of a body will be a combination of its free vibration and that due to the periodic disturbance. When the free vibration has died away, the steady-state vibration is of the same frequency as the disturbance.

20.2 Damped vibrations Fig. 20.1 shows an elastic system represented by a mass m supported by a spring of stiffness S, the motion being opposed by a damper which exerts a damping force c per unit velocity; c is called the *damping coefficient*. The equation of motion is given by

$$m\frac{d^2x}{dt^2} = -Sx - c\frac{dx}{dt}$$

The negative signs arise because the restoring force is opposite in direction to the displacement and the damping force is opposite in direction to the velocity.

This may be written as

$$\frac{d^2x}{dt^2} + 2\mu\frac{dx}{dt} + \omega^2 x = 0 \qquad (20.1)$$

where $2\mu = \dfrac{c}{m}$ and $\omega^2 = \dfrac{S}{m}$.

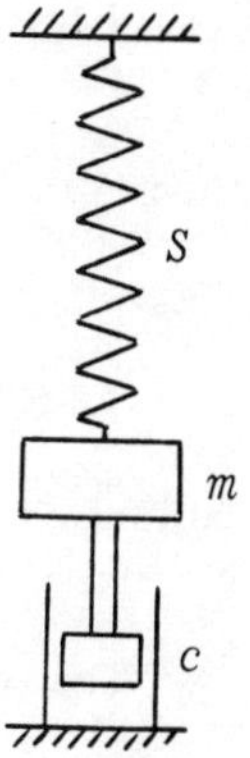

Fig. 20.1

If μ is equal to or greater than ω, the mass, when disturbed, will slowly return to its equilibrium position without oscillation but if μ is less than ω, the mass will overshoot its equilibrium position and then oscillate with decreasing amplitude until it eventually comes to rest.

The solution for this case is *

$$x = e^{-\mu t}\{A\cos\sqrt{\omega^2 - \mu^2}\,t + B\sin\sqrt{\omega^2 - \mu^2}\,t\} \qquad (20.2)$$

The periodic time $T = \dfrac{2\pi}{\sqrt{\omega^2 - \mu^2}}$ \qquad (20.3)

and the frequency $n = \dfrac{\sqrt{\omega^2 - \mu^2}}{2\pi}$ \qquad (20.4)

* See Appendix

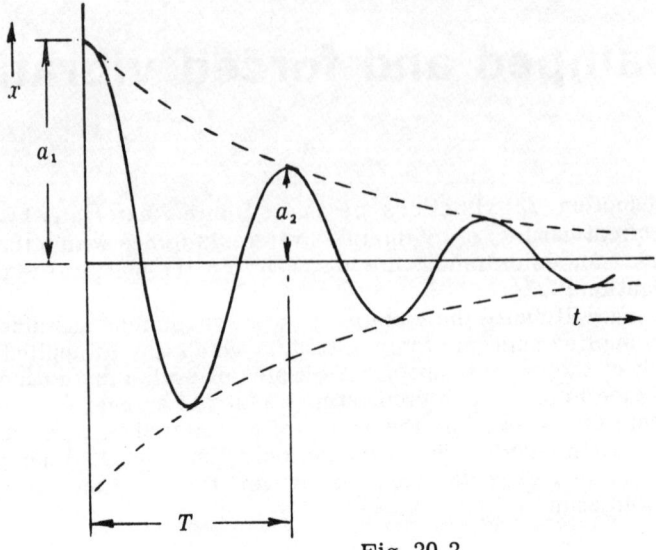

Fig. 20.2

Fig. 20.2 illustrates the motion represented by equation (20.2)

If $x = a_1$ when $t = 0$

then $x = a_2$ when $t = T = \dfrac{2\pi}{\sqrt{\omega^2 - \mu^2}}$

$$\therefore \frac{a_1}{a_2} = \frac{e^0 (A \cos 0 + B \sin 0)}{e^{-\mu T}(A \cos 2\pi + B \sin 2\pi)} = e^{\mu T} \quad (20.5)$$

so that $\log_e \dfrac{a_1}{a_2} = \mu T = \dfrac{2\pi\mu}{\sqrt{\omega^2 - \mu^2}}$ \quad (20.6)

The term $\dfrac{2\pi\mu}{\sqrt{\omega^2 - \mu^2}}$ is known as the *logarithmic decrement*.

For angular vibrations, in which a rotor of moment of inertia I is supported by a shaft of torsional stiffness q and is subjected to a damping torque c per unit angular velocity, the equation of motion is

$$I \frac{d^2\theta}{dt^2} = -q\theta - c \frac{d\theta}{dt}$$

which can be written

$$\frac{d^2\theta}{dt^2} + 2\mu \frac{d\theta}{dt} + \omega^2 \theta = 0 \quad (20.7)$$

where $2\mu = \dfrac{c}{I}$ and $\omega^2 = \dfrac{q}{I}$

The solution is similar to that for linear vibrations.

DAMPED AND FORCED VIBRATIONS

20.3 Forced vibrations Fig. 20.3 shows a system represented by a mass m supported by a spring of stiffness S, the mass being subjected to a harmonic disturbing force $P\cos pt$. The equation of motion is given by

$$m\frac{d^2x}{dt^2} = -Sx + P\cos pt$$

or $\quad \dfrac{d^2x}{dt^2} + \omega^2 x = \dfrac{P}{m}\cos pt \quad$ where $\omega^2 = \dfrac{S}{m}\quad$ (20.8)

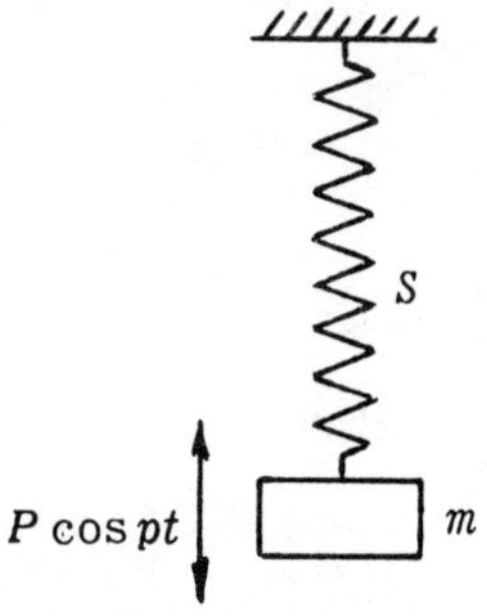

The solution is *

$$x = A\cos\omega t + B\sin\omega t + \frac{P\cos pt}{m(\omega^2 - p^2)}$$

Fig. 20.3

The first two terms (i.e. the complementary function) represents the free vibration of the body, which dies out, leaving

$$x = \frac{P\cos pt}{m(\omega^2 - p^2)}$$

(i.e. the particular integral) to represent the steady-state or sustained vibration. This is a harmonic motion of frequency $\frac{p}{2\pi}$ Hz and amplitude

$$a = \frac{P}{m(\omega^2 - p^2)} \quad (20.10)$$

If p is less than ω, the body oscillates in phase with the disturbing force but if p is greater than ω, the body oscillates 180° out of phase with the disturbing force and the amplitude then becomes $\dfrac{P}{m(p^2 - \omega^2)}$.

When $p = \omega$, the amplitude becomes infinite and *resonance* occurs.

The force transmitted to the support is the sum of the weight of the body and the dynamic force due to the vibration.

Hence, $\quad$ maximum force on support $= W + aS$

$$= mg + \frac{PS}{m(\omega^2 - p^2)} \quad (20.11)$$

For the corresponding angular case of a rotor of moment of inertia I supported by a shaft of torsional stiffness q and subjected to a harmonic torque $\tau\cos pt$, the equation of motion is

$$I\frac{d^2\theta}{dt^2} = -q\theta + \tau\cos pt$$

or $\quad \dfrac{d^2\theta}{dt^2} + \omega^2\theta = \dfrac{\tau}{I}\cos pt \quad$ where $\omega^2 = \dfrac{q}{I} \quad (20.12)$

* See Appendix

20.4 Forced damped vibrations Fig. 20.4 shows the combination of forcing and damping considered in the preceding articles.

This leads to the equation of motion

$$m \frac{d^2x}{dt^2} = -Sx - c \frac{dx}{dt} + P \cos pt$$

or
$$\frac{d^2x}{dt^2} + 2\mu \frac{dx}{dt} + \omega^2 x = \frac{P}{m} \cos pt \qquad (20.13)$$

The complementary function is *

$$x = e^{-\mu t} \{ A \cos \sqrt{\omega^2 - \mu^2} \, t + B \sin \sqrt{\omega^2 - \mu^2} \, t \}$$

as in equation (20.2).

This motion dies away, leaving the steady-state vibration represented by the particular integral, which is

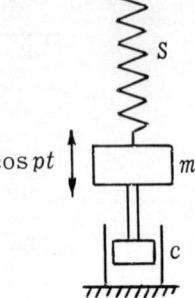

Fig. 20.4

$$x = \frac{P \cos(pt - \alpha)}{m \sqrt{4\mu^2 p^2 + (\omega^2 - p^2)^2}} \qquad (20.14)$$

where
$$\alpha = \tan^{-1} \frac{2\mu p}{\omega^2 - p^2} \qquad (20.15)$$

This sustained motion has a frequency $\frac{p}{2\pi}$ Hz and amplitude

$$a = \frac{P}{m \sqrt{4\mu^2 p^2 + (\omega^2 - p^2)^2}} \qquad (20.16)$$

lagging behind the disturbing force by the *phase angle* α.

When $p = \omega$, the amplitude is not infinite, as in undamped vibrations, but is equal to $\frac{P}{2\mu mp}$ and the maximum amplitude does not occur when $p = \omega$ but when $\frac{da}{dp} = 0$.

If the amplitude of the disturbing force P is constant, regardless of the frequency of application p, the amplitude of the mass varies in the forms shown in Fig. 20.5, depending upon the ratio μ/ω. In most practical cases, however, such as out-of-balance rotors or piston inertia, the magnitude of P is proportional to p^2 and this leads to curves of the form shown in Fig. 20.6. For both types of disturbing force, the effect of damping is small except in the region of resonance, i.e. when $p = \omega$.

The variation of phase angle α with frequency p is shown in Fig. 20.7.

For forced damped angular vibrations, the corresponding equation of motion is

$$I \frac{d^2\theta}{dt^2} = -q\theta - c \frac{d\theta}{dt} + \tau \cos pt$$

or
$$\frac{d^2\theta}{dt^2} + 2\mu \frac{d\theta}{dt} + \omega^2 \theta = \frac{\tau}{I} \cos pt \qquad (20.17)$$

* See Appendix

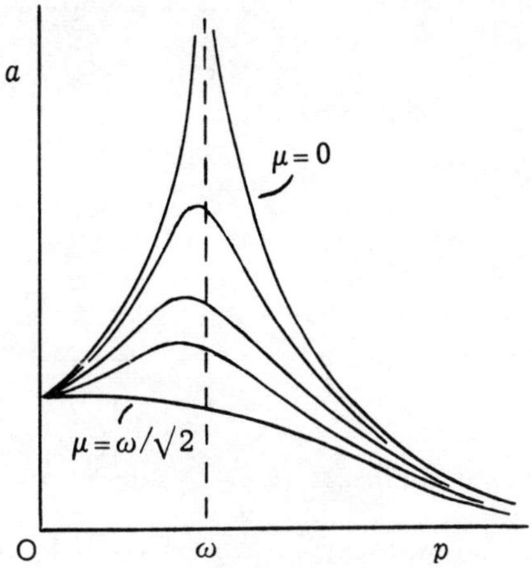

DAMPED AND FORCED VIBRATIONS

P constant

Fig. 20.5

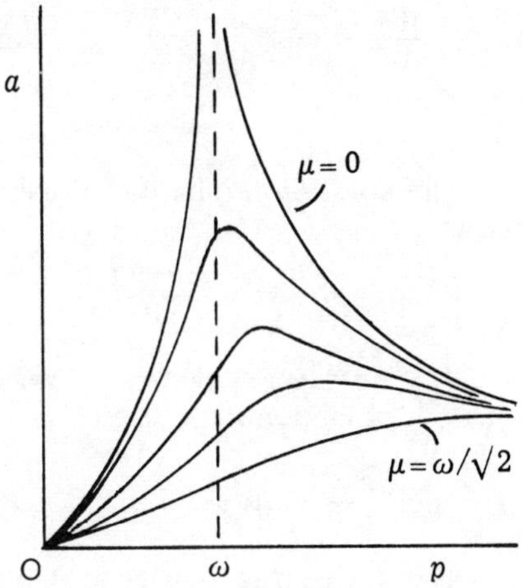

$P \propto p^2$

Fig. 20.6

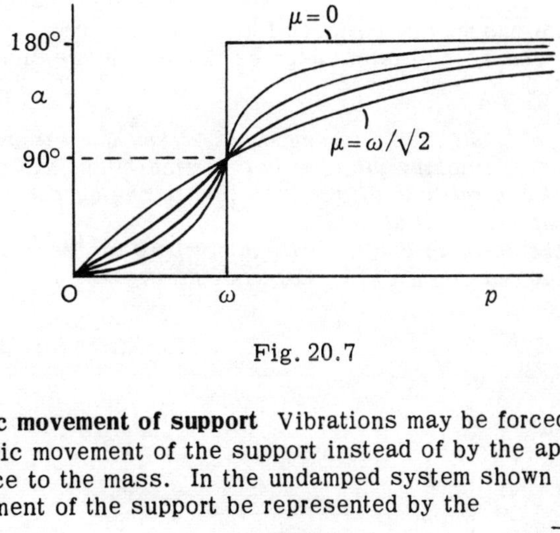

Fig. 20.7

20.5 Periodic movement of support Vibrations may be forced upon a system by the periodic movement of the support instead of by the application of a harmonic force to the mass. In the undamped system shown in Fig. 20.8, let the movement of the support be represented by the equation

$$y = h \cos pt$$

Then change of spring length $= x - y$

Therefore restoring force $= S(x - y)$

The equation of motion is therefore

$$m \frac{d^2 x}{dt^2} = -S(x - y)$$

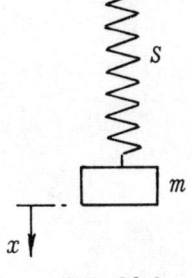

Fig. 20.8

or $\quad \dfrac{d^2x}{dt^2} + \omega^2 x = \omega^2 y \quad$ where $\omega^2 = \dfrac{S}{m}$

$$= \omega^2 h \cos pt \qquad (20.18)$$

The solution is similar to that for equation (20.8), giving

$$x = \dfrac{\omega^2 h \cos pt}{\omega^2 - p^2} \qquad (20.19)$$

For the damped system shown in Fig. 20.9, the equation of motion is

$$\dfrac{d^2x}{dt^2} + 2\mu \dfrac{dx}{dt} + \omega^2 x = \omega^2 h \cos pt \qquad (20.20)$$

The solution is similar to that for equation (20.13), giving

$$x = \dfrac{\omega^2 h \cos(pt - \alpha)}{\sqrt{4\mu^2 p^2 + (\omega^2 - p^2)^2}} \qquad (20.21)$$

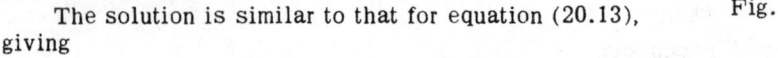

Fig. 20.9

Similar equations may be obtained for angular motion.

Amplitude curves for different degrees of damping are similar to those shown in Fig. 20.5.

1. *A machine of mass 100 kg is supported on springs which deflect 20 mm under the load. The machine vibrates in a vertical plane and a dash-pot is fitted to reduce the amplitude of free vibration to one quarter of its initial value in two complete oscillations.*

Calculate the damping coefficient and compare the frequencies of the damped and undamped vibrations of the system.

$$T = 2\pi \sqrt{\dfrac{g}{\delta}} = \dfrac{2\pi}{\omega} \quad . \quad . \quad . \quad \text{from Art. (18.2 \& 18.3)}$$

$$\therefore \omega^2 = \dfrac{g}{\delta}$$

$$= \dfrac{9 \cdot 81}{0 \cdot 020} = 490 \cdot 5 \ (\text{rad/s})^2$$

$$\dfrac{a_1}{a_2} = e^{\mu T} \quad . \quad . \quad . \quad . \quad . \quad . \quad \text{from equation (20.5)}$$

Similarly $\quad \dfrac{a_2}{a_3} = e^{\mu T}$

$$\therefore \dfrac{a_1}{a_3} = \dfrac{a_1}{a_2} \times \dfrac{a_2}{a_3} = e^{2\mu T} = 4$$

$$\therefore \mu T = \dfrac{2\pi \mu}{\sqrt{\omega^2 - \mu^2}} = \log_e 2$$

$$\therefore 2\pi \mu = 0 \cdot 693 \ 1 \sqrt{490 \cdot 5 - \mu^2}$$

DAMPED AND FORCED VIBRATIONS

$$\therefore \mu = 2.43 = \frac{c}{2m} \quad . \quad . \quad . \quad \text{from equation (20.1)}$$

$$\therefore c = 2.43 \times 2 \times 100 = 486 \text{ N s/m}$$

$$\text{Frequency of damped vibrations} = \frac{\sqrt{\omega^2 - \mu^2}}{2\pi} \quad . \quad \text{from equation (20.4)}$$

$$= \frac{\sqrt{490.5 - 2.43^2}}{2\pi}$$

$$= 3.505 \text{ Hz}$$

$$\text{Frequency of undamped vibrations} = \frac{\omega}{2\pi}$$

$$= \frac{\sqrt{490.5}}{2\pi}$$

$$= 3.525 \text{ Hz}$$

2. *A uniform bar of mass m and length l is hinged at one end while the other end is carried by a spring of stiffness S so that in the rest position, the bar is horizontal. Half-way along the bar, a dash-pot is attached which produces a damping force c per unit velocity. Obtain an expression for the periodic time of oscillation about the hinge.*

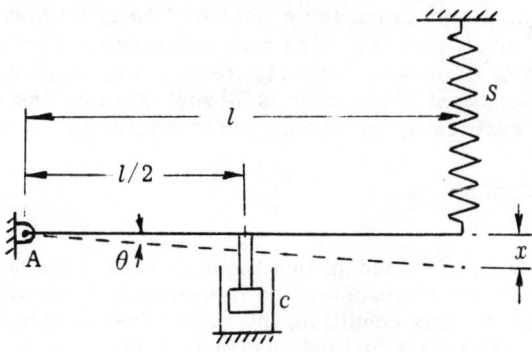

Fig. 20.10

Referring to Fig. 20.10, $I_A = \dfrac{ml^2}{3}$

Let the free end be depressed a distance x and released.

Restoring force in spring $= Sx$

$\therefore$ moment about A $= Sx \times l$

Velocity of dash-pot $= \frac{1}{2}\dfrac{dx}{dt}$

∴ damping force $= \dfrac{c}{2}\dfrac{dx}{dt}$

∴ moment about A $= \dfrac{c}{2}\dfrac{dx}{dt}\times\dfrac{l}{2} = \dfrac{cl}{4}\dfrac{dx}{dt}$

The equation of motion is therefore

$$I\dfrac{d^2\theta}{dt^2} = -Slx - \dfrac{cl}{4}\dfrac{dx}{dt}$$

i.e. $\quad \dfrac{ml^2}{3}\times\dfrac{1}{l}\dfrac{d^2x}{dt^2} + \dfrac{cl}{4}\dfrac{dx}{dt} + Slx = 0 \quad$ since $\quad x = l\theta \quad$ and $\quad \dfrac{d^2x}{dt^2} = l\dfrac{d^2\theta}{dt^2}$

i.e. $\quad \dfrac{d^2x}{dt^2} + 2\mu\dfrac{dx}{dt} + \omega^2 x = 0 \quad$ where $\quad 2\mu = \dfrac{3c}{4m} \quad$ and $\quad \omega^2 = \dfrac{3S}{m}$

$$\therefore T = \dfrac{2\pi}{\sqrt{\omega^2 - \mu^2}} \qquad \text{from equation (20.3)}$$

$$= \dfrac{2\pi}{\sqrt{\dfrac{3S}{m} - \left(\dfrac{3c}{8m}\right)^2}}$$

$$= \dfrac{16\pi m}{\sqrt{192\,Sm - 9c^2}}$$

3. *A light helical spring carries a mass of 5 kg at its lower end and the upper end is attached to a pin which moves in a vertical path with S.H.M., with a total stroke of 40 mm. When the frequency of oscillation is 200 cycles/min, the total movement of the mass is 30 mm. Find (a) the stiffness of the spring, (b) the maximum spring force, (c) the natural frequency of the system.*

From equation (20.19), $\quad x = \dfrac{\omega^2 h \cos pt}{\omega^2 - p^2}$

From Fig. 20.5, it is evident that the amplitude of the mass can only be less than that of the disturbance if the frequency is above the critical, i.e. if $p > \omega$. Thus, for this condition, the mass is oscillating 180° out of phase with the disturbance and the amplitude is given by

$$a = \dfrac{\omega^2 h}{p^2 - \omega^2} \qquad \ldots \qquad \text{(As in Art. 20.3)}$$

i.e. $\quad 15 = \dfrac{\omega^2 \times 20}{\left(200\times\dfrac{2\pi}{60}\right)^2 - \omega^2}$

from which $\quad \omega^2 = 188 \ (\text{rad/s})^2$

Hence stiffness $\quad S = \omega^2 m$

$\qquad\qquad\qquad\qquad = 188\times 5 = \underline{940 \text{ N/m}}$

DAMPED AND FORCED VIBRATIONS

As the mass and disturbance are 180° out of phase,

$$\text{maximum extension of spring} = \frac{0 \cdot 04 + 0 \cdot 03}{2} = 0 \cdot 035 \text{ m}$$

$$\therefore \text{maximum spring force} = 0 \cdot 035 \times 940 + 5 \times 9 \cdot 81$$

$$= \underline{82 \text{ N}}$$

$$\text{Frequency } n = \frac{\omega}{2\pi} = \frac{\sqrt{188}}{2\pi} = \underline{2 \cdot 18 \text{ Hz}}$$

4. *A machine, fixed to the floor of a workshop, produces a static deflection of 2 mm immediately under the machine. When the machine is working, there is an unbalanced mass which produces a vertical alternating force whose frequency is equal to the speed of the machine shaft. When the speed is 240 rev/min, the amplitude of the forced vibration of the floor is 1·25 mm. If the floor is assumed to be elastic and damping is neglected, what will be the amplitude of the forced vibration when the speed is 480 rev/min?*
At what speed will resonance occur?

$$\omega^2 = \frac{g}{\delta} = \frac{9 \cdot 81}{0 \cdot 002} = 4\,905 \text{ (rad/s)}^2 \quad . \quad . \quad . \quad \text{as in Ex. 1}$$

From equation (20.10), $a = \dfrac{P}{m(\omega^2 - p^2)}$

The inertia force on the unbalanced mass will be proportional to the square of the machine speed,

i.e. $\qquad\qquad\qquad P = kp^2$

When the machine speed is 240 rev/min, $p = 240 \times \dfrac{2\pi}{60} = 8\pi$ rad/s

When the machine speed is 480 rev/min, $p = 16\pi$ rad/s

Therefore, at 240 rev/min, $1 \cdot 25 = \dfrac{k(8\pi)^2}{m[4\,905 - (8\pi)^2]}$ \qquad (1)

At 480 rev/min, $\qquad a = \dfrac{k(16\pi)^2}{m[4\,905 - (16\pi)^2]}$ \qquad (2)

Therefore, from equations (1) and (2)

$$\frac{a}{1 \cdot 25} = \frac{4\,905 - 64\pi^2}{64\pi^2} \times \frac{256\pi^2}{4\,905 - 256\pi^2}$$

from which $\qquad\qquad a = \underline{8 \cdot 98 \text{ mm}}$

At resonance, $\qquad p = \omega = \sqrt{4\,905} = 70$ rad/s

$$\therefore \text{speed} = 70 \times \frac{60}{2\pi} = \underline{669 \text{ rev/min}}$$

5. *A machine, mounted on elastic supports, is free to vibrate vertically. The machine has a mass of 40 kg and rotor out-of-balance effects are equivalent to 2 kg at 150 mm radius. Resonance occurs when the machine is run at 621 rev/min, the amplitude of vibration at this speed being 45 mm. Determine the amplitude when running at 500 rev/min and find the angular position of the out-of-balance mass when the machine is at its highest position during vibration.*

Since the amplitude at resonance is not infinite, the vibration is damped and from equation (20.16),

$$a = \frac{P}{m\sqrt{4\mu^2 p^2 + (\omega^2 - p^2)^2}}$$

where $\quad P = mp^2 r = 2p^2 \times 0.15 = 0.3 p^2$

At resonance, $\quad p = \omega = 621 \times \dfrac{2\pi}{60} = 65$ rad/s

and $\quad a = \dfrac{P}{2\mu m p}$

i.e. $\quad 0.045 = \dfrac{0.3 \times 65^2}{40 \times 2\mu \times 65}$

$\therefore \quad 2\mu = 10.84$ rad/s

At 500 rev/min, $\quad p = 500 \times \dfrac{2\pi}{60} = 52.36$ rad/s

$\therefore \quad a = \dfrac{0.3 \times 52.36^2}{40\sqrt{10.84^2 \times 52.36^2 + (65^2 - 52.36^2)^2}}$

$\quad = 0.013$ m $\quad$ or $\quad \underline{13\text{ mm}}$

From equation (20.15), the phase angle,

$$\alpha = \tan^{-1}\frac{2\mu p}{\omega^2 - p^2}$$

$\quad = \tan^{-1}\dfrac{10.84 \times 52.36}{65^2 - 52.36^2}$

$\quad = \underline{21°}$

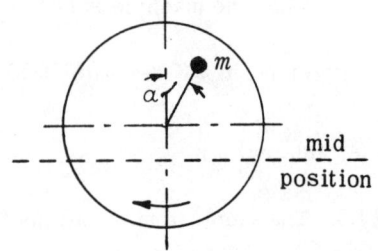

Fig. 20.11

Fig. 20.11 shows the rotor and the position of the out-of-balance mass when the machine is in the highest position.

6. *A rotor attached to the lower end of a vertical rod is observed to make one oscillation in ½ s and the amplitude of the second oscillation is half that of the first. If the top of the rod is now compelled to make angular oscillations of period 2 s and amplitude 5°, find the amplitude of the rotor oscillations.*

DAMPED AND FORCED VIBRATIONS

As in equation (20.5), $\quad \dfrac{\phi_1}{\phi_2} = e^{\mu T}$

i.e. $\quad 2 = e^{\mu \times \frac{1}{2}}$

from which $\quad \mu = 1{\cdot}386 \text{ rad/s}$

From equation (20.3), $\quad T = \dfrac{2\pi}{\sqrt{\omega^2 - \mu^2}}$

$\therefore \ \tfrac{1}{2} = \dfrac{2\pi}{\sqrt{\omega^2 - 1{\cdot}386^2}}$

from which $\quad \omega^2 = 159{\cdot}1 \ (\text{rad/s})^2$

If the angular amplitude of the periodic disturbance is γ, then, from equation (20.21), the amplitude of the rotor is given by

$$\phi = \dfrac{\omega^2 \gamma}{\sqrt{4\mu^2 p^2 + (\omega^2 - p^2)}}$$

If the periodic time is 2 s, $\quad p = \dfrac{2\pi}{2} = \pi \text{ rad/s}$

$\therefore \ \phi = \dfrac{159{\cdot}1 \times 5°}{\sqrt{4 \times 1{\cdot}386^2 \pi^2 + (159{\cdot}1 - \pi^2)^2}}$

$\quad = \underline{5{\cdot}33°}$

---

7. A mass suspended from a spring is subjected to damping proportional to the velocity. The frequency of damped vibrations is 1·5 Hz and the amplitude decreases to 20% of its initial value in one complete vibration. Determine the frequency of free undamped vibrations of the system. (*Ans.*: 1·55 Hz)

8. A mass of 14 kg is suspended from a spring of stiffness 8·4 kN/m and due to damping, the amplitude of the vibration diminishes to 1/10 of its original value in two complete vibrations. Find the frequency of vibration and the value of the damping coefficient. (*Ans.*: 3·83 Hz; 123·6 Ns/m)

9. A mass of 2·4 kg suspended from a spring is pulled downwards and released. The subsequent motion is controlled by a viscous damper such that the ratio of the first downward displacement to the third is 4 : 1 and five vibrations are completed in 4 s. Find the stiffness of the spring and the damping coefficient. (*Ans.*: 150 N/m; 4·16 Ns/m)

10. A disc of moment of inertia 0·6 kg m² is attached to one end of a shaft of stiffness 5 N m/rad, the other end of the shaft being fixed. Find the frequency of vibration of the disc when it is subjected to a damping torque of 0·3 N m s/rad and also the ratio of successive amplitudes on the same side of the equilibrium position. (*Ans.*: 0·458 Hz; 1·726)

11. A flywheel of mass 10 kg makes rotational oscillations under the control of a torsion spring of stiffness 4 N m/rad. Calculate the radius of gyration of the flywheel if the periodic time of oscillation is 2·5 s.

When a viscous damper is fitted to the system, the ratio of successive amplitudes on the same side is 0·1. Find the periodic time of the damped oscillation. (*Ans.*: 0·252 m; 2·665 s)

12. A mass of 35 kg is supported by a spring of stiffness 25 kN/m and is acted upon by a disturbing force of amplitude 40 N and frequency 5 Hz. Find the amplitude of forced vibration. (*Ans.*: 4·185 mm)

13. A mass of 100 kg is suspended from a spring of stiffness 4·5 kN/m. The upper end of the spring is given S.H.M. in a vertical direction by means of a crank 6 mm long. Determine the total vertical movement of the mass and also its maximum velocity when the crank is driven at 20 rad/s. (*Ans.*: 1·52 mm; 15·2 mm/s)

14. An undamped vibrating system is excited by a sinusoidal force of constant amplitude but variable frequency. The amplitude of vibration is 25 mm at 10 Hz and this decreases continuously to 2·5 mm at 20 Hz. Determine the critical frequency of the system and the static deflection of the load. (*Ans.*: 8·165 Hz; 3·73 mm)

15. An undamped vibrating system, excited by a sinusoidal force of constant magnitude but variable frequency, has a total travel of 100 mm at 3 Hz and 50 mm at 9 Hz and the critical frequency lies between these two rates. Determine the natural frequency of the system and the magnitude of the vibrating mass if the stiffness of the suspension is 7 kN/m. (*Ans.*: 5·7 Hz; 5·38 kg)

16. An engine rests on an elastic foundation which deflects 0·85 mm under the dead load. Find the frequency of free vertical vibration.

If the engine has a mass of 1·25 t and when running at 450 rev/min there is an out-of-balance force of this frequency and amplitude 2·4 kN, find the amplitude of the forced vibration. Find also the maximum force exerted on the foundation. (*Ans.*: 17·15 Hz; 0·206 mm; 15·22 kN)

17. A mass of 2 kg vibrating on a spring of stiffness 15 kN/m is subjected to a damping force of 7 N s/m. Find the amplitude of the forced vibration of the mass when it is acted upon by a periodic disturbing force of $25 \cos 100t$ N. (*Ans.*: 4·95 mm)

18. A machine of mass ½ t stands on a floor which deflects 6 mm under the load. The floor exerts a damping force which is proportional to the velocity of motion and of magnitude equal to 15% of the critical damping force. When the machine runs at 500 rev/min, the amplitude of vibration of the floor under the machine is 5 mm. Calculate the maximum value of the disturbing force within the machine. (*Ans.*: 3·19 kN)

19. A mass of 100 kg is suspended from a spring of stiffness 10 kN/m and the motion of the mass is damped by a frictional resistance proportional to the velocity and of value 220 N s/m.

If the top of the spring vibrates vertically with an amplitude of 2·5 mm and frequency 20 Hz, find the amplitude of forced vibration of the mass. (*Ans.*: 0·015 9 mm)

20. An engine of mass 240 kg is mounted on a spring support. When the engine is not running, free vertical vibrations are subject to damping forces which reduce the amplitude by 20% during each complete oscillation, the frequency of these oscillations being 9 Hz.

When the engine is running at 480 rev/min, there is an out-of-balance harmonic force of amplitude 200 N. Find the amplitude of steady-state vibration at this speed. (*Ans.*: 1·183 mm)

21. An unbalanced engine of mass 200 kg is mounted on an elastic support and an increase of speed from 600 rev/min to 900 rev/min trebles the amplitude of the forced vibration. Find the damping coefficient required for a damping device which will reduce the amplitude of vibration at 600 rev/min by 10%. (*Ans.*: 30·4 kN s/m)

APPENDIX

Differential Equations

Differential equations of the type

$$\frac{d^2x}{dt^2} + 2\mu\frac{dx}{dt} + \omega^2 x = f(t)$$

may be written in the form

$$\frac{d^2x}{dt^2} + 2\mu\frac{dx}{dt} + \omega^2 x = 0 + f(t)$$

and the solution consists of two parts:

(a) the value of x which satisfies the equation

$$\frac{d^2x}{dt^2} + 2\mu\frac{dx}{dt} + \omega^2 x = 0,$$

this being called the *complementary function*;

(b) the value of x which satisfies the equation

$$\frac{d^2x}{dt^2} + 2\mu\frac{dx}{dt} + \omega^2 x = f(t),$$

this being called the *particular integral*.

(a) *Complementary function*

(i) $$\frac{d^2x}{dt^2} + \omega^2 x = 0$$

The solution is $\quad x = A\cos\omega t + B\sin\omega t$

(ii) $$\frac{d^2x}{dt^2} + 2\mu\frac{dx}{dt} + \omega^2 x = 0$$

The solution is $\quad x = e^{-\mu t}(A\cos\sqrt{\omega^2-\mu^2}\,t + B\sin\sqrt{\omega^2-\mu^2}\,t)$

(b) *Particular integral*

(i) $$\frac{d^2x}{dt^2} + \omega^2 x = C$$

The solution is $\quad x = \dfrac{C}{\omega^2}$

(ii) $\quad \dfrac{d^2x}{dt^2} + 2\mu \dfrac{dx}{dt} + \omega^2 x = C$

The solution is $\quad x = \dfrac{C}{\omega^2}$

(iii) $\quad \dfrac{d^2x}{dt^2} + \omega^2 x = C \cos pt$

The solution is $\quad x = \dfrac{C \cos pt}{\omega^2 - p^2}$

(iv) $\quad \dfrac{d^2x}{dt^2} + 2\mu \dfrac{dx}{dt} + \omega^2 x = C \cos pt$

The solution is $\quad x = \dfrac{C \cos(pt - \alpha)}{\sqrt{4\mu^2 p^2 + (\omega^2 - p^2)^2}}$

where $\quad \alpha = \tan^{-1} \dfrac{2\mu p}{\omega^2 - p^2}$